INTELLIGENT STRUCTURES — 2

MONITORING AND CONTROL

Proceedings of the International Workshop on Intellligent Systems held in Perugia, Italy, 27-29 June 1991.

INTELLIGENT STRUCTURES — 2

MONITORING AND CONTROL

Edited by

Y.K. WEN

University of Illinois at Urbana-Champaign, U.S.A

ELSEVIER APPLIED SCIENCE
LONDON and NEW YORK

ELSEVIER SCIENCE PUBLISHERS LTD
Crown House, Linton Road, Barking, Essex IG11 8JU, England

Sole Distributor in the USA and Canada
ELSEVIER SCIENCE PUBLISHING CO., INC.
655 Avenue of the Americas, New York, NY 10010, USA

WITH 23 TABLES AND 168 ILLUSTRATIONS

British Library Cataloguing in Publication Data

International Workshop on Intelligent Systems
(2nd: 1991: Perugia, Italy)
Intelligent structures 2 ; monitoring and control.
I. Title II. Wen, Y. R.
624.176

ISBN 1-85166-757-1

Library of Congress CIP data applied for

Printed in Great Britain at The Alden Press, Oxford

Preface

Construction is the largest industry in the world. Yet technology to monitor and maintain the health of building and structural systems under natural and man-made loadings has only begun to receive the attention that it deserves. Recently there have been fast developments in analytical/numerical modeling of the structural systems including uncertainty in both loading and resistance. At the same time, new technologies and hardware have been invented which make real time sensing, monitoring and control of structural systems possible and feasible. To promote exchange of information, research cooperation and introduction of new ideas and concepts, a trilateral US/Italy/Japan Workshop/Seminar on Intelligent Systems was held in Perugia, Italy, 27-29 June, 1991. Investigators from US, Italy, Japan and other Western European countries gave presentations of papers and participated in group discussion. The topical areas of interest were:

1 Assessment of state of system: damage and effect of uncertainty, methods of knowledge-based expert systems, and neural networks.

2 Measurement of system performance: sensor techniques and computer-aided adaptive systems.

3 Control and isolation of system: dynamic experiments, system design, and algorithm development.

The presentations and summaries of group discussions are included in this book. It represents a summary of the state-of-art of this new and fast growing research area and also suggests directions for further research. This is the second International Workshop* of this nature. The first one was held in Taipei, Taiwan in July, 1990.

Y.K. Wen

Acknowledgements

The Workshop/Seminar was organized and coordinated by the Scientific Committee; the members are F. Casciati, H. Kameda, M. Shinozuka and Y.K. Wen. The Local Organizing Committee was chaired by L. Faravelli. The Workshop/Seminar was supported by the International Program of the National Science Foundation through Grant INT-9100511 to the University of Illinois at Urbana. The program official is H. Stolberg. The interest and encouragement of and support from S.C. Liu, H.J. Lagorio and K.P. Chong of N.S.F. are greatly appreciated. Financial support was also provided by the University of Perugia and the University of Pavia, Italy. Their support is gratefully acknowledged.

Contents

Measurement of System Performance and Adaptive Systems

Structural Control

Summary of Group Discussion and Recommendations

Assessment of State of Systems

NEW DEVELOPMENTS IN NONDESTRUCTIVE EVALUATION

KEN P. CHONG* and JOHN B. SCALZI
PROGRAM DIRECTORS
National Science Foundation
1800 G Street, N.W.
Room, 1108
Washington, D.C. 20550 USA

ABSTRACT

Due to environmental, man-made and natural hazards, infrastructure of public works in the world is deteriorating. Nondestructive evaluation (NDE) is essential for condition assessment for repair, retrofit, rehabilitation, and replacement (if needed) to insure the public safety. In this paper an overview of nondestructive evaluation projects, including a new initiative and four new awards on Quantitative NDE for Large Structural Systems, is presented.

KEYWORDS: Non-destructive evaluation; structures, condition assessment

INTRODUCTION

The world invests about US $1,430 billion annually in the construction of structures, including: buildings (residential, commercial, industrial, etc.), civil works, and utilities, such as highways, water and sewer structures, railroads, and transmission lines. Construction is the largest industry in the world amounting to 10% of the world's gross domestic product.

Structures are generally the most expensive investments/assets in any country. In addition

---

Note: Part of this paper is taken from "Overview of Nondestructive Evaluation Projects and Initiative at NSF" by K. P. Chong, J. B. Scalzi and O. W. Dillon in Intelligent Structures, Elsevier, London, 1990.

*to present paper

structures have long service life compared with any other kinds of commercial product, and are rarely replaceable once they are erected. Yet the feedback and controls on the "state of health" of structures are practically non-existent (compared with say the much less costly automobiles). Nondestructive evaluation is an essential part of this feedback and monitoring system for infrastructures.

During the past several years, many reports have been published concerning the sad state of deterioration of the nation's public works, such as bridges, roadways, water and sewer systems, ports, harbors, airports, and buildings of all types. According to the 1988 National Research Council Report on "Building for Tomorrow" estimates of public infrastructure amounted to $409 trillion in 1984, and growing rapidly. The U.S. Interstate System, built in the 1950's needs repair, so do most of the state and county secondary roads. Some 42% of all U.S. highway bridges are inadequate and half of school buildings are 50 years or older. In the U. S. alone, recent bridge collapses include:

- Mianus River Bridge - Connecticut, 1986, 3 deaths
- Schoharie Creek - New York, 1987, 10 deaths
- Hatchie River - Tennessee, 1989, 8 deaths
- Miamis Town River - Ohio, 1989, 2 deaths

According to David Aschauer (Chief Economist of U.S. Federal Reserve Bank - Chicago), a 1% increase in public capital causes 1/3% rise in productivity. From 1971 to 1985 U.S. productivity growth fell 1.2%, out of which 1% was attributed to neglect of the infrastructure. If the present infrastructure deterioration continues, the Department of Transportation estimates that the U.S. economy will miss out on:

- 3.2% growth in GNP
- 5.9% disposable income
- 2.2% employment
- 2.7% productivity

The infrastructure ages and deteriorates with time. The deterioration is mostly a result of aging of the materials, excessive use, overloading, climatic conditions, lack of sufficient maintenance, and difficulties encountered in proper inspection methods. All of these factors contribute to the obsolescence of the structural system as a whole. As a result, repair, retrofit, rehabilitation, and replacement become necessary to insure the safety of the public.

THE PROBLEM

In order to evaluate the safety of a structure, it is essential to determine the remaining strength of the total system by a performance analysis based upon the strength and inspection data of the materials and the structural system.

A visual inspection process will provide data with respect to the external physical condition of the material such as the degree of corrosion and general deterioration. By past experience and judgement an engineer will evaluate the remaining life of the material and the contribution to the safety of the structure.

Strength determinations may be made by sampling the material and testing it in a laboratory. However, this method destroys some of the structure which subsequently must then be repaired. The repair is usually noticeable and disruptive, as a result, many engineers would prefer a nondestructive technique.

THE SOLUTION

Although there are many techniques for nondestructive evaluation (NDE) currently available for metals which are used in defense and manufactured products, there are only a few which are available for the construction materials such as wood, concrete, masonry, and steel. These current methods for construction materials are not very reliable and in some instances they can only be used on a comparative qualitative basis. Instruments specifically designed for construction materials and with greater accuracy than those currently available are required in order to quantitatively evaluate the safety of the structure and to estimate its remaining life adequately.

NSF ACTIVITY IN NDE

The NSF Division of Mechanical and Structural Systems is encouraging basic research to investigate the requirements to develop suitable instruments for nondestructive evaluation of strength properties and other physical conditions for construction materials. The Program Directors in the Division are encouraging multi-disciplinary activities involving electrical, mechanical, chemical, computer, electronic and structural engineers, as suitable for the project to perform group or team efforts in the research projects. Unfortunately, this "encouragement" should not be interpreted as large sums of new money for this approach. Rather this is our personal assessment with which all of us agree.

NSF is also encouraging close cooperation and active participation by the industries involved in the manufacture of nondestructive equipment and/or instrumentation. By joint efforts viable products can result which will be of great help to engineers responsible for the safety and longevity evaluation of structural systems.

As a source of information on the type of questions which the engineering community would like to have answered by the use of nondestructive evaluation techniques, NSF supported two workshops to outline specific needs of the profession. One was held in February, 1988, at the University of Southern California, Los Angeles, and the Proceedings are titled, "Nondestructive Evaluation for Performance of Civil Structures." The contact person is Dr. Sami F. Masri. The other workshop was held in June, 1989, at the National Institute of Standards and Technology, at Gaithersburg, Maryland, titled International Workshop on Sensors and Measurement Techniques for Assessing Structural Performance," organized by Dr. Richard D. Marshall and his associates.

CENTERS AND OTHER NDE ACTIVITIES

In addition to NDE research done at MIT, Illinois, Cornell, West Virginia, Virginia Tech, New Jersey Institute of Technology, Georgia Tech, etc., the following centers concentrate on various aspects of NDE.

I. Johns Hopkins Center for NDE (R. Green)

- Heavy on experiment
- Advanced sensors
- Artificial intelligence
- Life cycle management - include fracture, fatigue
- Measurement science for microelectronics
- Process control

The closest project to large structural system is the infrared television for wall thinning of buried pipelines.

II. Iowa State Center for NDE (D. Thompson)

- NSF Industry/University CRC
- Modeling efforts emphasized
- Advanced ultrasonic NDE
- Advanced electromagnetic NDE
- X-Ray & thermal NDE imaging
- NDE for materials properties
- NDE for advanced composites
- Ultrasonic stress measurement

III. National Institute of Standards and Technology

- Transducer standards
- Pulse-echo testing of concrete
- Automated manufacturing
- Metals, plastics, ceramics, paints, etc.

IV. NASA - Langley NDE Center (J. Heyman)

- Space structures
- Aging aircrafts
- Imbedded sensors
- Smart structures

NSF INITIATIVE

Believing that "an ounce of prevention is better than a pound of cure," and that NSF should play a leading role in addition to be acting as a catalyst in the "Quantitative Nondestructive Evaluation for Large Structural Systems" in civil engineering, a NSF initiative was issued in April 1990. The deadline for receipt of proposals was set at June 1, 1990. The initiative was sponsored by the following four programs, totalling $1M, over a two year period. Program directors involved are listed below.

Dr. John B. Scalzi - Structures and Large Systems

Dr. Ken P. Chong - Structures and Building Systems

Dr. Oscar Dillon - Solid Mechanics

Dr. Thomas Zimmie - Geotechnical, Geomechanics

Academic and research institutions in the United States with engineering, research and educational programs were invited to submit proposals. The contents of the proposal must clearly relate to the research topics described in the Programs of the Division of Mechanical and Structural Systems of the National Science Foundation.

Projects were encouraged to be multidisciplinary in nature involving investigators from one or more departments at a single institution or from several institutions wherever the required expertise may be located. It was envisioned that projects may include researchers from the following engineering and science disciplines: structural, mechanical, material science, electrical, geotechnical, computer science, chemical or chemistry, electronic and others as appropriated. A team or group effort was strongly encouraged to achieve the desired result and final product of research. Contributions of all the participants must be combined into a single integrated proposal which conformed to the usual requirements and limitations of a proposal length. Each researcher must have a scope of work which contributes to the entire project.

Industry collaboration is highly desirable and may be in the form of participating personnel, loan of equipment, gifts, supplies of various kinds, and/or financial contributions. The academic institution must be the prime research management organization.

A panel of experts have been convened to evaluate the proposals on a competitive basis. The investigators and the synergistic benefits derived, as well as how these interactions are to be organized and performed, are examined.

SPECIFIC TOPICS OF THE NSF INITIATIVE

Research efforts are to be directed toward increasing fundamental knowledge of quantitative nondestructive evaluation of construction materials and their application to large structural systems of wood, concrete, masonry and steel. The data developed for equipment may be for the material properties alone or for the fabricated forms as total systems. Some data may be used to determine how each system or material deteriorates with time and excessive loads. The data and prototype resulting from the research should be of such a nature that the resulting equipment may be useful for practitioners, and can be produced as a viable portable and economical commercial product.

SUGGESTED RESEARCH TOPICS OF INTEREST

The following topics of interest were published in the NSF Initiative.

- Determination of wood properties - above and below ground
- Determination of concrete properties - above and below ground
- Determination of masonry properties
- Determination of steel safety remaining as a result of corrosion
- Determination of initial cracking in steel and concrete, including welds and rebars due to fatigue, fabrication, or design faults

- o Rate of deterioration in steel or concrete
- o Early warning of impending failure
- o General inspection of the degree of deterioration in steel, concrete, masonry, soil retaining systems and wood
- o Determination of the effective length of piles of wood, steel, or concrete, and the amount of erosion or scour in foundations
- o Methodology for predicting the remaining safe life of large structural systems

PROJECTS SUPPORTED BY THE INITIATIVE

The recently awarded projects are briefly described below. The projects are in four basic areas of interest to the Structures, Solid Mechanics, Geomechanics, and Building Systems programs of the National Science Foundation.

1. Fiber Optic Sensors for Strength Evaluation and Early Warning of Impending Failure in Structural Components; Dr. Edward G. Nawy, Mohamad H. Maher - Rutgers University, The State University of New Jersey

The objective of this research is to develop and verify by experimentation the optical fiber strain and stress sensor systems for on-line, accurate, and real time monitoring of critical components of structural systems under service load conditions. The application of these sensor systems will permit easier inspection of structures for impending failures.

A series of experiments are to be undertaken on scale models of reinforced and prestressed concrete structural beams, rolled structural steel sections, and beam to column connections.

2. Nondestructive Methods of Assessing Damage in Reinforced Concrete Structures; Drs. Kenneth H. Stokoe, Jose M. Roesset and James O. Jirsa - University of Texas-Austin

The objective of this research is to study nondestructive methods based on the propagation and attenuation of stress waves for evaluating (1) the degree of damage to structural and foundation elements, (2) the location and extent of that damage, and (3) the extent and quality of repair. One of the key thrusts of this research will be the integrated study of both body and surface stress waves. The techniques to be used will be selected in terms of the type of structural elements and their accessibility for use of nondestructive test equipment. The rapidly-developing technique based on surface wave measurements will be used where there is access to an exposed surface of the structural element which has dimensions similar in magnitude to the depth of the element being investigated. The Spectral Analysis of-Surface-Waves (SASW) method will be used.

3. <u>Micromagnetic Surface Studies of Materials for Nondestructive Evaluation</u>, Dr. David Jiles - Iowa State University,
Dr. W. A. Theiner - University of Saarland, Federal Republic of Germany

This project is a joint research effort into the use of magnetic surface studies for nondestructive evaluation between the Center for NDE at Iowa State University and the Fraunhofer Institute for Zerstorungfreie Prufverfahren (NDT) at the University of Saarland in West Germany. The objective of the research is to identify alternating current magnetic measurements which can be made on the surfaces of constructional steel components to evaluate their mechanical condition. Such methods, in which the penetration of the magnetic field is restricted to the surface, will be useful for determining the mechanical state of the surface as a result of shot peening, grinding, surface hardening and residual surface stress. It is also known that in fatigue failure, the fatigue crack initiates at the surface and then propagates into the bulk of the material. Monitoring of the surface condition is therefore critical in identifying the onset of failure in the material as a whole.

4. <u>Fundamentals of Embedded Optical Fibers In Structural Concrete;</u> Drs. K. S. Kim, T. F. Morse - Brown University

The thrust of this project will be to combine the knowledge and expertise of faculty members from solid mechanics, materials science, civil engineering, and optical sensing to obtain a more quantitative description of the response of structural concrete to external loads. By embedding optical fibers in a concrete structure during the fabrication process, it is possible to sense, through a variety of techniques, the local internal state of the structure. This research will aid in the prediction of the behavior of such materials, will point toward directions for the improvement of structural concrete, and can be applied to the retrofitting of existing structures with novel sensors.

NSF SUPPORTED WORKSHOP ON NDE

o "NDE for performance of Civil Structures," Sam Masri (USC), 2/88.

o "International Workshop on Sensors and Measurement Techniques for Assessing Structural Performance" Richard Marshall (NIST), 6/89.

o "Colloquium on Facility Monitoring and for Infrastructure Management" (Building Research Board/NRC).

o "NDE of Construction Materials and Large Structural Systems," (Colorado-Boulder), 10/90.

o "NSF-ATLSS Workshop on Sensor and Signal Processing for Structural Control" (NSF), 2/91.

SUMMARY AND CONCLUSION

An overview of non-destructive evaluation projects at NSF and a discussion on the new NSF initiative and projects on the "Quantitative Nondestructive Evaluation for Large Structural Systems" are presented. The authors hope that this paper will act as a catalyst, sparking interest and further research in quantitative nondestructive evaluation of civil engineering structures. This paper reflects the personal views of the authors, not necessarily those of the National Science Foundation.

SIMULATION OF CONDITIONAL RANDOM FIELDS
—A Basis for Regional Seismic Monitoring for Urban Earthquake Hazards Mitigation—

HIROYUKI KAMEDA, and HITOSHI MORIKAWA

Urban Earthquake Hazard Research Center
Disaster Prevention Research Institute
Kyoto University
Gokasho, Uji, Kyoto 611, Japan

ABSTRACT

A first step of the theory of conditional random fields is developed, which will be a basis for stochastic interpolation of earthquake motions in seismic monitoring of urban regions. First the significance of the development of regional seismic monitoring systems is discussed with some practical examples. On this basis, further research needs that motivate this study are made clear. Next, a preliminary discussion is made on stochastic interpolation of earthquake ground motion, which is a main issue for optimization of a regional seismic monitoring system. The main part of this paper presents a method of simulation of conditional random fields involving deterministic time functions, which is to be used as a major tool for stochastic interpolation of earthquake motions. Closed form solutions are derived for conditional probability density functions for harmonic amplitudes and phase angles of interpolating motions conditioned by a given number of deterministic time functions through coherence functions. Numerical examples for the analytical results as well as those for simulated interpolating motions are presented.

INTRODUCTION

General Remarks

Protection of urban regions against severe earthquakes is a crucial issue. Tremendous developments in seismic hazard assessment, and design and construction method for earthquake resistant structures in the past few decades have made major contributions to making urban regions much safer than before under seismic environments. While these pre-quake measures are of essential importance for upgrading earthquake preparedness of urban regions, we have learned from recent urban earthquakes such as the 1985 Mexico City Earthquake and the 1989 Loma Prieta Earthquake that existing developed urban regions inevitably contain elements with high seismic vulnerability that cannot be covered directly by those achievements of modern earthquake engineering. Buildings constructed under old design codes, many non-engineered small buildings that are subject to diverse levels of workmanship are constantly threats to human lives. For lifeline systems that constitute large regional networks of underground pipelines, poor soil conditions which are truly hazardous under strong earthquake shaking are often unavoidable.

This notion of the problem already indicates that pre-quake measures, as important it is, are not capable of solving all issues of comprehensive seismic protection of urban regions. Importance of emergency operations is then recognized as a key issue for safety in intra-quake (during the earthquake shaking) and post-quake periods. Emergency operation problems are related to search and rescue, information management under emergency, physical and functional restoration of lifeline systems, rehabilitation of damaged residential areas, etc., that are to be continued on to post-earthquake reconstruction. This is an area where close research cooperation should be made by experts in engineering and social sciences. It should be emphasized that much more contributions can be made from engineering fields to the improvement of emergency response capability of urban regions. In this regard, regional seismic monitoring which has motivated the study presented in this paper is one of promising concepts for future engineering developments in this area.

Basic Concepts of Regional Seismic Monitoring

The concept of regional seismic monitoring is closely connected with information management and automatic system control under earthquake emergency. It is a new and practically important issue to make quick and accurate estimation of damage over an entire urban region immediately after an earthquake. This is particularly important as basic information for effective organization of search and rescue activities, and efficient management of restoration works for lifeline systems. It is also a useful measure for detecting strong seismic motions before they reach urban regions and send alarm information so that automatic shut-down systems are operated at critical facilities such as chemical plants, high-speed railways, etc.

Typical systems for regional seismic monitoring can be realized by installing seismometers at various appropriate locations of urban regions, and make it an automatic telemetered system combined with data processing packages to generate useful information required for emergency response activities and other system operations. Some of such systems already exist or going to be realized for emergency management of lifeline systems as explained in the following. It will be seen that the part of data processing is very important and what type of intelligence it should be equipped with depends on its purpose. It is, therefore, required to establish a technology of regional seismic monitoring systems that can be used widely for urban earthquake hazards reduction through a comprehensive development regarding its physical equipment and optimum layout as well as its level of intelligence. Research needs and their orientations are being discussed by some Japanese researchers [1),2),3)] .

Examples of Existing Regional Monitoring Systems

The two monitoring systems introduced below have been chosen as they can be regarded as "regional" systems as well as "intelligent" systems. Both of them have been realized for a limited use for specific lifeline systems. However, there is a possibility that they will be prototype systems for further development of more comprehensive seismic protection systems for urban regions.

Gas supply system disaster control (Tokyo Gas Co.[4)]) — Fig.1 schematically illustrates the disaster control system built in for the city gas supply network of Tokyo Gas Co. It has thirty-one seismometers installed at governor stations within its service area extending over the Tokyo Metropolitan Region. Peak ground accelerations and their spectrum intensities are telemetered to the Headquarters Supply Center, where decision is made if remote-controlled service shut-down should be operated for safety purposes.

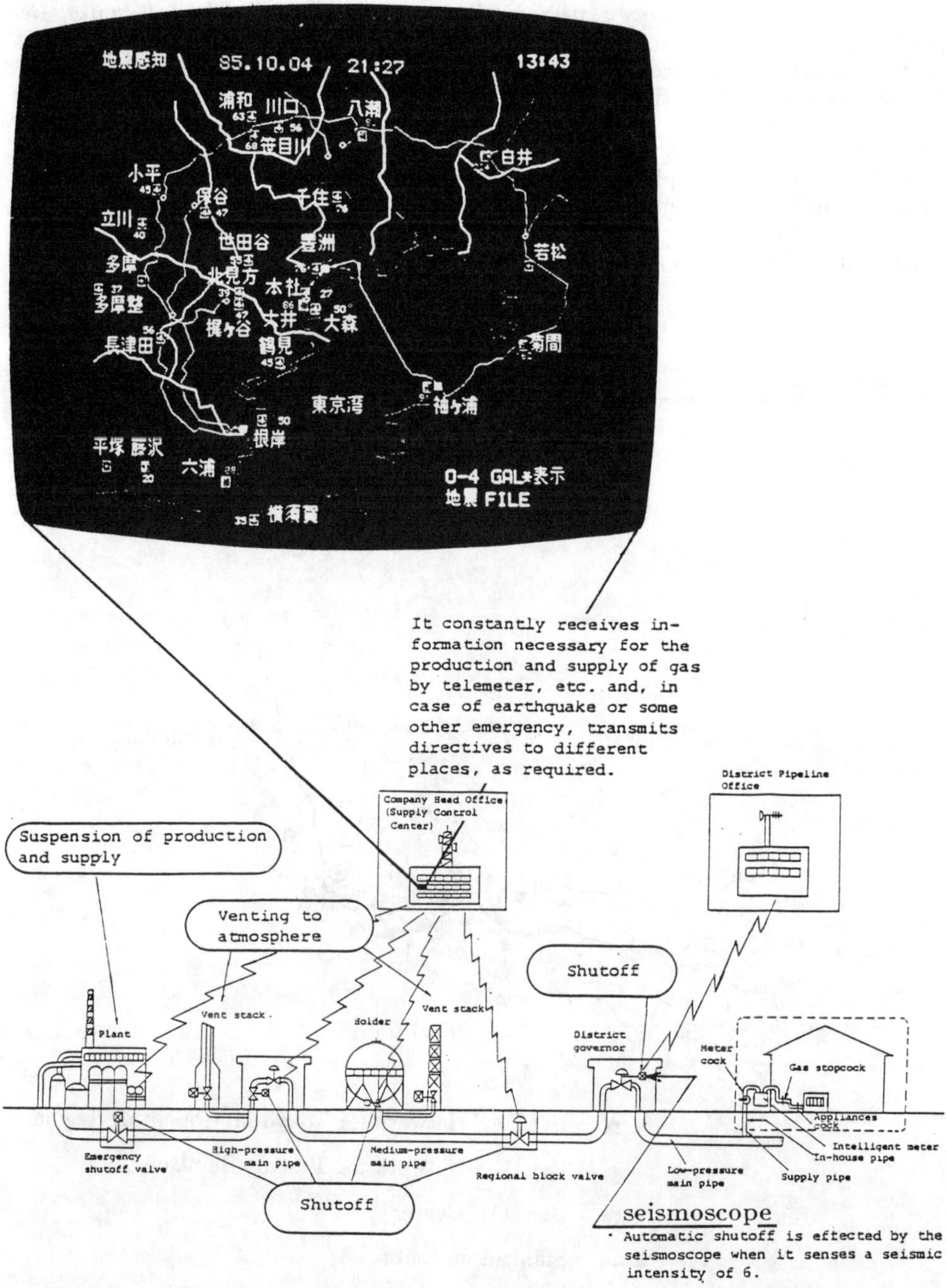

Fig.1 Earthquake monitoring and emergency operation of a city gas supply system (Tokyo Gas Co.[4])

Intelligent earthquake warning system for high-speed railways (Railway Technical Research Institute[5),6)]*)* — Besides conventional earthquake warning seismometer systems which cover main railway routes in Japan, a new earthquake warning system UrEDAS (Urgent Earthquake Detection and Alarm System) has been developed by the Railway Technical Institute, Japan. It is capable of identifying the magnitude and location of earthquakes detected at its intelligent seismometer stations (UrEDAS stations) by data processing of the P-wave phase record. This information is telemetered to the control center and judgment is automatically made and emergency brakes are applied to all trains operated in the affected zone before the strong motion phase reaches. Locations of the existing UrEDAS stations are shown in Fig.2.

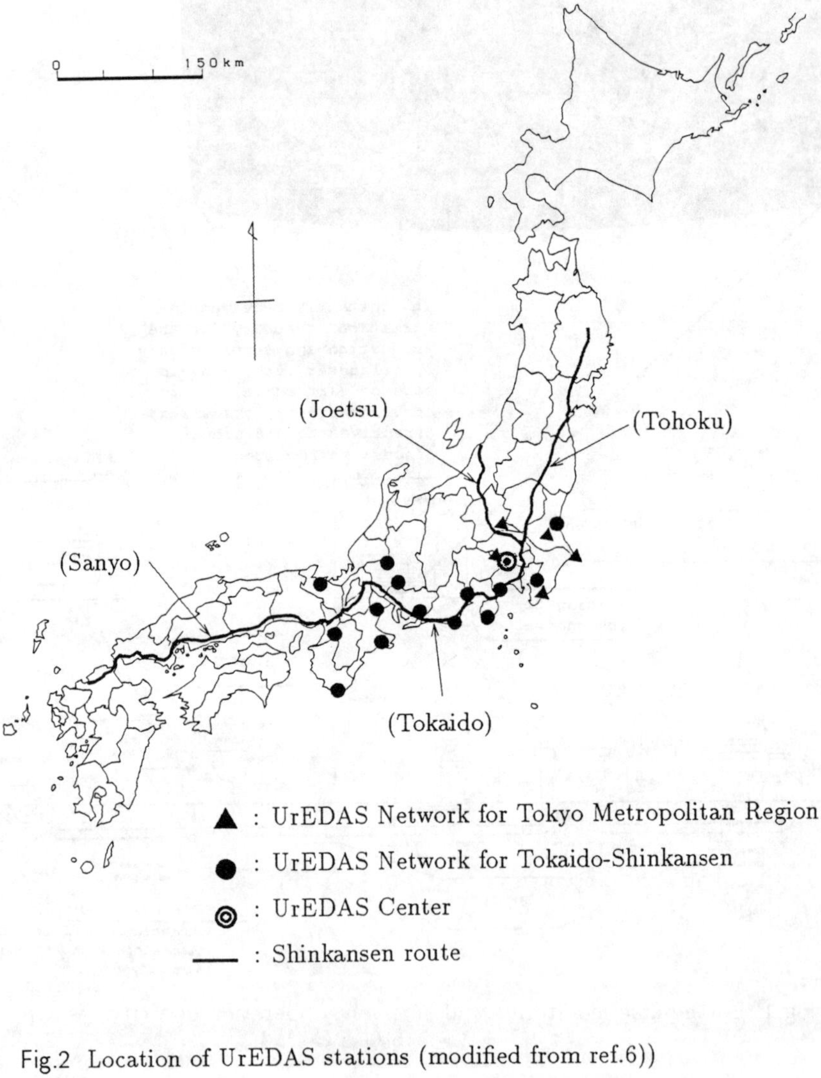

Fig.2 Location of UrEDAS stations (modified from ref.6))

Scope of This Study
As previously mentioned, regional seismic monitoring systems are used mainly to aid quick (sometimes automatic) emergency operations immediately following (or at the moment of) earthquake occurrence to save urban regions from catastrophic effects of the earthquake. Fig.3 shows the role of such systems in the earthquake operations for lifeline systems[3] arranged in a logarithmic time scale.

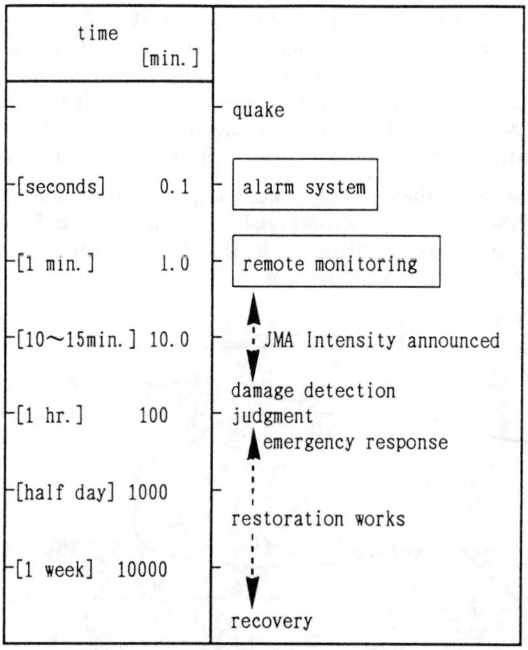

Fig.3 Roles of regional seismic monitoring systems in intra- and post-earthquake operations of lifeline systems (translated from Ref.3))

It should be noted that many research efforts are needed to establish methodologies for optimum system development under financial constraint. Takada[3] made preliminary discussion on the factors to affect the locations selection of monitoring seismometer stations for lifeline systems. In any case, it is required that the distribution of ground motions over a region must be estimated through interpolation from actual ground motions recorded at discrete monitoring stations. It should also be noted that this type of estimation involves various sources of uncertainty. For these reasons, it has been decided in this study to approach the general problem of regional seismic monitoring from the aspect of stochastic interpolation of earthquake ground motions. The theory of conditional random field presented in the main part of this paper has been developed as one of major tools to be used for this purpose.

PRELIMINARY DISCUSSION OF STOCHASTIC INTERPOLATION OF EARTHQUAKE GROUND MOTIONS

For realizing an optimum regional seismic monitoring system under a financial constraint, an effective method should be developed for interpolation of ground motions in space on the basis of the records obtained at discrete locations. The method should also be capable of quantitative assessment of estimation errors caused by various sources of uncertainty: this type of information is needed in risk-cost-benefit optimization of the system. For these reasons, the subject is addressed herein in terms of stochastic interpolation of earthquake ground motions.

The stochastic interpolation of earthquake ground motion should be dealt with from two aspects; i.e., (i) physically, it is a problem of seismic wave propagation over an area where known records are available at discrete locations, and (ii) from a view point of decision under uncertainties coming from imperfect site information and random nature of the seismic wave itself, it is a problem of conditional random field involving deterministic time functions at discrete locations. Schematical illustration is given in Fig.4. A theoretical basis should be laid for the development of optimum monitoring systems by incorporating research outputs from these two problem areas.

This paper presents a method of simulation of conditional random field which was developed as a first step of comprehensive studies of stochastic interpolation of earthquake ground motions whose ultimate goal is to contribute to the development of optimum regional seismic monitoring systems.

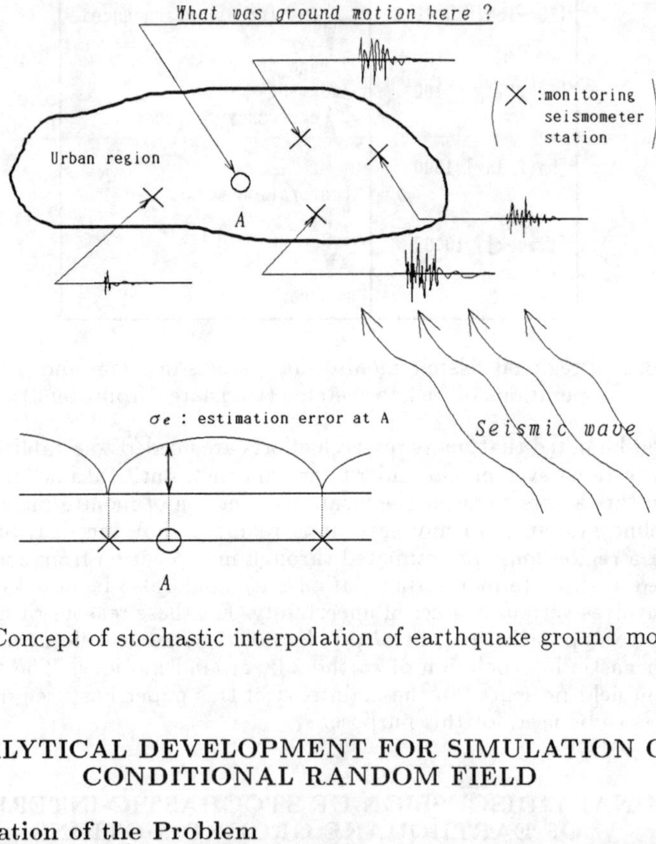

Fig.4 Concept of stochastic interpolation of earthquake ground motions

ANALYTICAL DEVELOPMENT FOR SIMULATION OF CONDITIONAL RANDOM FIELD

Basic Formulation of the Problem

Theory of random fields[7),8)] is an established area with various extended developments and applications including earthquake motions[9)–11)]. Stochastic interpolation of earthquake ground motions requires its extension to conditional random fields. A method of simulating interpolating earthquake ground motions was recently proposed by Kawakami [12]. With his own engineering view, Kawakami's method is a deterministic method based on optimization of harmonic amplitudes of interpolating motions under the constraint of given deterministic motions and coherency. As it is essential to develop a method of "stochastic" interpolation from the foregoing discussion, this study deals with the problem on a probability basis.

Under the condition that deterministic time functions $u_1(t), u_2(t), \cdots, u_{n-1}(t)$ are obtained (recorded) at *n-1* locations, a method is developed for simulating a random time function $U_n(t)$ at the *n*-th location. This corresponds to a problem of estimating a random earthquake ground motion at a specified site on the basis of *n-1* records obtained at separate monitoring seismometer stations.

As a first step of the study, stationary random processes are dealt with herein. The deterministic functions $u_1(t), u_2(t), \cdots, u_{n-1}(t)$ are treated as realizations, respectively, of random processes $U_1(t), U_2(t), \cdots, U_{n-1}(t)$. It is assumed that the power spectral densities of $U_1(t), U_2(t), \cdots, U_n(t)$ and coherence functions among them are specified.

The development is achieved by deriving the probability density functions of harmonic amplitudes and phase angles of $U_n(t)$ conditioned by those for realized $u_1(t), u_2(t), \cdots, u_{n-1}(t)$. Closed form solutions for these functions are derived analytically. On this basis, the simulation of $U_n(t)$ is made possible.

Conditional Probability Distribution of Harmonic Amplitudes and Phase Angles of $U_n(t)$

Let $U_n(t)$ be represented by $U_n(t \mid u_1(t), u_2(t), \cdots, u_{n-1}(t))$ in the sense that it is a random process conditioned by the deterministic time functions $u_1(t), u_2(t), \cdots, u_{n-1}(t)$. Then we have[13]

$$U_n(t \mid u_i(t)\ ;\ i = 1, 2, \ldots, n-1) = \sum_k C_{nk} \cos(\omega_k t - \Phi_{nk}) \tag{1}$$

in which C_{nk} and Φ_{nk} are, respectively, random amplitude and random phase angle at frequency ω_k.

The deterministic time functions $u_i(t)\ ;\ i = 1, 2, \ldots, n-1$ are represented in a similar manner as

$$u_i(t) = \sum_k \tilde{c}_{ik} \cos(\omega_k t - \tilde{\phi}_{ik})\ ;\ i = 1, 2, \ldots, n-1 \tag{2}$$

in which $\tilde{c}_{ik}$ and $\tilde{\phi}_{ik}$ are regarded as realized values, respectively, of random amplitude C_{ik} and random phase angle Φ_{ik} at frequency ω_k corresponding to a random process $U_i(t)$.

Then the conditional probability density function of C_{nk} and that of Φ_{nk} for the interpolating random process, $U_n(t \mid u_i(t)\ ;\ i = 1, 2, \ldots, n-1)$ are analytically derived as[13]

$$\begin{aligned} & f_{C_{nk}|C_{1k},\Phi_{1k},\cdots,C_{n-1\ k},\Phi_{n-1\ k}}(c_{nk} \mid \tilde{c}_{1k}, \tilde{\phi}_{1k}, \cdots, \tilde{c}_{n-1\ k}, \tilde{\phi}_{n-1\ k}) \\ & = c_{nk} A_{k\ 2n-1\ 2n-1} \cdot \mathrm{I}_0\left(c_{nk}\sqrt{\tilde{P}_k^2 + \tilde{Q}_k^2}\right) \\ & \cdot \exp\left[-\frac{1}{2}\left(c_{nk}^2 A_{k\ 2n-1\ 2n-1} + \frac{\tilde{P}_k^2 + \tilde{Q}_k^2}{A_{k\ 2n-1\ 2n-1}}\right)\right] \end{aligned} \tag{3}$$

$$f_{\Phi_{nk}|C_{1k},\Phi_{1k},\cdots,C_{n-1\ k},\Phi_{n-1\ k},C_{nk}}(\phi_{nk}\mid \tilde{c}_{1k},\tilde{\phi}_{1k},\cdots,\tilde{c}_{n-1\ k},\tilde{\phi}_{n-1\ k},\tilde{c}_{nk})$$
$$=\frac{1}{2\pi}\frac{1}{\mathrm{I}_0\left(\tilde{c}_{nk}\sqrt{\tilde{P}_k^2+\tilde{Q}_k^2}\right)}\cdot\exp\left[-\tilde{c}_{nk}R_k\right] \tag{4}$$

in which

I_0 : 0-th Bessel function of an imaginary argument of first kind

$$\tilde{P}_k=\sum_{i=1}^{n-1}\tilde{c}_{ik}\sqrt{\mathcal{S}_k}\cos\mathcal{T}_k,\quad \tilde{Q}_k=\sum_{i=1}^{n-1}\tilde{c}_{ik}\sqrt{\mathcal{S}_k}\sin\mathcal{T}_k,\quad R_k=\sum_{i=1}^{n-1}\tilde{c}_{ik}\sqrt{\mathcal{S}_k}\cos(\mathcal{T}_k-\phi_{nk})$$

$\mathcal{S}_k=A^2_{k\ 2i-1\ 2n-1}+A^2_{k\ 2i-1\ 2n},\ \mathcal{T}_k=\tilde{\phi}_{ik}+\theta_{ink}$
$\mathbf{\Lambda}_k^{-1}=[A_{kij}]$
$\mathbf{\Lambda}_k$: covariance matrix of $C_{ik},\Phi_{ik}\ (i=1,\ldots,n)$

It can be proved that if $U_1(t),U_2(t),\cdots,U_{n-1}(t)$ are Gaussian processes $U_n(t\mid u_i(t)\ ;\ i=1,2,\ldots,n)$ is also Gaussian. It is also proved that in a special case of no coherency between the motion at the n-th location and the motion at any other locations Eqs.(3) and (4) coincide with the Rayleigh distribution and the uniform distribution, respectively.

Numerical Results

For numerical illustration, the values for Eqs.(3) and (4) have been obtained. The form of power spectrum and that of coherency have been employed from Refs.14) and 15). Their governing parameters are: T_P = predominant period, $f_P=(1/T_P)$ = predominant frequency, v = mean speed of wave propagation, x_0 = distance between discrete locations, and α = distortion constant ($\alpha=0$: perfect coherency, $\alpha\to\infty$: no coherency).

Fig.5 shows the conditional probability density function of harmonic amplitudes for a case of $U_2(t\mid u_i(t)\ ;\ i=1)$; i.e., a case of a motion conditioned by a single deterministic motion $u_1(t)$. In this figure, μ stands for a mean value. Likewise, Fig.6 shows the conditional distribution of harmonic phase angles in which $P_2-\bar{P}_1-\theta_{12}$ stands for $\Phi_2-\tilde{\phi}_1-\theta_{12}$, where $\tilde{\phi}_1$ = deterministic value for Φ_1 , and $\theta_{12}=2\pi\cdot f_P\cdot x_0/v$ = mean phase lag between locations 1 and 2 at frequency f_P .

Fig.7 shows the conditional mean values of the harmonic amplitude C_2 for various values of conditioning amplitude $\tilde{c}_1$ as plotted against the distance between the two locations. Figs.8 and 9 show similar plots of the standard deviations of the harmonic amplitude C_2 and the phase angle Φ_2 , respectively. Observe that these values converge to the cases of no coherency (shown by dotted lines) as the distance x_0 increases, at various rates depending on the conditioning amplitude $\tilde{c}_1$, frequency f , distortion constant α , etc.

DIGITAL SIMULATION OF INTERPOLATING MOTIONS

Random numbers that conform with Eqs.(3) and (4), are generated, and are used as sample values for C_{nk} and Φ_{nk} in Eq.(1). Then the resulting time functions can be regarded as samples from interpolating stochastic motions at location n conditioned by the deterministic motions at the other *n-1* locations. Examples of such motions are shown in Figs.10–12.

Fig.10 shows the effect of coherency, denoted by the distortion constant α , for a case conditioned by a single deterministic time function. For the case of $\alpha=0$, perfect coherency, all simulated motions are identical with the given deterministic motion except for time shifts due to phase lag. The simulated motions get more independent with increasing values of α and distance x_0 .

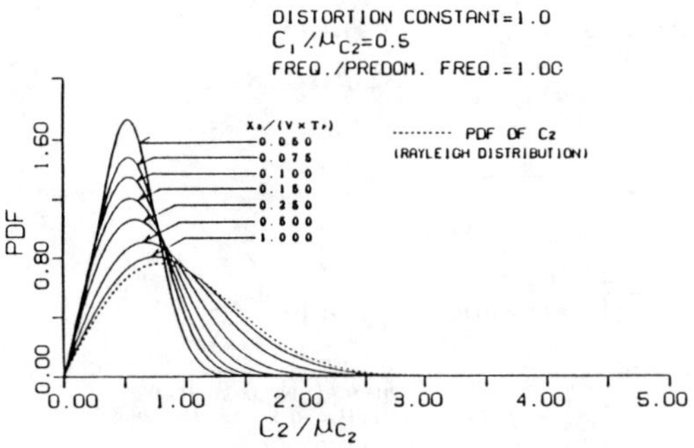

(a) $\tilde{c}_1/\mu_{C_2} = 0.5$

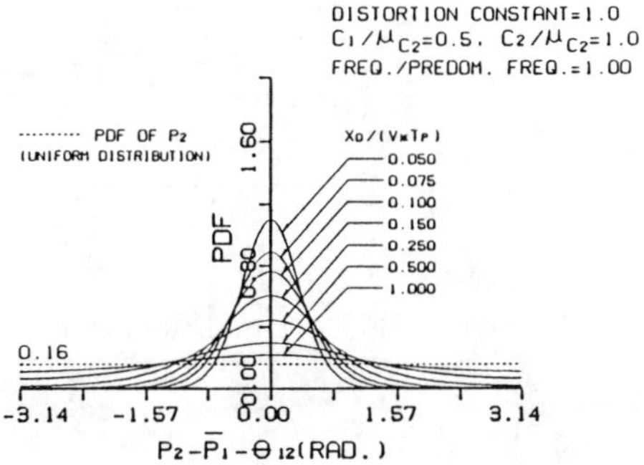

(a) $\tilde{c}_1/\mu_{C_2} = 0.5$

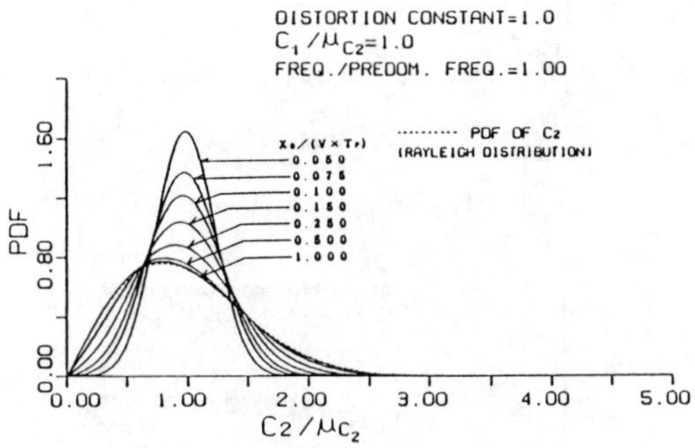

(b) $\tilde{c}_1/\mu_{C_2} = 1.0$

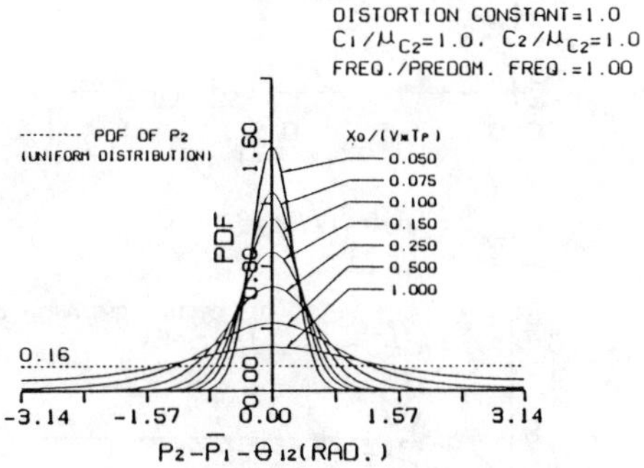

(b) $\tilde{c}_1/\mu_{C_2} = 1.0$

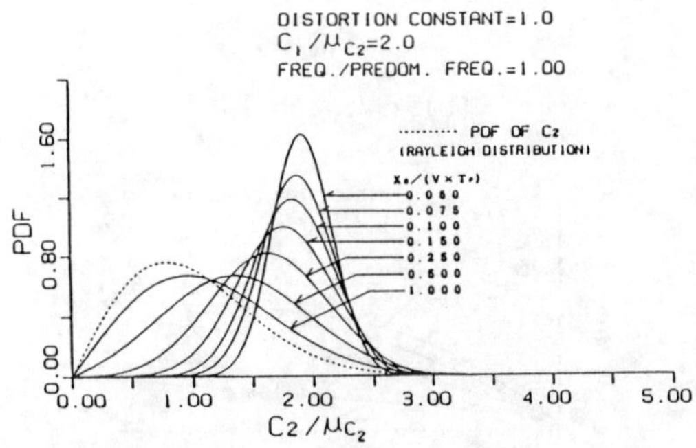

(c) $\tilde{c}_1/\mu_{C_2} = 2.0$

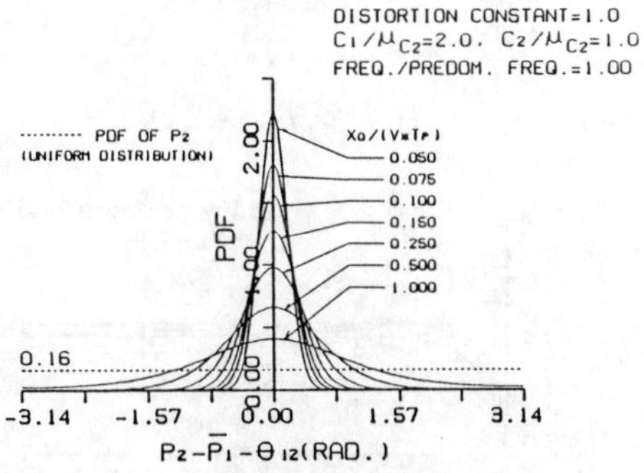

(c) $\tilde{c}_1/\mu_{C_2} = 2.0$

Fig.5 Conditional probability density function of harmonic amplitude

Fig.6 Conditional probability density function of harmonic phase angle

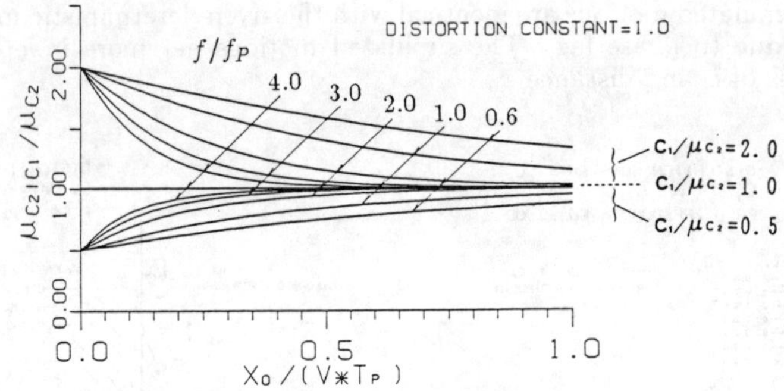

Fig.7 Conditional mean value of harmonic amplitude

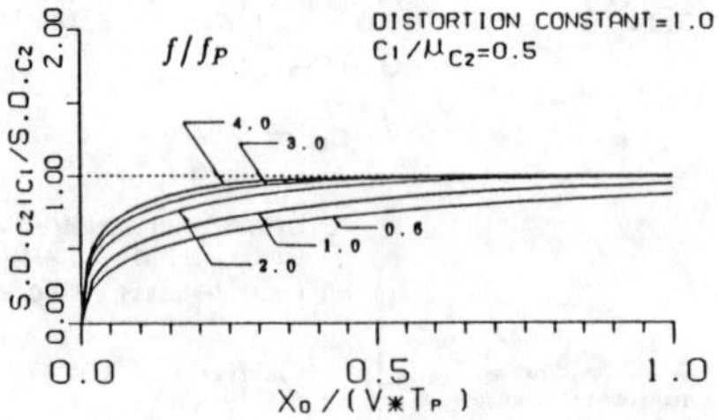

(a) $\tilde{c}_1/\mu_{C_2} = 0.5$

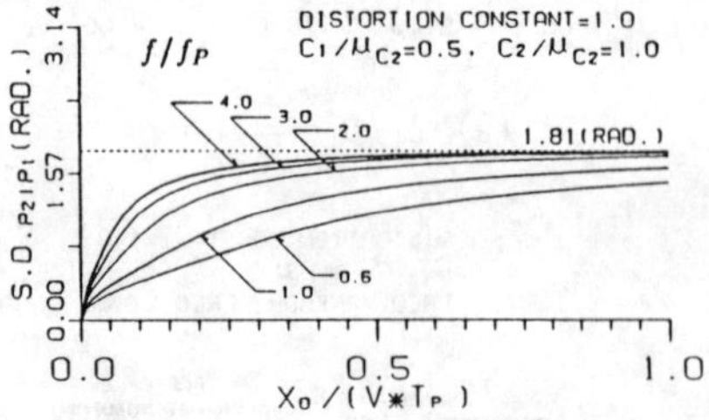

(a) $\tilde{c}_1/\mu_{C_2} = 0.5$

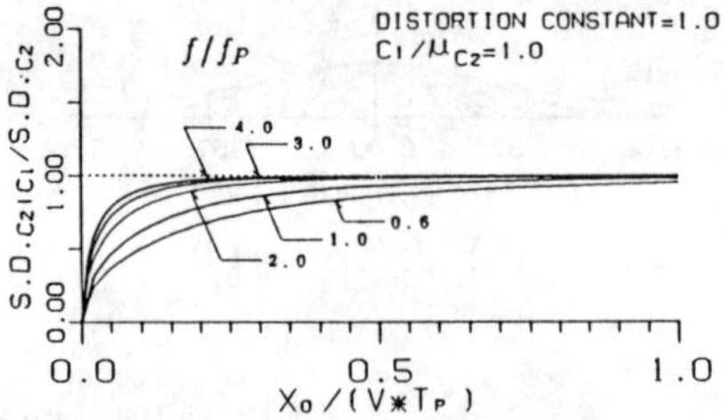

(b) $\tilde{c}_1/\mu_{C_2} = 1.0$

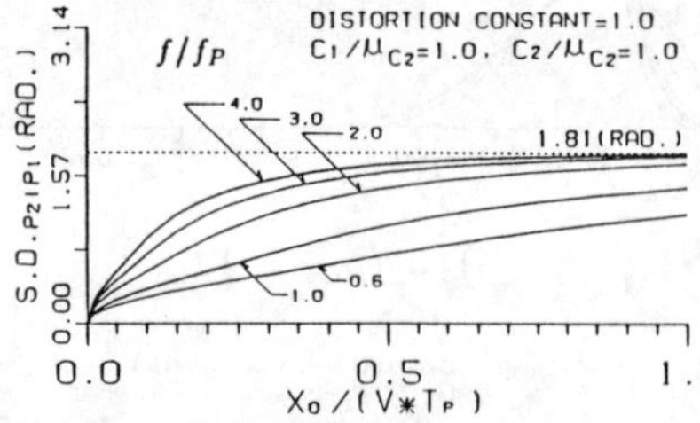

(b) $\tilde{c}_1/\mu_{C_2} = 1.0$

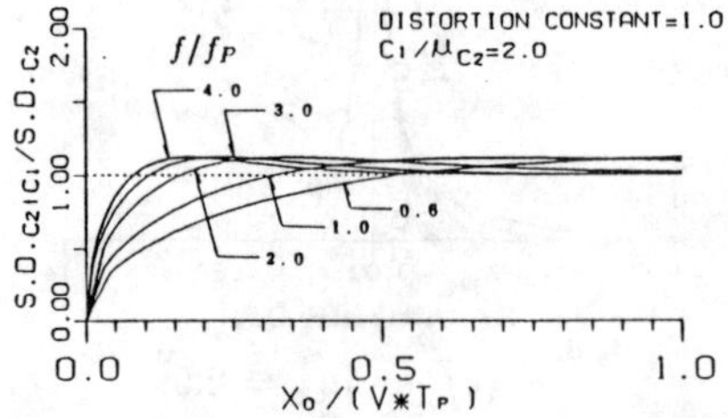

(c) $\tilde{c}_1/\mu_{C_2} = 2.0$

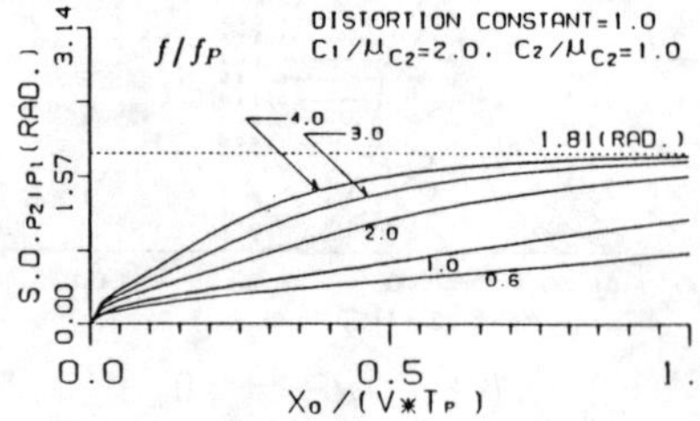

(c) $\tilde{c}_1/\mu_{C_2} = 2.0$

Fig.8 Conditinal standard deviation of harmonic amplitude

Fig.9 Conditional standard deviation of harmonic phase angle

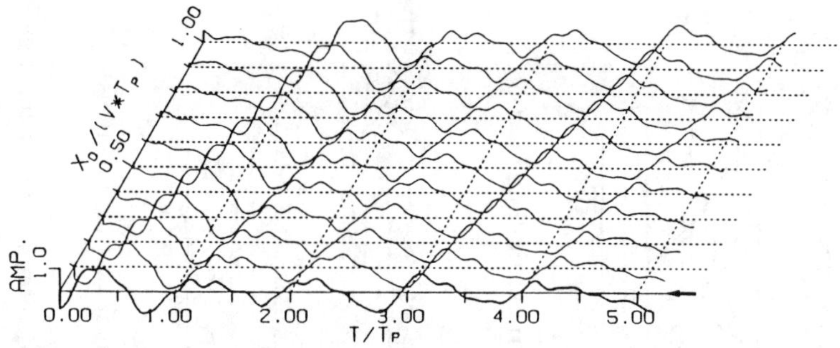

(a) distortion constant $\alpha = 0.0$

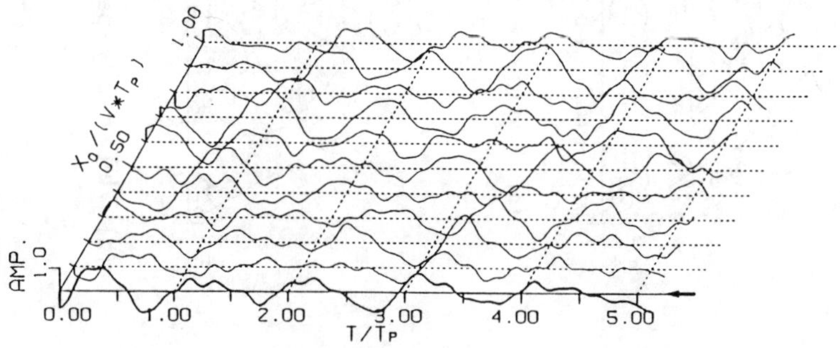

(b) distortion constant $\alpha = 1.0$

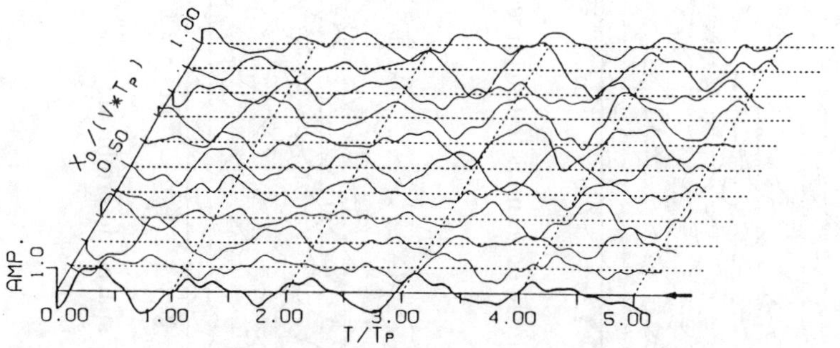

(c) distortion constant $\alpha = 2.0$

deterministic time function
simulated time functions

Fig.10 Simulated time functions conditioned by single deterministic time functions

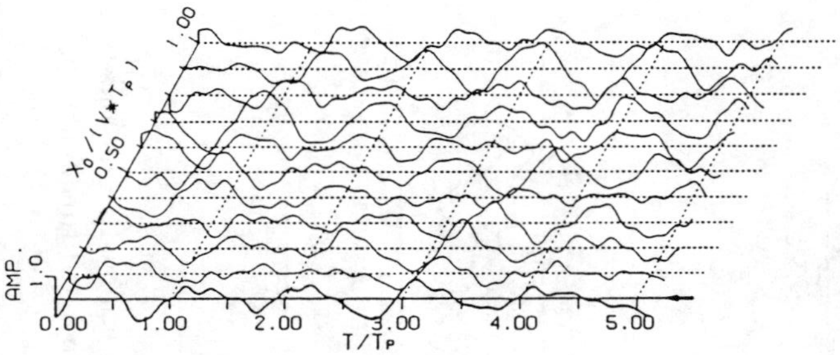

(a) Single deterministic time function

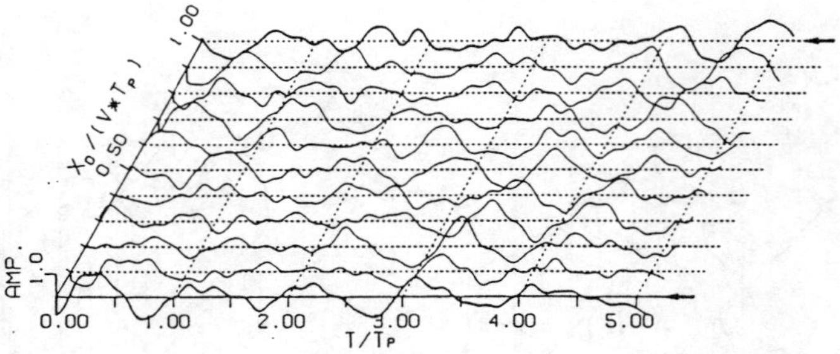

(c) Two widely spaced deterministic time functions

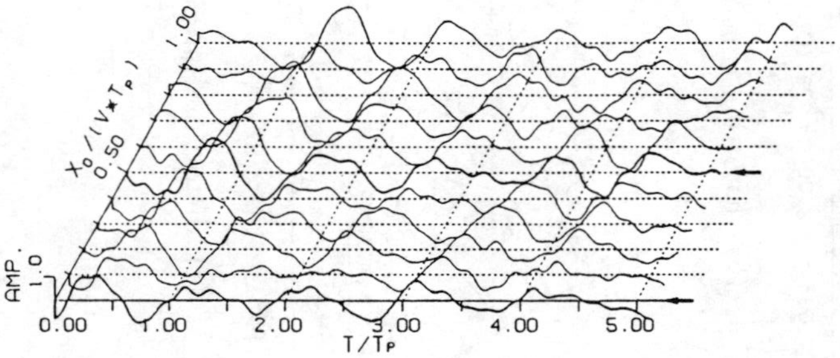

(b) Two closely spaced deterministic time functions

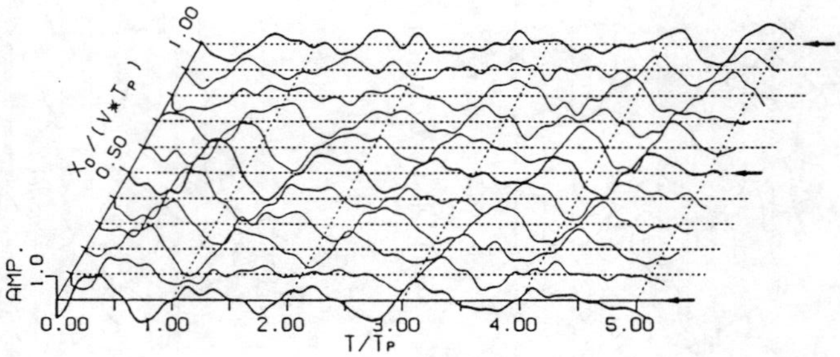

(d) Three closely spaced deterministic time functions

deterministic time function
simulated time functions

Fig.11 Simulated time functions conditioned by one, two, or three deterministic time functions (distortion constant $\alpha = 1.0$)

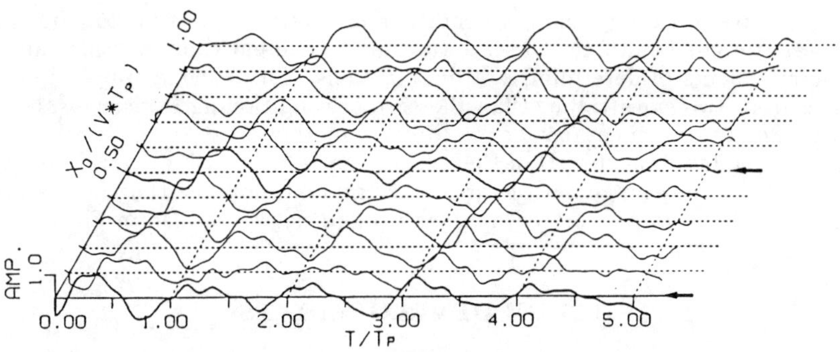

(a) Two closely spaced deterministic time functions

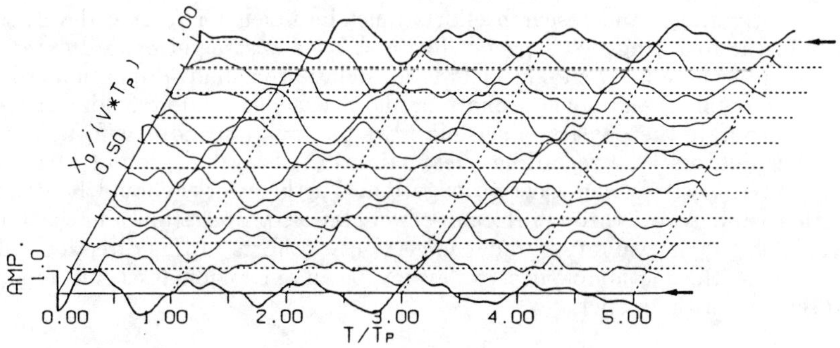

(b) Two widely spaced deterministic time functions

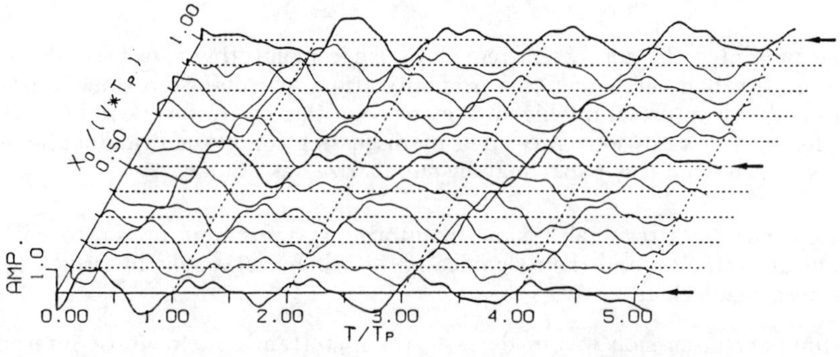

(c) Three closely spaced deterministic time functions

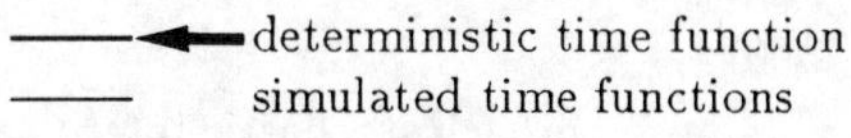

Fig.12 Simulated time functions conditioned by two or three identical deterministic time functions with phase lag (distortion constant $\alpha = 1.0$)

Fig.11 compares various cases of conditioning deterministic time functions whose number varies from one to three. Observe that more coherent time functions are generated for closer spacing of deterministic time functions. The same is true in the case of conditioning from two sides rather than the case of conditioning from one side. While Fig.11 used different time functions for conditioning motions at separate locations, Fig.12 uses identical time functions for all of these locations. Although Fig.12 will be less realistic than Fig.11 in the view point of earthquake ground motions, the above statement regarding the coherency of simulated motions is more clearly verified in Fig.12.

FUTURE DEVELOPMENTS

The results presented herein demonstrate reasonable properties that should characterize stochastic interpolation of earthquake ground motions. It is noted, however, the study is still in an early stage, and research efforts must be taken for further developments, particularly in the following two aspects. The first issue is straightforward extension of the work presented herein. It should include formulation in multi-dimensional domain, first-passage probability, and nonstationary nonhomogeneous random fields. The second issue is to incorporate the physics of seismic wave propagation. Some efforts have been taken by other authors[11),16)] regarding shear vibration of layered ground with random media and source to site shear wave propagation. Further works should be conducted to deal with seismic waves through irregular boundaries, propagation and scattering of surface waves and their evaluation under uncertainty. Efforts in this direction will contribute greatly to the methodologies to develop optimum regional seismic monitoring systems discussed in this paper.

CONCLUSIONS

This paper presented the concept of regional seismic monitoring as a feasible engineering measure for urban earthquake hazards mitigation, and discussed some issues to be covered to make them useful intelligent systems. On this basis, a work to lay a theoretical basis for such systems were presented by discussing conditional random fields. The conclusions drawn from this paper may be summarized as follows.

(1) Basic concepts of regional seismic monitoring systems were presented with some practical examples, and their significance in urban earthquake hazards mitigation has been made clear.

(2) Preliminary discussion was made as to the importance of stochastic interpolation of earthquake ground motions in optimum planning and design of regional seismic monitoring systems. Key issues to be studied for its development were also discussed.

(3) As a step to approach stochastic interpolation of earthquake ground motions from probabilistic aspects, basic formulation of conditional random field was presented. It was achieved by deriving conditional probability distributions of random harmonic amplitudes and random phase angles of the interpolating motion at a certain location, conditioned by given time functions specified for other separate locations.

(4) Numerical results were presented for these developments, which demonstrated the appropriateness of the work. On this basis, items for future studies were indicated.

REFERENCES

1. Kameda, H., "Seismic safety of modern urban regions," *Journal of the Japan Society of Civil Engineers*, Vol.74, No.6E, June 1989, pp.26–28, (in Japanese).
2. Ohta, Y., and Okada, S., "Value-added strong motion observation network, —Proposal and its significance—," *Proc. 26th Natural Disasters Science Symposium*, Sep. 1989, pp.42–43, (in Japanese).
3. Takada, S., Takaya, T., Fuke, T., and Fukui, S., "An estimating method of predicted seismic ground motions and an arrangement of remote monitoring seismographs for early detection of earthquake damages," *Proc. 21th JSCE Earthquake Engineering Symposium-1991*, Earthquake Engineering Committee, JSCE, to be held in July 1991, (in Japanese).
4. Supply Control Center, Tokyo Gas Co., "Disaster control," June 1991.
5. Nakamura, Y., "On the urgent earthquake detection and alarm system (UrEDAS)," *Proc. 9th World Conference on Earthquake Engineering*, Vol.VII, Aug. 1988, pp.673–678.
6. Track & Structures Research Group, Railway Technical Research Institute, "UrEDAS for high speed railways," *Track and Structures Research News*, No.8, Feb. 1991, pp.1–2, (in Japanese).
7. Vanmarcke, E., *Random Fields*, MIT Press, 1983.
8. Shinozuka, M., "Stochastic fields and their digital simulation," *Stochastic Methods in Structural Mechanics*, Hermes Publishing Co.
9. Harada, T., and Shinozuka, M., "Stochastic analysis of ground response variability for seismic design of buried lifeline structures," *Proc. 7th Japan Earthquake Engineering Symposium*, Dec. 1986, pp.595–600.
10. Deodatis, G., and Shinozuka, M., "Digital simulation of seismic ground motion using stochastic wave theory," *Proc. 9th World Conference on Earthquake Engineering*, Vol.II, Aug. 1988, pp.801–806.
11. Harada, T., and Fugasa, T., "Seismic response characteristics of stochastic ground," *Proc. 8th Japan Earthquake Engineering Symposium*, Dec. 1990, pp.427–432.
12. Kawakami, E., "Simulation of space-time variation of earthquake ground motion including a recorded time history," *Structural Engineering/Earthquake Engineering*, Proc. JSCE, No.410/I-12, Oct. 1989, pp.435–443, (in Japanese).
13. Morikawa, H., and Kameda, H., "Simulation of conditional random fields involving given time functions," *UEHR Report*, No.E8, Urban Earthquake Hazard Research Center, DPRI, Kyoto Univ., April 1991, (in Japanese).
14. Kameda, H., "Engineering application of stochastic earthquake motion models with non-linear soil amplification," *Trans. 9th SMiRT*, Vol.A, Aug. 1987, pp.327–336.
15. Kawakami, H., and Sato, Y., "Effect of distortion of seismic waves on ground strain," *Proc. 9th World Conference on Earthquake Engineering*, Vol.II, Aug. 1988, pp.477–482.
16. Deodatis, G., Shinozuka, M., and Papageorgiou, A., "Frequency-wave number spectra from seismic sources in a half-space," *Proc. 9th World Conference on Earthquake Engineering*, Vol.I, Aug. 1988, pp.447–456.

SEISMIC DAMAGE AND DAMAGE-CONTROL DESIGN FOR RC FRAMES

MASANOBU SHINOZUKA
AHMET S. CAKMAK

Department of Civil Engineering and Operations Research
Princeton University
Princeton, NJ 08544 USA

ABSTRACT

A review of three seismic damage models, that were more recently developed, is performed. The first damage model derives a local damage index in the form of a modified Miner's rule. This index is then used to develop the automated damage-control design procedure so that the design, while satisfying the necessary building design code, is optimized in the sense that it spatially distributes the damage as uniformly as possible. The second damage model considers maximum deformation and absorbed hysteretic energy. The local damage index of this model is a simple linear combination of maximum displacement and total hysteretic energy dissipation. On the other hand, the third damage model derives a global damage index base on the evolution of the natural period of a time-varying linear system equivalent to nonlinear system for a series of non-overlapping time windows.

INTRODUCTION

In the classical structural design for static loads, the structural safety is attained by keeping the stresses well below the material yield limit. This simple definition of safety is too conservative for the evaluation of the state of a structural system after earthquakes. Even moderate earthquakes may produce yielding in some of the structural members without producing a dangerous situation. Since inception of the earthquake engineering practice both the expected performance of structures subjected to earthquakes and the state of damage after actual seismic events have been characterized using indicators other than the stress level. The width and distribution of cracks, the interstory or global drift and ductility ratio have been used as engineering measures of seismic damage.

The maximum interstory drift has been shown to be a good measure of the extent of nonstructural damage such as those for cladding, partition, etc. [1]. On the other hand, the ductility ratio (maximum to yielding displacement), reflects the extent of inelastic deformation in the structure [2]. Other measures of damage which are functions of the maximum displacement have also been introduced [3–5]. As these measures are extremes or functions of extremes of the response, their statistics for a given excitation can be evaluated using known analytical procedures such as those based on the random vibration method or determined empirically using response time histories or response spectra.

That the above measures do not consider the effect of cumulative damage indicate that application would be limited to situations where such effects are unimportant. Recent models mostly aim at including these cumulative effects. The effect of cumulative damage to RC members and structures subjected to seismic and other loads has been studied by many researchers. A large number of analytical models that intend to describe measures of damage one way or another, were proposed for steel and RC structures, for example, by Yao and Munse [6], Stephen and Yao [7], Banon et al. [8–10], Gosain et al. [11], Hwang and Scribner [12], Darwin and Nmai [13], Lybas and Sozen [14], Bertero and Bresler [15], Blejwas and Bresler [15], Roufaiel and Meyer [16], Park et al. [18,19], Mizuhata et al. [20], Chung et al. [21], Casciati and Faravelli [22] and DiPasquale and Cakmak [23]. The reader is referred to Chung et al. [24] for a comprehensive review of these and other damage models.

In this paper, (1) the two damage models developed by Park et al. [18,19] and Chung et al. [21] and the global damage model introduced by DiPasquale and Cakmak [23] are reviewed, and (2) a damage control design based on the work by Chung et al [21,24] is demonstrated with an example.

MODEL BASED ON MODIFIED MINER'S RULE

The material presented in this section was primarily taken from Chung et al [25], where the study by Chung et al [21,24] was further refined utilizing the experimental results produced by Fujikake et al [26] so that, for example, the model for the pinching effect more realistically reflects the empirical evidence.

Under load reversals, the stiffness of a reinforced concrete member experiences progressive reduction due to concrete cracking and bond deterioration of the steel-concrete interface. In addition, reinforced concrete members experience strength degradation under cyclic loading beyond the yield load level. Figure 1 shows this hysteretic model characterized by five different kinds of branches so as to represent various hysteretic behaviors of reinforced concrete members during cyclic loading.

Meanwhile, when load reversal occurs within the elastic range in the presence of high shear force, open shear cracks will initially permit the transfer of shear cracks through the dowel bar, leading to rather low stiffness. After closing such cracks, however, aggregate interlock and shear friction cause a significant increase in member stiffness. To reflect this behavior caused by shear force, the point (M_n^+, ϕ_n^+), will be introduced with the following coordinates:

$$\phi_n^+ = \phi_r^- \frac{\overline{(EI)}}{\overline{(EI)} - (EI)_e} \qquad\qquad M_n^+ = (EI)_e \phi_n^+ \tag{1}$$

where

$$\overline{(EI)} = \frac{\overline{M}_x}{\overline{\phi}_x^+ - \phi_r^-} \tag{2}$$

The coordinates of the crack-closing point can be expressed as [24]

$$M_p^+ = \alpha_p M_n^+ \qquad\qquad \phi_p^+ = \alpha_p \phi_n^+ \tag{3}$$

where

$$\alpha_p = \begin{cases} 0 & \text{if } \frac{a}{d} \le 1.5 \\ \sqrt{0.4\frac{a}{d} - 0.6} & \text{if } 1.5 < \frac{a}{d} \le 4.0 \\ 1 & \text{if } \frac{a}{d} > 4.0 \end{cases}$$

and $\frac{a}{d}$ = shear span ratio, $a = \frac{l}{2}$ in general, and l = clear span length. A "pinching factor" α_p is such that $\alpha_p = 1$, if the shear effect is negligible, and $\alpha_p = 0$ if the shear effect completely controls the load-deformation behavior.

The extent of damage at each node is often expressed in terms of a damage index, which remains zero as long as the member's yield capacity is not exceeded, and is assigned a value of 1.0 when the member fails. This representation thus requires a definition of failure and a mechanism of strength deterioration for each load cycle. First, it is recognized that the failure moment level is related to the member's actual strength reserve or residual strength, which is a function of the load history and maximum experienced curvature level ϕ. The failure moment was proposed as a function of the normalized curvature level, Fig. 2, given by

$$M_{fi} = M_f \cdot \frac{2\Phi_i}{\Phi_i + 1.0} \tag{4}$$

where M_{fi} = failure moment for given curvature level ϕ_i, M_f = failure moment for monotonic loading, $\Phi_i = \phi_i/\phi_f$ = curvature ratio and ϕ_f = failure curvature for monotonic loading. As reported earlier [21,24], strength deterioration is initiated as soon as the yield load level is exceeded, and accelerates as the critical load level is reached. As shown in Fig. 3, the amount of strength deterioration given load level ΔM is represented by

$$\Delta M = \Delta M_f \left(\frac{\phi - \phi_y}{\phi_f - \phi_y} \right)^{\omega} \tag{5}$$

where ΔM_f = moment capacity (strength) reduction in a single load cycle at failure curvature and ϕ_y, ϕ_f = yield and failure curvature, respectively. Reproduction of quasi-static experimental load-deformation curves could give the values between 1.5 and 2.5 for parameter ω, which must be a function of the confinement ratio, longitudinal steel ratio, axial force, etc.

Taking into consideration the relationship between maximum displacement and dissipated energy, the strength deterioration rate and number of load cycles to failure, the damage index D_c is expressed by means of the modified Miner's Rule:

$$D_c = \sum_i \left(\alpha_i^+ \frac{n_i^+}{N_i^+} + \alpha_i^- \frac{n_i^-}{N_i^-} \right) \tag{6}$$

where i = indicator of different displacement or curvature levels, $N_i = (M_i - M_{fi})/\Delta M_i$ = number of cycles up to curvature level i to cause failure, ΔM_i = strength drop in one load cycle up to curvature level i, n_i = number of cycles up to curvature level i actually applied, α_i = damage modifier, and +,- = indicator of loading sense. The damage modifier α_i introduced above reflects the effect of the loading history. For positive moment loading, α_i is defined as

$$\alpha_i^+ = \frac{\frac{1}{n_i^+} \sum_{j=1}^{n_i^+} k_{ij}^+}{\overline{k_i^+}} \cdot \frac{\phi_i^+ + \phi_{i-1}^+}{2\phi_i^+} \tag{7}$$

where

$$k_{ij}^+ = \frac{M_{ij}^+}{\phi_i^+} \tag{8}$$

is the stiffness during the j-th cycle up to load level i and,

$$\overline{k_i^+} = \frac{1}{N_i^+} \sum_{j=1}^{N_i^+} k_{ij}^+ \tag{9}$$

is the average stiffness during N_i^+ cycles up to load level i and

$$M_{ij}^+ = M_{i1}^+ - (j-1)\Delta M_i^+ \tag{10}$$

is the moment reached after j cycles up to load level i. For negative loading, the damage accelerator is defined similarly [21,24].

In passing, it is mentioned that the question whether deformation capacity should be measured in terms of inelastic curvature or inelastic rotation is not entirely settled. In this connection, an extensive collection of rotation capacity data is reported by Casciati and Faravelli [22] which includes 310 tests performed at sixteen different laboratories in twelve countries. These data refer to a wide variety of beam spans, loading conditions, tension and compression reinforcement ratios, tie spacing, cross-section dimension, axial loading, shear span, etc. Out of the entire data set, the following relationship emerged: $\theta'_M = A + B\tau' + \epsilon$ where $\theta'_M = ln\ \theta_M$ and $\tau' = ln\ \tau$ with θ_M and τ representing the mid-span rotation capacity and neutral axis depth, respectively. It came out that $B = 0.876$, while A and ϵ are normally distributed random variables with mean values —4.235 and 0 and variances 0.361^2 and 0.548^2 respectively.

MODEL BASED ON MAXIMUM DISPLACEMENT AND TOTAL HYSTERETIC ENERGY DISSIPATION

A model based on consideration of the displacement and hysteretic energy dissipation has been recently proposed for reinforced and masonry structures [18,19]. The damage is measured by an index which is a simple linear combination of the maximum displacement and the total hysteretic energy dissipation,

$$D = \frac{\delta_{max}}{\delta_u} + \frac{\beta}{Q\delta_u}\int dE \tag{11}$$

in which δ_{max} is the maximum deformation, δ_u is the ultimate deformation under a static load, Q is the yielding strength, $\int dE$ is the total hysteretic energy dissipation, β is a nonnegative constant. The positive value of D represents different degrees of damage and $D > 1$ represents collapse. This model allows a quantitative estimate of the degree of damage to the structural members.

Under external excitations such as earthquake ground motion, the damage index can be evaluated numerically if response records are available. This estimation, however, requires a proper model for the structural system which can go into the nonlinear inelastic range and an accurate and reliable method of system identification.

In predicting damage of structural members due to future earthquakes, damage index statistics caused by earthquake can be evaluated based on recently developed random vibration method [27]. Because of the variability in material properties, workmanship and construction methods, the ultimate capacity of a structural member against collapse is highly uncertain. The capacity D_c in terms of the index, D, is therefore a random quantity. From analysis of data of a large number of RC members tested in the US and Japan, it was found [28] that the variability of D_c can be approximately described by a lognormal variate with mean value of 1.0 and standard deviation of 0.54. This provides the information needed for assessment of the uncertainty of member capacity against collapse. The collapse probability can be evaluated as that of the capacity D_c being exceeded by the damage caused by the earthquake [19,28].

For detailed analysis of seismic damage, the linear combination of the maximum displacement D_1 and energy dissipation D_2 may not serve as a good measure of the extent of damage. Other forms of damage index D, such as $D = \sqrt{(1.1D_1^{0.38})^2 + D_2^2}$, have been proposed [22].

A GLOBAL DAMAGE MODEL BASED ON THE LENGTHENING IN THE FUNDAMENTAL PERIOD

Most of the research in the field of damage assessment has dealt with the analysis of simple structural elements. The application of these techniques to the post earthquake analysis of existing structures is limited by the scarcity of data available in the field regarding the history of stress and strain for single elements. In order to obtain such detailed information a great number of instruments must be installed on the structure. Furthermore, when a certain degree of redundancy is present in the structure, local features of damage may not be very important in determining what, if any, repair actions should be carried out. Some average or global measures of damage may, in this case, be as useful to engineers as detailed information.

Global damage indices that give information about the state of complex structual systems are generally defined as a weighted average of the local damage in the simple structural elements that form the system. A procedure which was advocated in [29] and used in damage prediction in [19,28] gives the system (global) damage index as a weighted average of those of the components. The total hysteretic energy dissipations in the structural elements are the weights used in this procedure. Comparisons of such a global damage index with observed damage of reinforced concrete and masonry structures [19,28] seem to have lent some credence to this procedure. A different approach for the definition of global damage indices was proposed by DiPasquale and Cakmak [23,30]. Knowing that when a structure is damaged, it tends to lose part of its stiffness and as a result of this, the natural period of oscillation lengthens; they defined a global damage index for reinforced concrete structures as the degree of stiffness degradation. This method of identification of damage has also been suggested by researchers [31]. This global damage index, named maximum softening is defined as:

$$\delta_M = 1 - \frac{(T_0)_{initial}}{(T_0)_{maximum}} \tag{12}$$

where $(T_0)_{initial}$ = natural period for undamaged frame, $(T_0)_{maximum}$ = maximum value of natural period for frame under seismic excitation. In the calculation of maximum softening, the nonlinear mechanical behavior of the structure was taken into account by evaluating the natural period of a time-varying linear system equivalent to the actual nonlinear system for a series of non-overlapping time windows [23,30]. They used a maximum likelihood criteria for the identification of the time varying system.

It was shown in [32,33] that the maximum softening, δ_M, defined for reinforced concrete structures, is a theoretically grounded global damage indice, which carries the necessary elastic and plastic damage (stiffness degradation by cracks and plastic deformation by the yielding of reinforcement bars) independent contributions (also, see table 1).

In order to calibrate the model, acceleration records from seismic simulation experiments performed at the University of Illinois at Urbana-Champaigne have been used. The seismic behavior of seven reinforced concrete structures has been analyzed, for a total of 25 tests [23,30,34,35,36,37]. The damage detection criterion proposed yields the probability that the structure has undergone damage after the seismic event or not. So, a fragility curve, relative to the structural serviceability limit state has been derived. Figure 4 shows the probability of the onset of seismic structural damage is computed as a function of a global damage index (maximum softening) which is based on equivalent modal parameters,

This criterion has been used for the damage analysis of shaking table experiments performed at the University of California at Berkeley [38] and strong motion records from the San Fernando (1971), Imperial Valley (1979) and Loma Prieta (1989) earthquakes [39]. In all cases considered, an excellent correlation has been found between the values of the parameter-based damage indices estimated from strong motion records and the onset of seismic structural damage (see figure 5 and table 2).

Since the post earthquake structural reliability is intended solely from the latest recorded damage value, the Markov assuption of maximum softening has been studied and it is justified for the maximum value if transition probability density fuction, whereas some deviation is observed for variance and higher order moments [40].

In order to compare maximum softening with other damage indicators, the response of three code-designed buildings frames subjected to artificially generated and recorded earthquakes has been obtained by using the computer code SARCF-II [41,50]. First, one two bay three storey building frame, one three bay four storey and another three bay four storey building frames are designed according to the ACI 318-83 [42] code , to resist the equivalent static lateral loads specified in the Uniform Building Code [43] for seismic zone 4. The details of the structures can be found in [41]. Then a total of 29 different earthquakes have been applied to the three structural models. Two types of input acceleration have been used: scaled versions of the N-S component of the acceleration history recorded at El Centro during the Imperial Valley Earthquake (1940), and artificially generated earthquakes using an ARMA model according to the method proposed by Ellis and Cakmak [44,45]. The parametric relationships between modeling parameters and physical variables were obtained from a set of strong motion accelerograms recorded from the following Japaneses earthquakes: Tokachi-Ochi (1968), Miyagiken-Oki (1978), Nihonkai-Chubu (1983) and Michoacan (1985) [44]. Single event earthquakes have been generated using this capability of the program SARCF-II. The input parameters and their values used in this case are: an initial peak time of two seconds, a duration of the earthquake of 20 seconds, a distance from the epicenter of 10 or 100 kilometers, different values of the earthquake magnitude, and a soil factor, γ_f, with a value of 0.10. In this numerical study, several damage indices and traditional measures of damage have been computed and compared to maximum softening. A new way of computing maximum softening by moving averages of the instantaneous natural period has been derived and it is oberved that system identification and moving averages methods are equivalent from practical point of view. Before graphic comparisons, simulation results are compared to experimental results of the tests done at UIUC by Sozen *et al.* Considering the uncertainties that exist in the description of the nonlinear behavior of reinforced concrete structures, a good approximation to the experimental values was obtained by using SARCF-II [41].

Among the traditional means of damage the maximum interstorey drift, the permanent interstorey drift and the maximum ductility ratio for beams and columns, have been computed. The interstorey drift is defined as the tangent of the angle between the original position and the deformed position of the columns. The interstorey drift has been used as a measure of damage by M.A. Sozen [1]. The permanent drift is the interstorey drift after the earthquake. The permannet drift has has been used as an indication of damage by J.E. Stephens and J.P.T. Yao [46 and 47]. The ductility ratio which has been computed is a curvature ductility ratio, defined as the ratio between the curvature and yield curvature.

The global damage indices that were studied are:

1. Energy average of Chung, Meyer and Shinozuka's Local Damage Index [21,24,41].
2. Weighted average of Chung, Meyer and Shinozuka's Local Damage Index, using the triangular weighting function [21,24,41].
3. Energy average of Park and Ang's Damage Index [18,19,41].
4. Maximum Softening defined by Equation (12).

The comparisons have been done by means of a series of plots that show the value of the maximum softening as a function of the other damage indicator [figures 6,7,8,9,10,11,12]. Each point in the graphs has been obtained from the nonlinear dynamic analysis of any of the building frames described in [41]. The filled circles represent the cases for which the second occurrence of the same earthquake produced collapse. However, the filled circles correspond to the damage indices that were computed for the first earthquake.

Whether or not the structure collapses in the second earthquake is considered to be an objective measure of the level of damage in the first earthquake, the best global damage index would be the one that can predict the collapse of the structure in the second earthquake. In that case, there should be a cutoff value of that damage index so that a collapse resulting from the second identical earthquake only occurs for values greater than that cutoff value.

Since it is possible to draw a horizontal line that seperates the squares and filled circles in all figures 6, 7, 8, 9, 10, 11 and 12, recalling that the serviceability limit state was related to an average value of maximum softening of 0.43, one concludes from these figures that for all cases, a value of the maximum softening larger than 0.43 implies a collapse in the second identical earthquake and that there is not such a value for the other damage indicators which seperate the structures that collapse in the second earthquake. The reader is referred to reference [41] for detailed information about the calculation of other damage indicators.

For the studied case, it is concluded that maximum softening predicts the structural capacity for future earthquakes more consistently than any of the traditional measures of damage and that maximum softening is a more robust global damage index than the weighted averages of the local damage indices [41]. It should be noted that local damage indices give important information at the local level, but for obtaining global damage, an appropriate weighting function has to be defined. The maximum softening avoids the definition of a weighting function.

The maximum softening and the limit state probability defined can therefore be used in the detection of seismic structural damage.

Figure 13 shows the smoothed time history of the natural period of a three-bay, four-story reinforced concrete building frame of figure 16 subjected to the North-South component of the May 1940 El Centro earthquake linearly scaled up so that it has a 1.0g peak ground acceleration. The natural periods were calculated based on the moving averages method described in [41]. They were then smoothed to produce Fig. 13 by averaging the values over the 2 sec. time window. In this case, the global damage index δ_M was estimated as 0.416.

AUTOMATED DAMAGE-CONTROL DESIGN

This section draws most of its contents from Chung et al [21,24,25] in order to describe the automated damage-control procedure.

Conventional seismic design, which is typically based on equivalent static lateral loads specified in an appropriate building code, cannot guarantee the absence of a concentration of damage in a few more vulnerable structural elements. Such damage has been experienced repeatedly in recent earthquakes. On the other hand, a structure which is shown to dissipate energy uniformly in its main components can be expected to survive an earthquake of given intensity with the least amount of damage possible. It is therefore desirable that a design procedure strive for uniform distribution of energy dissipation and damage. This goal is the objective of the automated damage-control procedure for reinforced concrete frame buildings developed herein.

The key components of this automated damage-control design procedure are (1) an algorithm to evaluate the computed damage distribution by comparing it with user-specified criteria and (2) a set of design rules which permit the automated modification of the structure such that improved performance is guaranteed.

Optimum Design Criteria

Structural designers normally rely on their experience when designing a structure to withstand seismic loads. They can fall back on both knowledge of rational principles of structural theory and intuition. However, the design task is complicated by the fact that a typical reinforced concrete frame is a highly redundant structure with intricate load-resistant mechanisms. In addition, the random nature of earthquake loading makes the design task more difficult.

There are basically two approaches one can follow to accumulate knowledge to be incorporated into an automated design algorithm. The first approach consists of interviews with experts, that is in our case, structural engineers experienced in earthquake-resistant design. This is the approach traditionally followed by developers of knowledge-based expert systems [48]. In our work, a different approach was followed. By performing parametric sensitivity studies [48], the necessary numerical experience along with physical insight into building frames was accumulated. The damage acceptance algorithm for the automated damage-control design method [49] contains the following components:

1. Damage (and plastic hinges) in columns is unacceptable, as required by the strong-column weak-beam concept. Specifically a damage index larger than $D^{allow}_{column} = 0.01$ will be flagged as unacceptable in any column except at the base of the first story and at the top of the top story.
2. The damage index D_{beam} at any beam node shall not deviate from the spatial average $\overline{D}_{beam}$ computed for all the beam nodes by more than a user-specified allowance δ (= 0.05, for example). Thus, individual beam elements may be flagged as having too much or too little damage.
3. The spatial average $\overline{D}_{beam}$ of all beam damage indices shall not exceed a user-specified target level D^{target}_{beam}, say, equal to 0.1. A small tolerance τ (= 0.05, for example) is allowed.

These are summarized as:

$$D_{column} \leq D^{allow}_{column} \tag{13}$$

$$\left|D_{beam} - \overline{D}_{beam}\right| \leq \delta \tag{14}$$

$$\left|\overline{D}_{beam} - D^{target}_{beam}\right| \leq \tau \tag{15}$$

The rules indicated above and summarized in Fig. 5 in the form of a flow chart are considered to form only a useful breakpoint for an automated damage-control design procedure, and are coded to the program SARCF-I which is a modified version of the program SARCF [50] with a few enhancements. The flow chart in Fig. 14 is consistent, however, with the particular optimum design procedure used in the present study. For design improvement, ten artificial earthquakes are used as described in "Generation of Artificial Earthquakes" below, and D_{beam} and D_{column} are the (ensemble) average of the damage indices corresponding to these ten earthquakes at each beam and column node.

In the future, one may wish to supplement the design rules synthesized from these studies with rules obtained from experts in the field. Also, if in the future such a situation arises that a particular spatial distribution of damage is more preferable than uniform distribution, SARCF-I can easily be modified to satisfy that preference.

Automated Damage-Control Design Procedure

If the damage index of at least one frame member is unacceptable, corrective action has to be taken, i.e., the design must be modified such that an improved performance in a reanalysis is guaranteed and convergence towards an acceptable design is assured.

1. For any beam element which showed an unacceptable level of damage in the original design, the longitudinal steel will be increased (or decreased) by 5%:

$$\Delta A^1_s = 0.05 \times A_s \times SIGN\left[D - D^{target}_{beam}\right] \tag{16}$$

where A_s = original amount of steel, D = amount of damage at a beam node in the original design, and D_{beam}^{target} is the target value of the damage index for beam nodes.

2. In a subsequent design iteration "i," the amount of steel in any beam with unacceptable damage is changed according to

$$\Delta A_s^i = \Delta A_s^{i-1} \times \frac{D^i - D_{beam}^{target}}{D^{i-1} - D^i} \tag{17}$$

where ΔA_s^i = additional (or deductible) amount of longitudinal steel for the element in question, ΔA_s^{i-1} = steel added (or deducted) at previous iteration, D^i and D^{i-1} = damage values in the i-th and (i-1)–th iteration, respectively.

3. If the mean beam damage index $\overline{D}_{beam}$ satisfies the requirement of Eq. 15, Eq. 17 will be applied only to beam elements which do not satisfy Eq. 14.

4. To adhere to the strong-column weak-beam concept, any column with unacceptable damage has to satisfy the requirement $M_y^{column} \geq 1.25 \times M_y^{beam}$, where M_y^{column} is the sum of the absolute yield moment of the columns considered, and M_y^{beam} is the sum of the absolute yield moments of the beams framing into the same joint. The reinforcing steel of the upper column will be linearly increased by the amount

$$\Delta A_s^i = \Delta A_s^{column} \times \frac{M_y^i - M_y^{i-1}}{\Delta M_y^{column}} \tag{18}$$

where the superscript i indicates the iteration number. ΔM_y^{column} denotes an increment of the yield moment of the column when the longitudinal steel of the corresponding column is increased by ΔA_s^{column}.

5. At any section of an element, the longitudinal steel ratio ρ should not be less than the minimum required by ACI 318–83; $\rho \geq \rho_{min} = 200/f_y$ where f_y is the yield strength of the longitudinal steel (psi), and should not be greater than the Code maximum, $\rho \leq \rho_{max} = (3/4)\rho_b + \rho' f_s'/f_f$ where ρ_b is the balanced steel ratio, ρ' is the compressive steel ratio and f_s' is the compressive steel stress. For any element which does not satisfy the above steel requirements, the member depth will be reduced (or increased) by

$$\Delta d = \frac{A_s \cdot (\rho_{min} - \rho)}{b \cdot (\rho_{min} \cdot \rho)} \quad \text{or} \quad \Delta d = \frac{A_s \cdot (\rho_{max} - \rho)}{b \cdot (\rho_{max} \cdot \rho)} \tag{19}$$

where b is the width of the element section in question.

The steel ratios of all the beams with unacceptable damage indices are either increased or decreased simultaneously. Thus, full use of the superposition principle is made. Design rule 5 is subject to practical constraints; however, since most preliminary frame designs are reasonable, the reinforcing limits are seldom violated.

Generation of Artificial Earthquakes

In this study, artificial ground acceleration histories, $x(t)$, are generated by multiplying an envelope function $s(t)$, and a stationary Gaussian process $g(t)$

$$x(t) = s(t) \times g(t) \tag{20}$$

Here, the envelope function is assumed to have a trapezoidal shape as shown in Fig. 15. Other envelope functions could easily be substituted. A Gaussian process $g(t)$ is assumed to have a Kanai-Tajimi power spectral density (PSD) function $S(\omega)$ as shown in Fig. 15.

$$S(\omega) = S_0 \times \frac{1 + 4\zeta_g^2 \left(\frac{\omega}{\omega_g}\right)^2}{\left[1 - \left(\frac{\omega}{\omega_g}\right)^2\right]^2 + 4\zeta_g^2 \left(\frac{\omega}{\omega_g}\right)^2} \tag{21}$$

where ω_g = characteristic ground frequency, ζ_g = predominant damping coefficient, S_0 = intensity measure. The Gaussian process $g(t)$ can be generated by the following expression [51]

$$g(t) = \sqrt{2}\sum_{k=1}^{N} \sqrt{G(\omega_k)\Delta\omega} \cdot \cos(\omega_k t - \phi_k) \tag{22}$$

where ϕ_k = random phase angle, uniformly distributed between 0 and 2π, $\omega_k = k\Delta\omega$, $G(\omega_k) = 2S(\omega_k)$ = one-sided power spectrum, and $\omega_* = N\Delta\omega$ = upper cutoff frequency.

To generate an artificial earthquake, Shinozuka [52] suggested a relationship between the intensity S_0 and peak ground acceleration (PGA). Since the standard deviation σ_g of $g(t)$ with PSD given in Eq. 21 is

$$\sigma_g = \left[\pi\omega_g\left(\frac{1}{2\zeta_g} + 2\zeta_g\right)\right]^{\frac{1}{2}} \times S_0^{\frac{1}{2}} \tag{23}$$

the peak ground acceleration can be written as

$$PGA = \alpha_g S_0^{\frac{1}{2}} \tag{24}$$

where

$$\alpha_g = p_g\left[\pi\omega_g\left(\frac{1}{2\zeta_g} + 2\zeta_g\right)\right]^{\frac{1}{2}} \tag{25}$$

with p_g indicating the peak factor, empirically assumed to be 3.0 in this study.

With the aid of Eq. 23, the value of S_0, depending on the intensity of the earthquake measures in terms of PGA can be determined. For firm soil conditions, the other two parameters in the Kanai-Tajimi spectrum are empirically given as $\omega_g = 9\pi$ (rad/sec) and ζ_g = 0.6. Under these conditions,

$$S_0 = 0.0827,\ 0.160 \text{ and } 0.639\ ft^2/sec^3 \tag{26}$$

respectively for PGA = 0.36g (1940 El Centro), 0.5g and 1.0g.

Because of the random nature of earthquake acceleration histories, it is more meaningful to consider structural response quantities (such as damage indices) formed as averages for an ensemble of sample earthquake histories, rather than responses to individual histories. Chung et al [25] determined that a sample size of 10 is reasonably adequate.

Spatial Damage Distribution

A three-bay four-story frame structure is designed as shown in Fig. 16 in accordance with the 1988 Uniform Building Code. This structure is then subjected to ten (10) artificial earthquakes of the type described in the preceding section with PGA = 1.0g and redesigned iteratively in the manner described earlier until the specified conditions of spatial distribution of damage are satisfied. Figure 8 shows the spatial distribution of the ensemble average of damage index D_e (Eq. 6) evaluated at each beam node with the (ensemble) average taken over the ten artificial earthquakes. The mean damage $\overline{D}$ = 0.079 and standard deviation σ_D = 0.084 shown in Fig. 17 indicate respectively the (spatial) average and standard deviation of these ensemble averages computed at all the beam nodes. Figure 18 shows the spatial distribution of the ensemble average of damage after twelve (12) design iterations to satisfy the requirements of Eqs. 13–15 with D_{column}^{allow} = 0.01, D_{beam}^{target} = 0.1, τ = 0.05, and δ = 0.05. It can be easily verified from Fig.18 that these requirements are met. The improved design (Fig.18) requires a total of reinforcing steel equal to 29,078 in^3 (15,010 in^3 for beams and 14,068 in^3 columns) as opposed to a total of 24,262 in^3 (14,774 in^3 for beams and 9,488 in^3 for columns) for the original design (Fig17). It is also noted that the improved design results in columns having larger member yield moments throughout the building than in the original design. It is further noted, however, that although beams in the lower stories of the improved design have larger member yield moments, beams in the higher stories have smaller values than the corresponding beams in the original design.

Evaluation of Damage-Control Design

The automated damage-control design presented here aims for uniform energy dissipation throughout the building frame, as measured by the individual member damage index. The usefulness of this design method is shown by demonstrating that the damage-control design reduces damage as measured by the global damage index under the same earthquake and, at the same time, it leads to a larger capacity accommodating more intensive earthquakes.

The original and improved designs are now subjected to ground acceleration histories obtained by linearly scaling the NS component of the 1940 El Centro earthquake so that $PGA = 0.2g, 0.4g, ..., 1.6g$. The global damage index δ_g defined in Eq. 12 is then computed for each time history for both designs. The results are plotted in Fig. 19 which indicates that at each PGA value considered, the damage-control design exhibits smaller global index values than the original design. This is in spite of the fact that the spatially averaged beam damage is increased from $\overline{D}_{beam} = 0.079$ for the original design to $\overline{D}_{beam} = 0.089$ for the improved design. Similar trends are observed when the Taft and Hachinohe earthquakes are used instead of the El Centro, although the results are not shown here.

Figure 20 plots the maximum damage index among the beam nodes in the frame as a function of PGA for both original and improved designs. If, for safety reasons, any node of the frame should not be stressed beyond the point where damage index D_e is, say, equal to 0.5, then the original design can accommodate the ground acceleration history with $PGA = 0.86g$, while the improved design with $PGA = 1.04g$, a capacity increase of 21% in terms of PGA. Again, similar results are obtained using the Taft and Hachinohe earthquakes.

As demonstrated by Chung et al [25], the characteristics of the Kanai-Tajimi-based artificial earthquakes and three natural earthquakes mentioned above are quite different from each other. In spite of this, the improved design upgraded the original design from the viewpoint of both attaining smaller global damage if these two designs were subjected to the same earthquake (Fig. 19) and increasing the capacity if they could accommodate the damage only up to the extent specified by the same damage index (Fig. 20). This does suggest that the proposed damage-control design appears effective regardless of the type of earthquake.

CONCLUSION

Among many possible models for local damage indices, the one based on the modified Miner's rule was used to develop a damage-control design procedure for RC frame buildings. The procedure requires iterative design improvements, satisfies the necessary building code requirements and at the same time produces an optimum design in the sense that the seismic damage measured in terms of this local damage index is spatially distributed as uniformly as desired. In fact, an example design indicates that the optimum design significantly enhances the seismic capacity of the frame building.

ACKNOWLEDGEMENT

In carrying out this work, the authors were supported by the National Center for Earthquake Engineering Research under contract number 89-3005 and 89-1104. The authors also acknowledge the support of the Princeton-Kajima Joint Research program.

REFERENCES

1. SOZEN, M.A., "Review of Earthquake Response of Reinforced Concrete Buildings with a View to Drift Contro", *State of the Art in Earthquake Engineering*, Istanbul, Turkey, 1981.
2. NEWMARK, N.M. AND ROSENBLUETH, E., "Fundamentals of Earthquake Engineering", Prentice-Hall, Englewood Cliffs, New Jersey, 1971.
3. OLIVERIA, C.S., "Seismic Risk Analysis of a Site and a Metropolitan Are". EERC, Report No. 75–3, University of California, Berkeley, 1975.
4. ALGAN, B.B., "Drift and Damage Consideration in Earthquake Resistant Design of Reinforced Concrete Building", Ph.D. Thesis, Department of Civil Engineering, University of Illinois at Urbana-Champaigne, 1982.
5. KRAWINKLER, H. AND ZOHREI, M., "Cumulative damage in steel structures subjected to earthquake ground motion", *Computers and Structures*, Vol. 16, No. 14, 1983.
6. YAO, J.T.P. AND MUNSE, W.H., "Low Cycle Fatigue Behavior of Mild Stee", Special Technical Publication No. 338, American Society for Testing and Materials, 1962.
7. STEPHEN, J.E. AND YAO, J.T.P., "Estimation of Innerstory Load-Deformation Relations for Damaged Structure", *Proc. 4th International Conference on Structural Safety and Reliability*, Kobe, Japan, 1985.
8. BANON, H., BIGGS, J.M. AND IRVINE, H.M., "Prediction of Seismic Damage in Reinforced Concrete Frame", Publication No. R80–16, Dept. of Civil Engineering, MIT, Cambridge, MA, May 1980.
9. BANON, H., BIGGS, J.M. AND IRVINE, H.M., "Seismic Damage in Reinforced Concrete Frame", *ASCE J. Structural Division*, Vol. 107, No. ST9, September 1981.
10. BANON, H. AND VENEZIANO, D., "Seismic Safety of Reinforced Concrete Members and Structure", *Earthquake Engineering and Structural Dynamics*, Vol. 10, 1982.
11. GOSAIN, N.K., BROWN, R.H. AND JIRSA, J.O., "Shear Requirements for Load Reversals on RC Members", *J. Structural Division*, Vol. 103, No. ST7, July 1977.
12. HWANG, T.H. AND SCRIBNER, C.F., "RC Member Cyclic Response During Various Loading", *ASCE J. Structural Division*, Vol. 103, No. ST7, July 1977.
13. DARWIN, D. AND NMAI, C.K., "Energy Dissipation in RC Beams Under Cyclic Load", *ASCE J. Structural Division*, Vol. 112, No. 8, August 1986.
14. LYBAS, J. AND SOZEN, M.A., "Effect of Beam Strength and Stiffness on Dynamic Behavior of Reinforced Concrete Coupled Walls", Civil Engineering Studies, SRS No. 444, University of Illinois at Urbana-Champaigne, July 1977.
15. BERTERO, V.V. AND BRESLER, B., "Design and Engineering Decision: Failure Criteria (Limit States), Developing Methodologies for Evaluating the Earthquake Safety of Existing Buildings", Report No. EERC-77–6, University of California at Berkeley, February 1977.
16. BLEJWAS, T. AND BRESLER, B., "Damageability in Existing Buildings", Report No. EERC-78–12, University of California, Berkeley, August 1979.
17. ROUFAIEL, M.S.L. AND MEYER, C., "Analysis of Damaged Concrete Frame Buildings", Report No. NSF-CEE-81–21359–1, Dept. of Civil Eng. and Eng. Mech., Columbia University, New York, 1983.
18. PARK, Y.J., ANG, A.H-S. AND WEN, Y-K., "Seismic Damage Analysis and Damage-Limiting Design of R/C Structures", *ASCE J. Structural Division*, Vol. 111, No. 4, April 1985.
19. PARK, Y.J., ANG, A.H-S. AND WEN, Y-K., "A Mechanistic Seismic Damage Model for Reinforced Concrete", *ASCE J. Structural Division*, Vol. 111, No. 4, April 1985.

20. Mizuhata, K., Sugiyama, K. and Shimizu, H., "Comparative Study on the Evaluation Method of Cumulative Damage of Reinforced Concrete Columns", Technical Report, Kinki Division, Japanese Architectural Society, Japan, 1987.

21. Chung, Y-S., Meyer, C. and Shinozuka, M., "Modeling of Concrete Damage", *ACI J.*, Vol. 86, No. 3, May-June 1989, pp. 259–271.

22. Casciati, F. and Faravelli, L., "Biaxial Damage Sssessment", Seismic Engineering Research and Practice, *ASCE Proc. of Struc. Cong. '89*, San Francisco, pp. 497–506.

23. DiPasquale, E. and Cakmak, A.S., "Detection and Assessment of Seismic Structural Damage", Technical Report NCEER-87–15, NCEER, Buffalo, NY, August 1987.

24. Chung, Y-S., Meyer, C. and Shinozuka, M., "Seismic Damage Assessment of Reinforced Concrete Members", Report No. NCEER-87–22, NCEER, Buffalo, NY, October 1987.

25. Chung, Y-S., Hatamoto, H., Shinozuka, M. and Meyer, C., "Seismic Safety Improvement by Damage-Control Desig", Draft Technical Report, Princeton University, June 1989.

26. Fujikake, K., Ohno, T. and Nishioka, T., "Experimental Study on Energy Absorption Capacity of Reinforced Concrete Fram", *Proc. Japan Society of Civil Engineers*, Vol. 8, No. 390, February 1988.

27. Wen, Y.K., "Methods of Random Vibration for Inelastic Structures", *Applied Mechanics Reviews*, Vol. 42, No. 2, 1989.

28. Kwok, Y.H., Ang, A.H-S. and Wen, Y-K., "Random Seismic Damage Analysis and Design of Unreinforced Masonry Structure", *Proc. 9th World Conf. on Earthquake Eng.*, Tokyo, Japan, 1988.

29. Akiyama, H., "Earthquake-Resistant Limit-State Design for Buildings", University of Tokyo Press, Tokyo, 1985.

30. D'Pasquale, E., Cakmak, A.S., "Detection of Seismic Structural Damage Using Parameter-Based Global Damage Indices", *Probabilistic Engineering Mechanics*, Vol. 5, No. 2, pp. 60-65.

31. Yao, J.T.P., "Safety and Reliability of Existing Structures", Pitman Publishing, London, 1985.

32. DiPasquale, E., Cakmak, A.S., "On the Relation Between Local and Global Damage Indices", Technical Report NCCEER-89-0034, National Center for Earthquake Engineering Research, State University of New York at Buffalo, 1989.

33. DiPasquale, E., Ju, J.W., Askar, A., and Cakmak, A.S., "Relation Between Global Damage Indices and Local Stiffness Degradation", *J. of Struc. Eng.*, Vol. 116, No. 5, pp. 1440-1456, 1990.

34. DiPasquale, E., Cakmak, A.S., "Identification of the Serviceability Limit State and Detection of Seismic Structural Damage", Report NCEER-88-0022, National Center for Earthquake Engineering Research, State University of New York at Buffalo, 1988.

35. Healey, T.J., Sozen, M.A., "Experimental Study of the Dynamic Response of a Ten Story Reinforced Concrete Frame with a Tall First Story", Rep. No. UILU-ENG-78-2012, SRS 450, University of Illinois at Urbana-Champaigne, 1978.

36. Abrams, D.P., Sozen, M.T., "Experimental Study of Frame-Wall Interaction in Reinforced Concrete Structures Subjected to Strong Earthquake Motions", Report No. SRS 460, UILU-ENG-79-2002, University of Illinois at Urbana-Champaigne, 1979.

37. Cecen, H., "Response of Ten Story, Reinforced Concrete Models Frames to Simulated Earthquakes", Ph.D. Dissertation, University of Illinois at Urbana-Champaigne, 1979.

38. Bertero, V.V., Aktan, A.E., Charney, F.A., Sause, R., "U.S.-Japan Cooperative Research Program: Earthquake Simulation Tests and Associated Studies of a 1/5th Scale

Model of a 7-Story Reinforced Concrete Test Structure", Report No. UCB/EERC-84/05, University of California at Berkeley, 1984.

39. Rodriguez-Gomez, S., Cakmak, A.S., and Shinozuka, M., "Damage Analysis of Simulated I-880 Structures Under the Loma Prieta Earthquake", *Soil Dyn. and Earth. Eng.* (to appear).

40. Nielsen, S.R.K., Cakmak, A.S., "Evaluation of Maximum Softening as a Damage Indicator for Reinforced Concrete Under Seismic Excitation", *Proc. of 1st Inter. Confrence on Computational Stochastic Mechanics* (to appear September 1991).

41. Rodriguez-Gomez, S., "Evaluation of Seismic Damage Indices for Reinforced Concrete Structures", M.Sc. Eng. Dissertation. Department of Civil Engineering and Operations Research, Princeton University, Princeton, NJ, 1990.

42. ACI Committee 318-83, "Building Code Requirements for Reinforced Concret", American Concrete Institute, 1983.

43. Uniform Building Code, "Earthquake Regulation", 1988.

44. Ellis, G.W., Cakmak, A.S., "Modeling Earthquakes in Seismically Active Regions Using Parametric Time Series Methods", Technical Report NCEER-87-0014, National Center for Earthquake Engineering Research, State University of New York at Buffalo, 1987.

45. Ellis, G.W., Cakmak, A.S., "Modelling Strong Ground Motion from Multiple Event Earthquakes", Technical Report NCEER-88-0042, National Center for Earthquake Engineering Research, State University of New York at Buffalo, 1988.

46. Stephens, J.S., Yao, J.P.T., "Damage Assessment Using Response Measurements", *ASCE J. Struc. Eng.*, Vol. 113, No. 4, pp. 787-801, "Theory of Evidence", *Structural Safety*, Vol. 1, pp. 107-121, 1987.

47. Toussi, S., Yao, J.P.T., "Hysteresis Identification of Existing Structures", *ASCE J. Eng. Mech.*, Vol. 109, No. 5, pp. 1189-1203, 1983.

48. ASCE Special Issue, "Expert Systems in Civil Engineerin", *ASCE J. Computing in Civil Eng.*, Vol. 1, No. 4, October 1987.

49. Chung, Y-S., Meyer, C. and Shinozuka, M., "Automated Seismic Design of Reinforced Concrete Building", Report No. NCEER-88-0024, NCEER, Buffalo, NY, October 1988.

50. Chung, Y-S., Meyer, C. and Shinozuka, M., "SARCF User's Guide, Seismic Analysis of Reinforced Concrete Frame", Report No. NCEER-88-0044, NCEER, Buffalo, NY, November 1988.

51. Shinozuka, M. and Wai, P., "Digital Simulation of Short-Crested Sea Surface Elevation", *J. Ship Research*, Vol. 23, No. 1, March 1979, pp. 76–84.

52. Shinozuka, M., Hwang, H. and Reich, M., "Reliability Assessment of Reinforced Concrete Containment Structures", *Nuclear Engineering and Design*, Vol. 80, 1984, pp. 247–267.

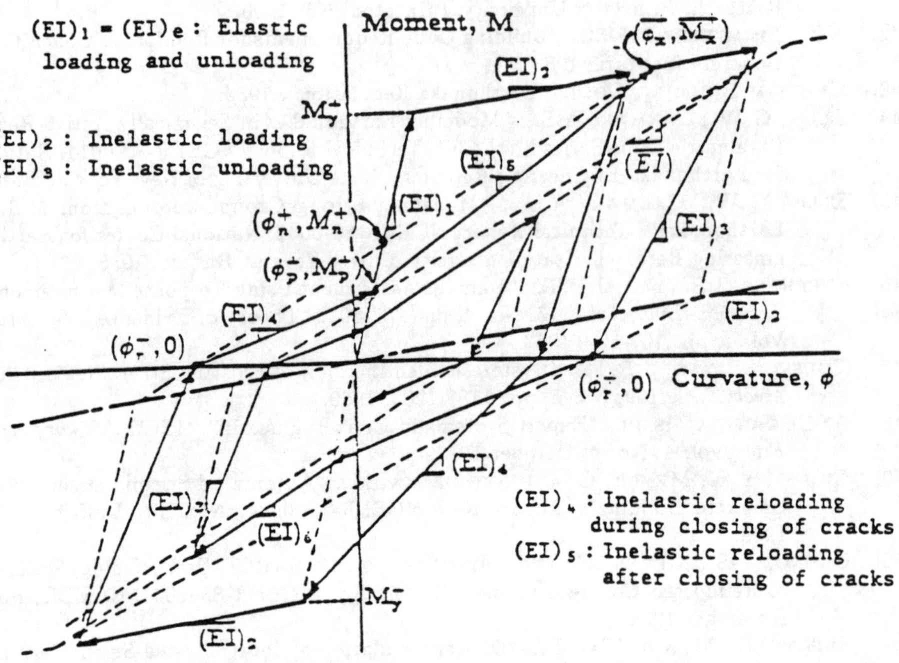

FIGURE 1. Typical Hysteretic Moment-Curvature Relationship

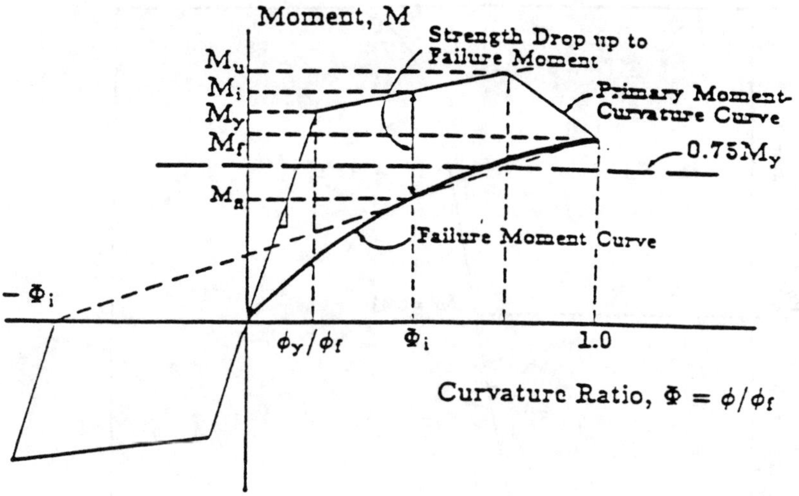

FIGURE 2. Definition of Failure

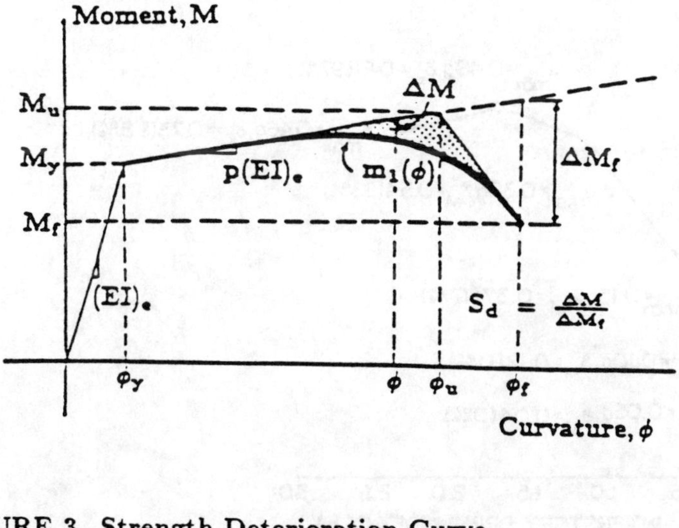

FIGURE 3. Strength Deterioration Curve

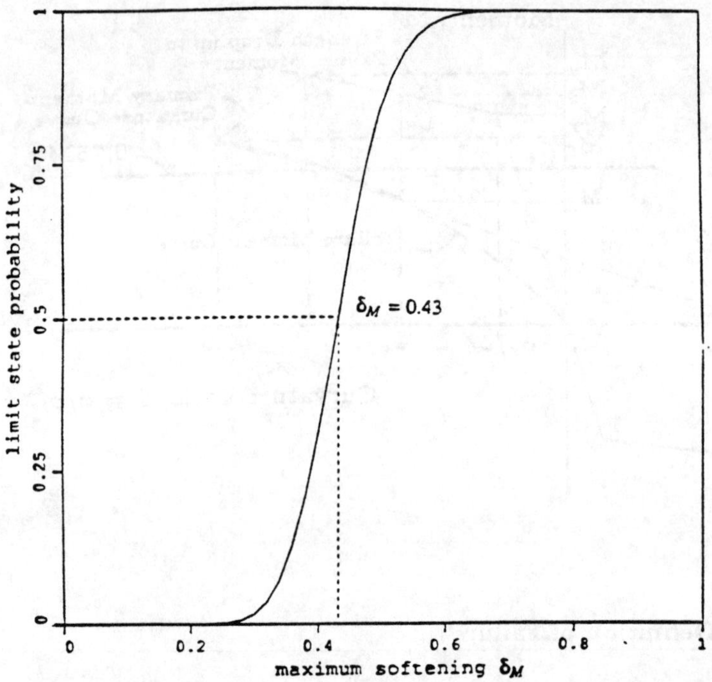

FIGURE 4. Fragility Curve, Structural Serviceability Limit State

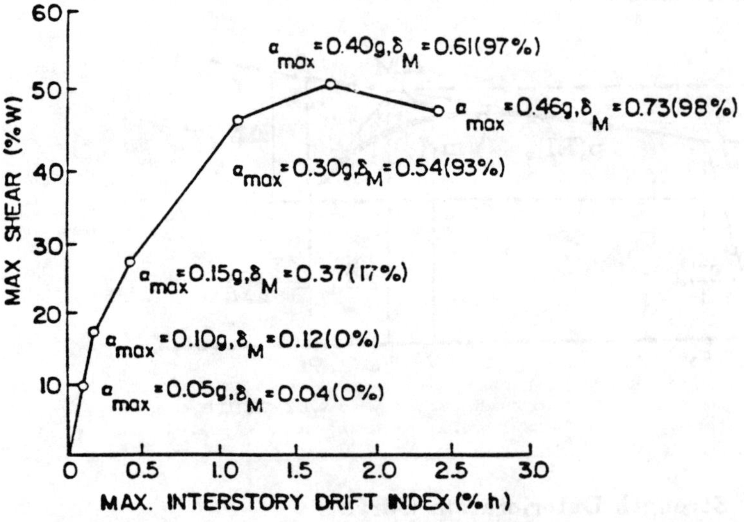

FIGURE 5. Detection of Structural Damage for the UCB Model (7-Storey, 1/5-Scale Reinforced Concrete)

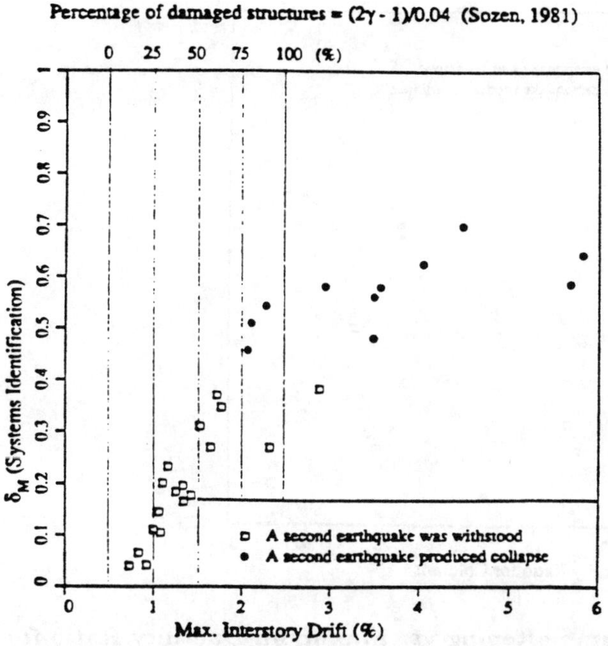

FIGURE 6. Maximum Softening vs. Maximum Interstorey Drift

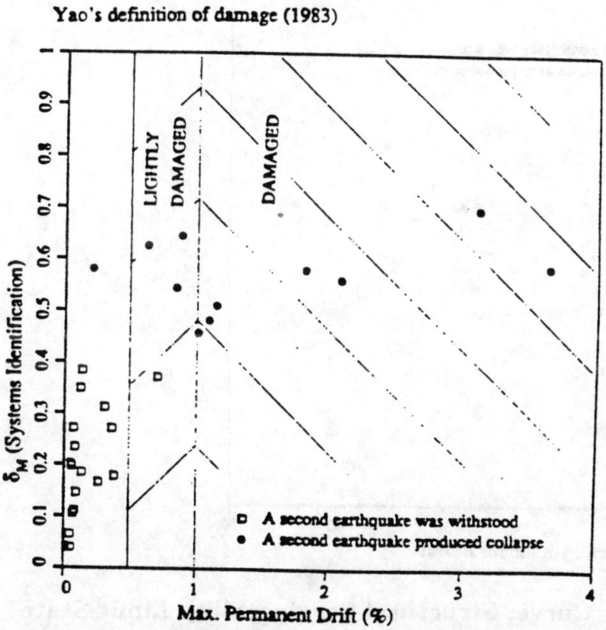

FIGURE 7. Maximum Softening vs. Maximum Final Drift

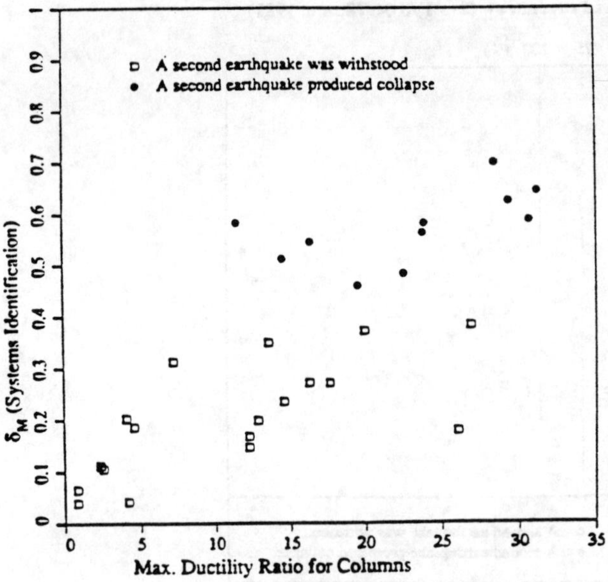

FIGURE 8. Maximum Softening vs. Maximum Ductility Ratio for Columns

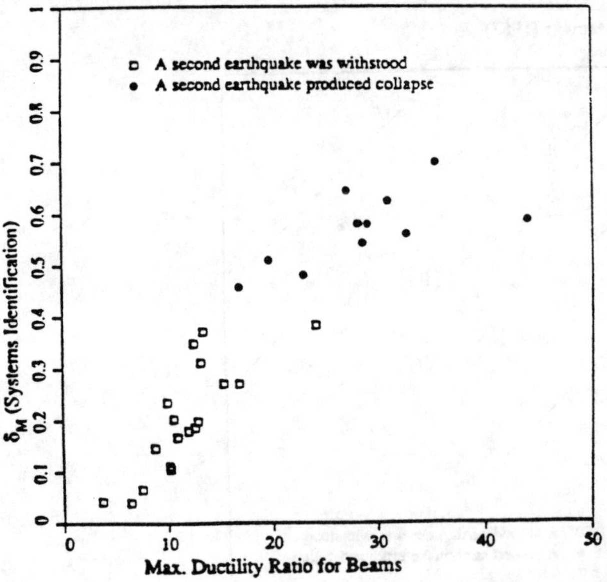

FIGURE 9. Fragility Curve, Structural Serviceability Limit State

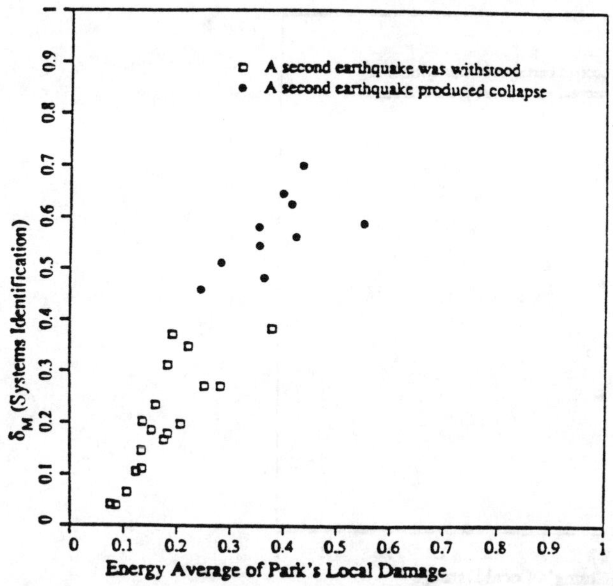

FIGURE 10. Maximum Softening vs. Energy Average of Park's Index

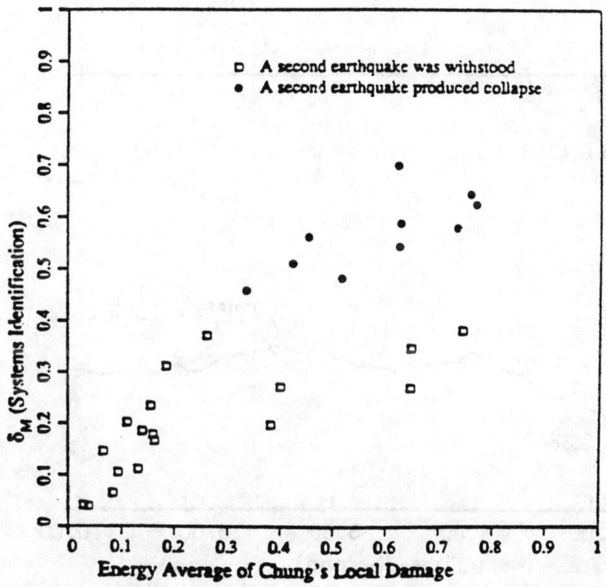

FIGURE 11. Maximum Softening vs. Energy Average of Chung's Index

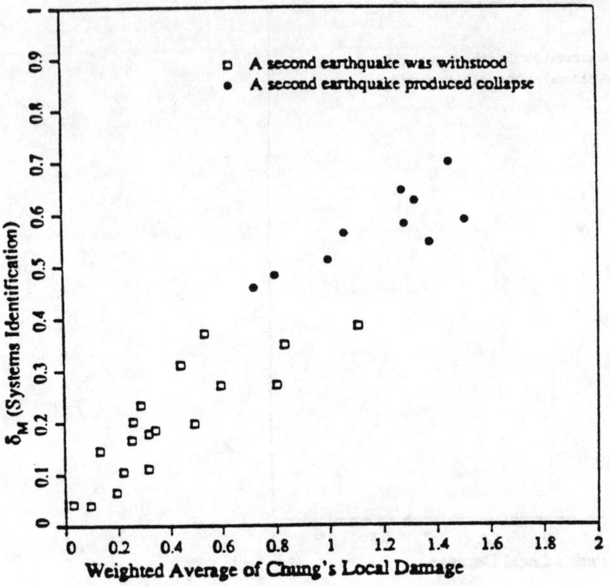

FIGURE 12. Maximum Softening vs. Chung's Weighted Avergage

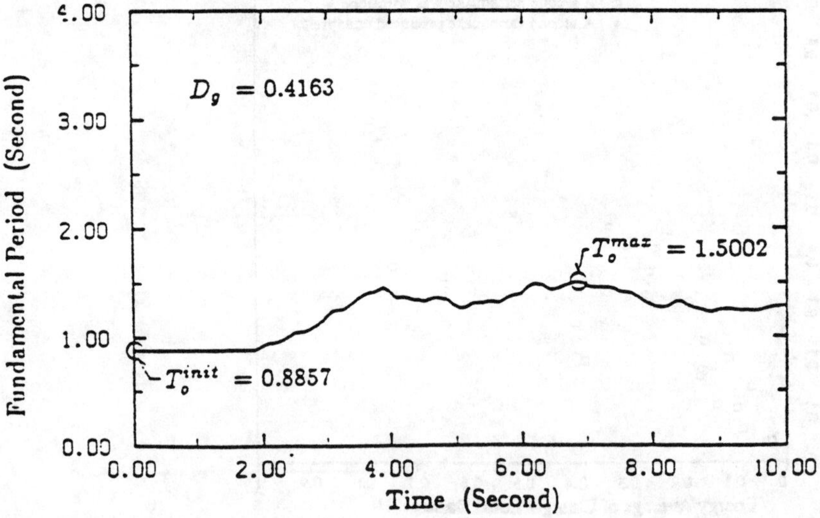

FIGURE 13. Smoothed Time History of Natural Periods for Three Bay Four Story Frame under the El Centro Earthquake Linearly Scaled to have PGA = 1.0g

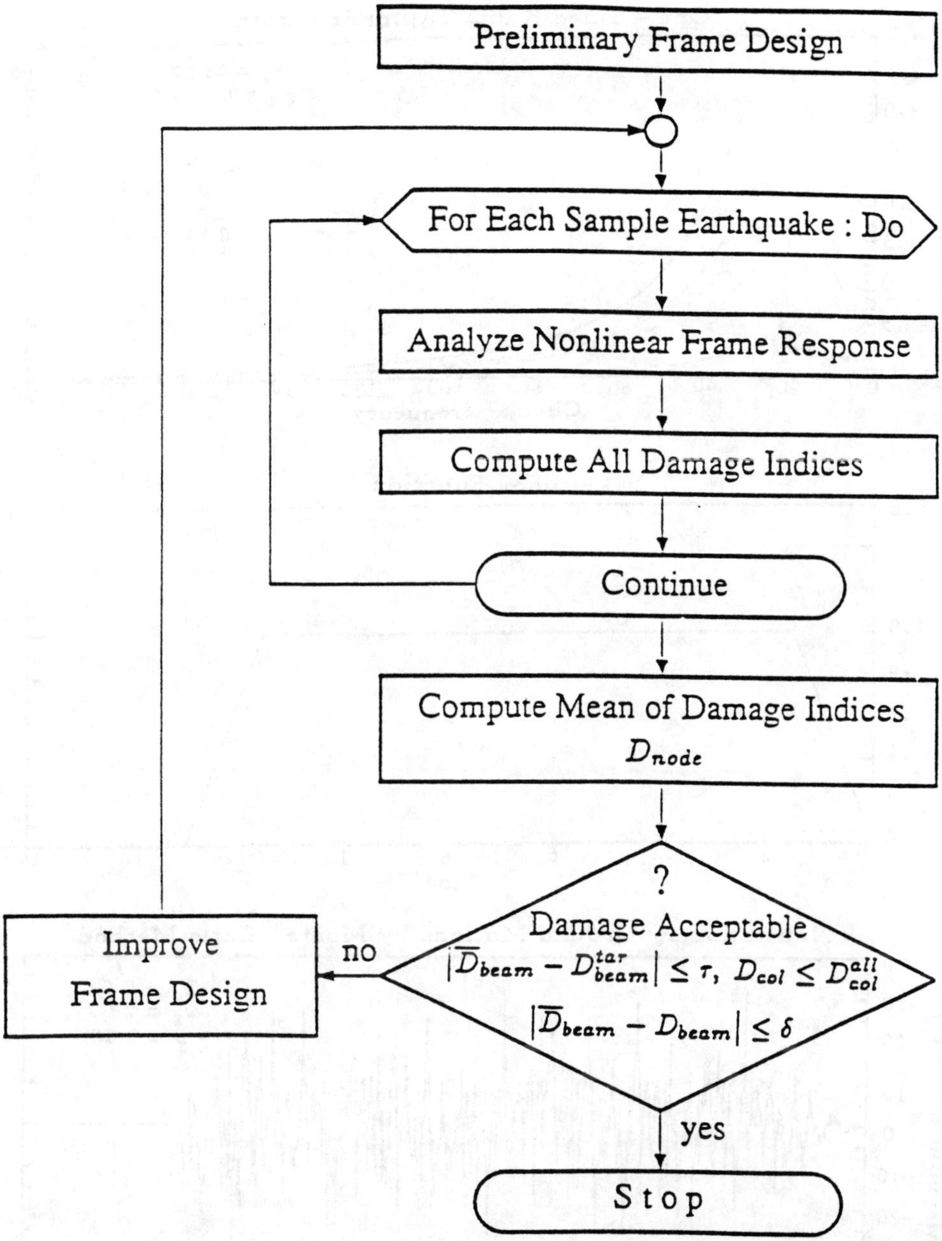

FIGURE 14. Automated Damage-Control Design

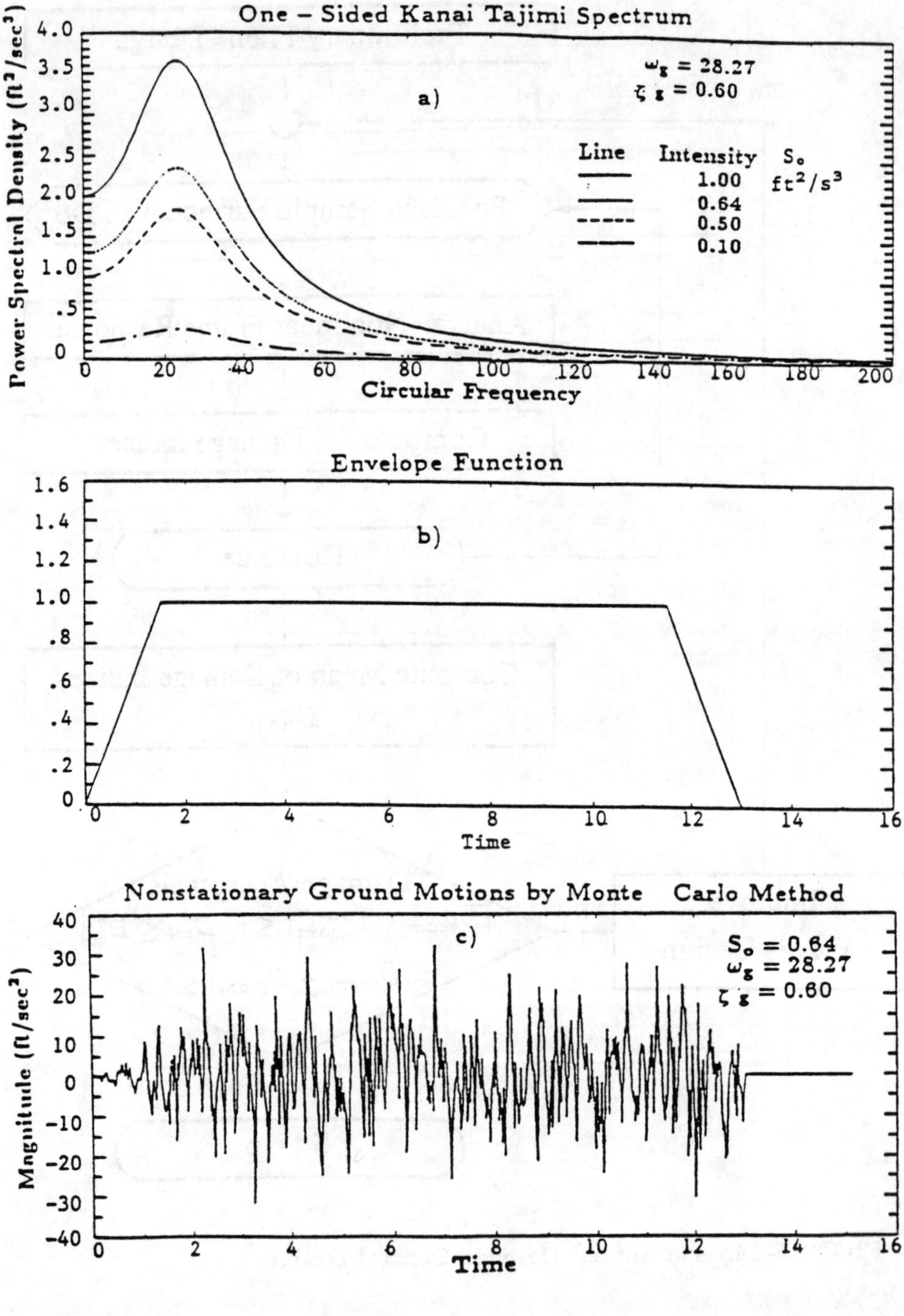

FIGURE 15. Simulation of Nonstationary Ground Acceleration Histories

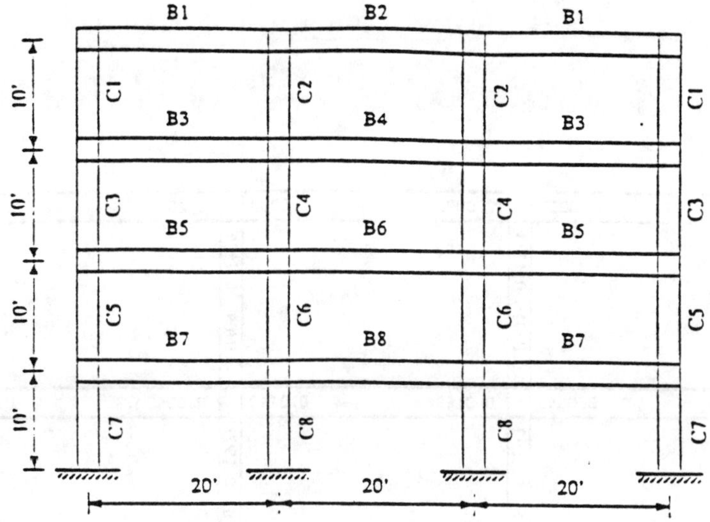

ELEVATION
Frame spacing 20'

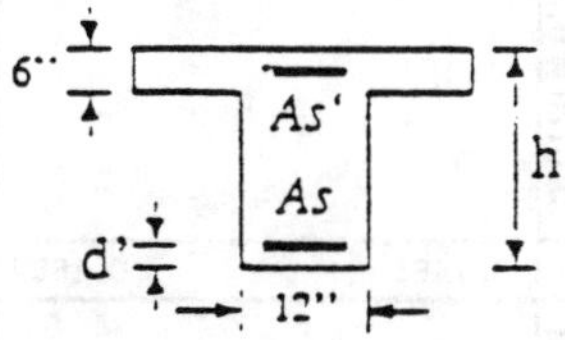

$f_{sy} = 60ksi$
$f'_c = 4ksi$

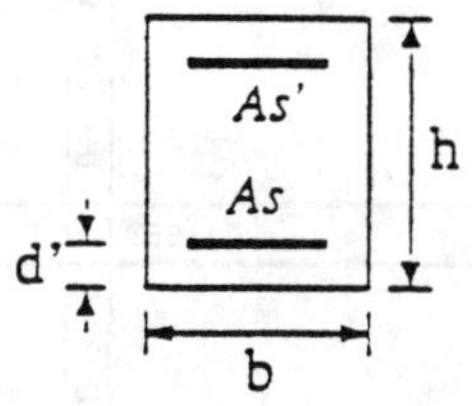

BEAM

COLUMN

	d'	As	As'	h
	(in)	(in²)	(in²)	(in)
B1,B2	2.0	2.280	2.280	20.0
B3,B4	2.0	2.622	2.622	20.0
B5,B6	2.0	2.622	2.622	22.0
B7,B8	2.0	2.736	2.736	22.0

	d'	As	As'	b	h
	(in)	(in²)	(in²)	(in)	(in)
C1,C2,C3	2.0	2.280	2.280	12.0	15.0
C4,C5	2.0	2.622	2.622	12.0	15.0
C6,C7	2.0	2.622	2.622	15.0	18.0
C8	2.0	2.736	2.736	15.0	18.0

FIGURE 16. Details of Three Bay Four Story Building Frame
(1 in = 2.54 cm; 1 kip = 4.448 kN)

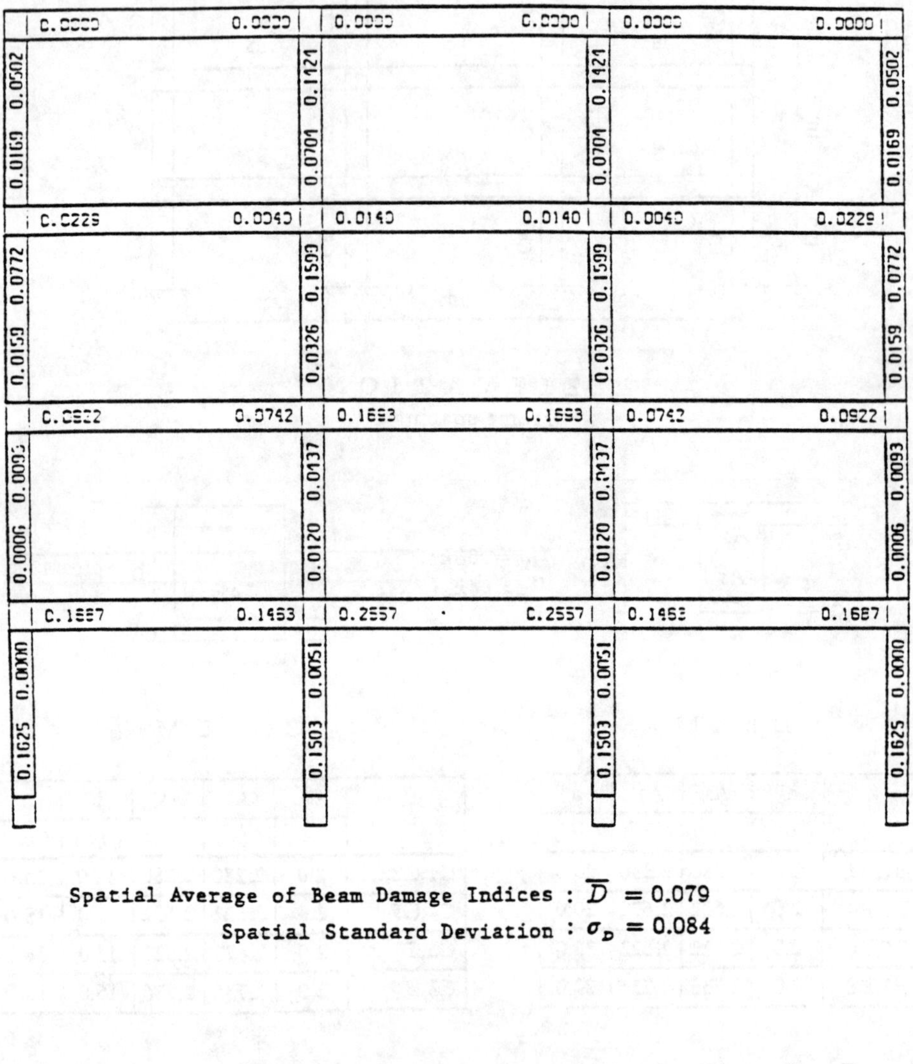

Spatial Average of Beam Damage Indices : $\overline{D} = 0.079$

Spatial Standard Deviation : $\sigma_D = 0.084$

FIGURE 17. Ensemble Average of Damage Indicies for Three Bay Four Story Frame

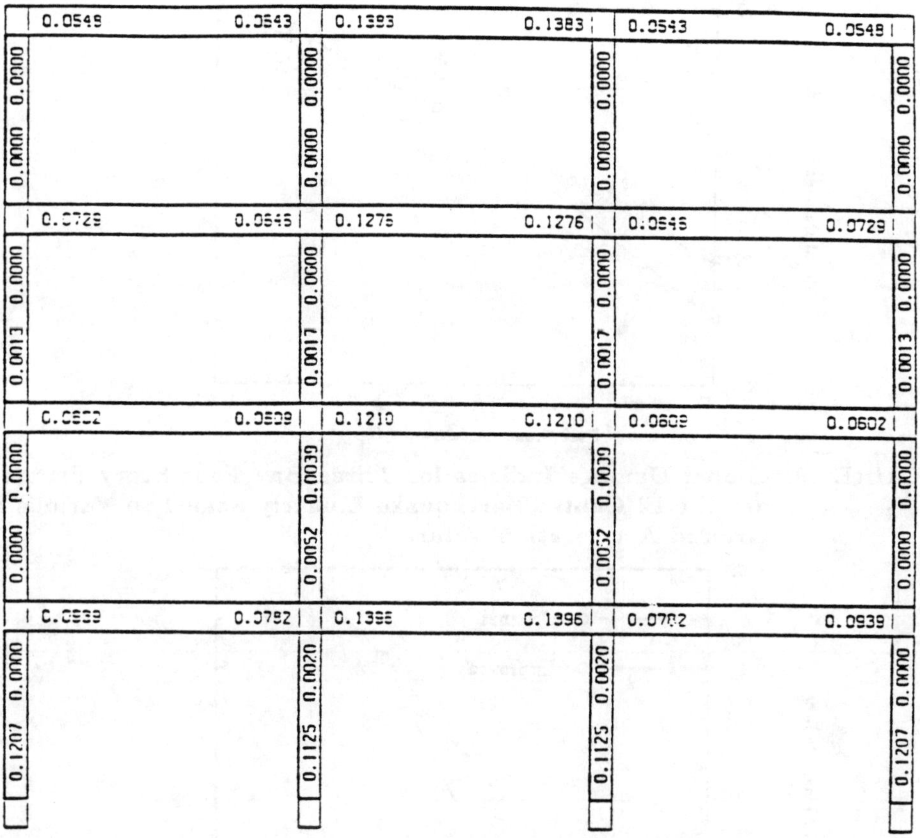

Spatial Average of Beam Damage Indices : $\overline{D} = 0.089$

Spatial Standard Deviation : $\sigma_D = 0.032$

FIGURE 18. Ensemble Average of Damage Indices for Three Bay Four Story Frame After 12 Design Iteration

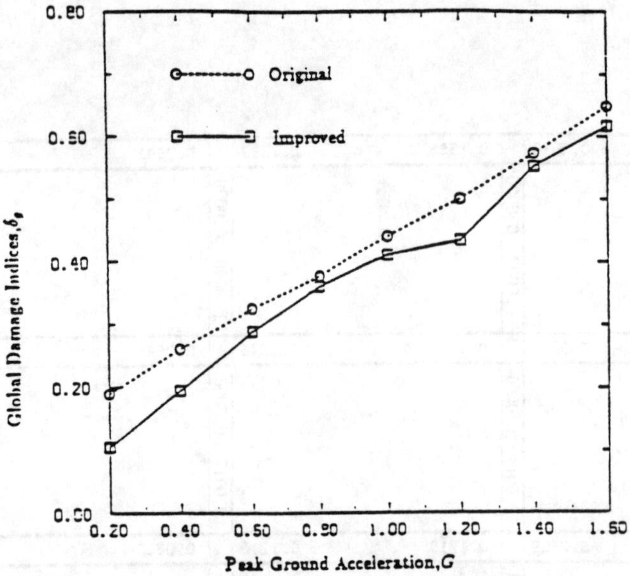

FIGURE 19. Global Damage Indicies for Three Bay Four Story Frame Under the El Centro Earthquake Linearly Scaled to Various Peak Ground Acceleration Values

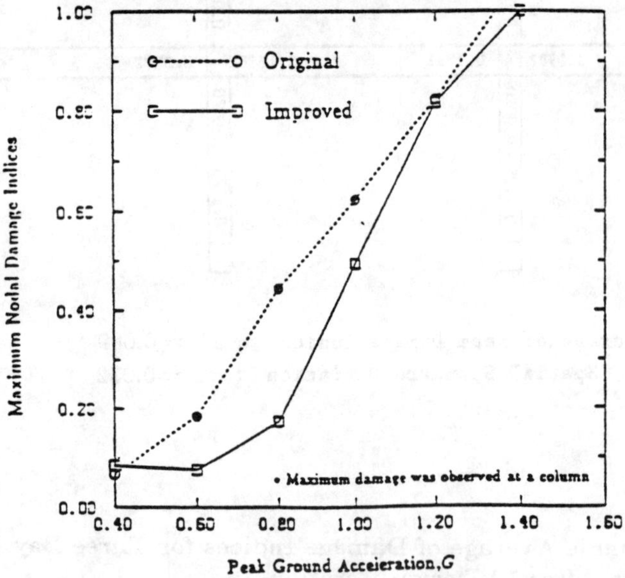

FIGURE 20. Maximum Nodal Damage Indicies for Three Bay Four Story Frame Under the El Centro Earthquake Linearly Scaled to Various Peak Ground Acceleration Values

TABLE 1. Damage Indices Based on Equivalent Linear Models

LOCAL PHENOMENON	MACROSCOPIC FEATURE	GLOBAL INDEX
Stiffness degradation	Cracks	Final softening $\delta_f = 1 - \frac{(T_0)^2_{initial}}{(T_0)^2_{final}}$
Plastic deformation	Yielding of reinforcement bars	Plastic softening $\delta_p = 1 - \frac{(T_0)^2_{final}}{(T_0)^2_{max}}$
Combined effect of plastic deformation and stiffness degradation	Onset of structural damage (service-ability limit state)	Maximum softening $\delta_M = 1 - \frac{(T_0)_{initial}}{(T_0)_{max}}$

TABLE 2. Damage Detection for Actual Buildings

SAN FERNANDO EARTHQUAKE (1971)

STRUCTURE	DAMAGE		*P*[damage]
611 6th Street	None	0.13	0
Sheraton Hotel	None	0.29	0
Milikan Library	None	0.32	0.03
Holiday Inn	Repaired	0.52	0.88
Bank of California	Repaired	0.81	1

IMPERIAL VALLEY EARTHQUAKE (1979)

STRUCTURE	DAMAGE		*P*[damage]
Imperial County (EW)	Demolition	0.61	0.97
Imperial County (NS)	Demolition	0.59	0.97

LOMA PRIETA EARTHQUAKE (1989)

STRUCTURE	DAMAGE		*P*[damage]
Double-deck structure on I-880	Demolition	0.62*	0.97

*Table 1. Type of the structure II [41].

Uncertainty of the Event and Fuzzy Approach

Shibata Heki, Professor-Dr.

Institute of Industrial Science
University of Tokyo
22-1, Roppongi 7, Minato-ku, Tokyo 106, JAPAN

Synopsis

§1 Introduction

This paper deals with the practical concept how to convert the uncertainty of events to fuzzy expression, and fuzzy expression to more engineering expression such as a failure probability using for the Seismic PSA study.

In this paper, the author is mainly discussing on the second part.

§2 Uncertainty in Earthquake Engineering

The author frequently discussed on uncertainty observed in the earthquake engineering, because the main reason of failures of various structures in industrial facilities came from the lack of knowledges on their failure mechanisms and behaviors. This problem is summarized as "Uncertainty". And also came from errors in the processes of design, fabrication, construction, inspection and maintenance. The author has been discussing both subjects. On "Un-certainty" he presented his paper(1) at the first Symposium on this topic last year in Maryland. Term "Uncertainty" is understood in various way. He has been observed the responses of a chemical complex model to natural earthquakes for last twenty years. The fluctuation of its model has been observed, then he found that the distribution of response factors was expressed by a normal distribution model. But not only on their parameters, but also distribution shape was fluctuated. To understand this fluctuation, there are two approaches. That is, a model based on its response to various input waves, and a subjective model without any interpretation. In the case where we employed a subjective model, we must develop the method to convert the subjective measuring result to probabilistic value of engineering parameters, for example, the membership function of the normal distribution. The practical approach should be developed. Only the concept will be discussed in Chapter 4.

§3 Human Error in Design

The second cause is human errors in their processes in design and others as mentioned above. The author discussed in various previous papers(2) and summarized in Ref. [3] with the method to prevent it.

§4 Flow of Analysis and Practice for Qualitative Measuring the Uncertainty Expressed by Fuzzy Expression

The flow of analysis how to analyze may be divided into two parts. One is from fuzzy expression to engineering expression, and another is from one fuzzy expression to other fuzzy expression. On the second subject the author presented it at the 6th International Seminar on Earthquake Prognostics in Berlin in this week. Its flow is shown in Fig. 1. The first part is going to present at the 2nd Japan Conference on Structural Safety and Reliability which will be held in this coming November in Tokyo. The following is the summary of the practice.

Against a engineering parameter like a computed stress, the designer judge which the structure is safe or unsafe. The expression on "safe" or "unsafe" is fuzzy. Usually we have a criteria like an allowable stress χ_0 as shown in Fig. 2. If the judgement is crispy like curve a, then the failure probability is just expressed by its fragility curve as shown in Fig. 3. However, the designer has some ambiguity for his judgement as the curve b in Fig. 2, then, the fragility curve should be modified. And in some other case, his judgement may be biased as the curve c by his error. Moreover, he may has a possibility to judge completely different way as the curve d. Last one is out of discussion here.

The evaluation of these three cases on its failure probability is summarized as follows:

i) crispy: curve a

$$f(\chi_0)$$

in this case

$$\mu(\chi_0) = 0 \text{ or } 1$$

ii) Type F, Fuzzy: curve b

$$[1-\mu(x)]*f(r)$$

iii) Type S, biased: curve c

$$[1-G_x(y)\ \mu(x)]*f(r)$$

In both cases, we assume $x \equiv r$.

For such calculations, $\mu(x)$ should be restricted. It was discussed in the previous papers including that of JCOSSAR-91. It may be referred in the presentation.

§5 CAE to Avoid Human Error

The author summarized his idea in Ref. [3]. Some idea was employed for the Aseismic Design of High Pressure Facility based on the Notice No.515 from the Ministry of International Trade and Industry in Japan in 1981, and it has been practically used for the governmental regulation of the design of plants through Japan.

§6 Concluding Remarks

The author has been working on how to measure the fuzzyness of the human operator in the plant. This concept may be applied to the measure of the judgement of designers as described in Chapter 4. He will discussed on this subject at the JCOSSAR-91, but it has been not developed enough. He feels more study to complete the series of the research necessary why structure, equipment and piping systems in industrial facilities were failed during past destructive earthquake, and how to eliminate the human errors which may cause catastrophic disaster of nuclear power plants under such seismic event.

§7 Reference

[1] SHIBATA, H., Uncertainty in Earthquake Engineering in Relation to Critical Facilities, *Preprint for 1st International Symposium on Uncertainty Modeling and Analysis*, *IEEE* (Maryland), Dec. 1990, 6pp.

[2] SHIBATA, H., Effect of Human Error for Structural Design under Seismic Loadings and its Evaluation, *Proc. of NSF Workshop on Civil Eng'g. Application of Fuzzy Sets* (Annarbor), Oct. 1986, p.45.

[3] SHIBATA, H., FUJII, H., TSUCHINO, H. and NAGAYA, Y., On a Concept of Aiding Anti-earthquake Design by a Computer for Industrial Critical Facilities, *Proc. of 9 WCEE*, V, March 1989, p.1013.

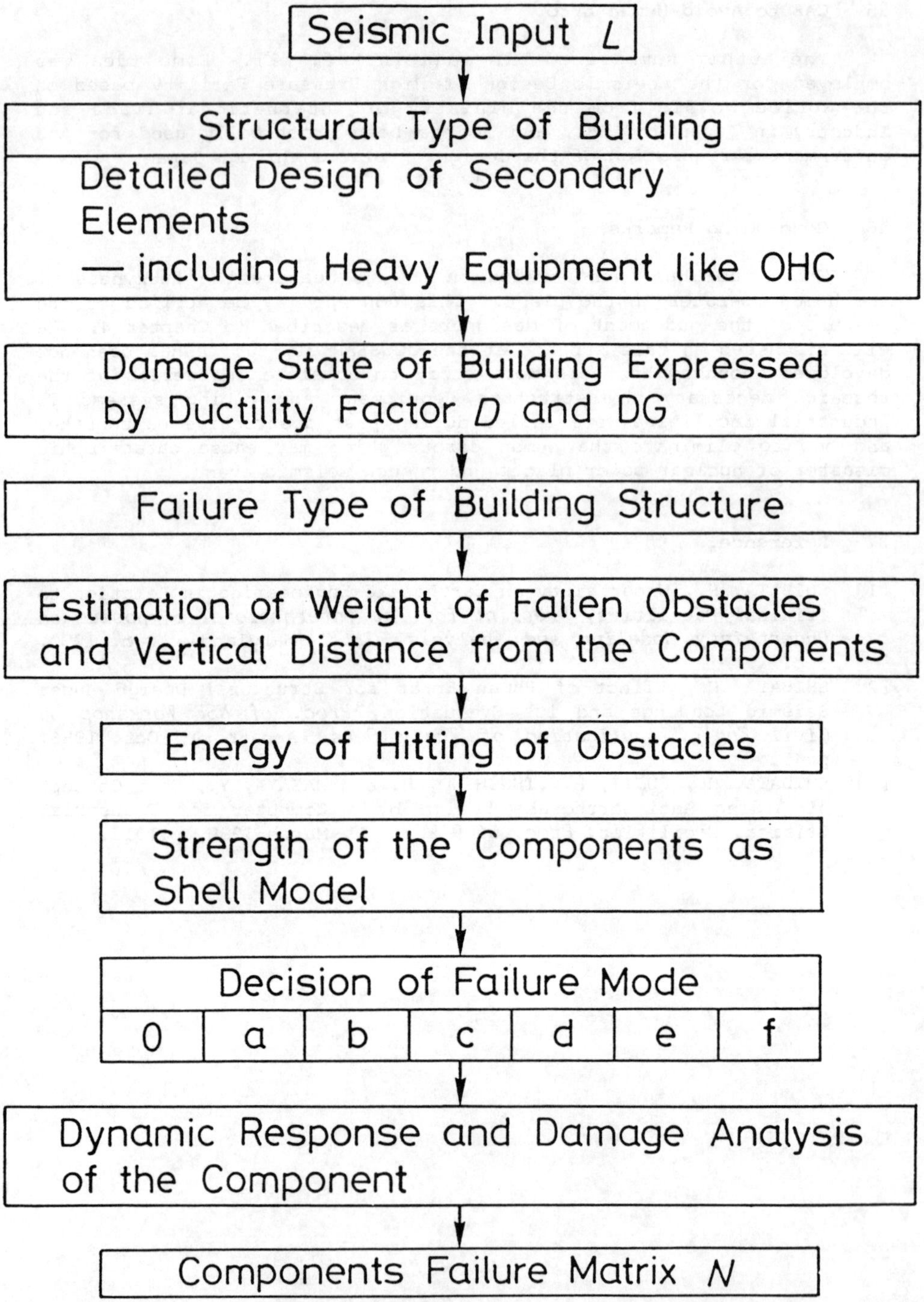

Fig.1 Process Flow of Design Analysis and Communication

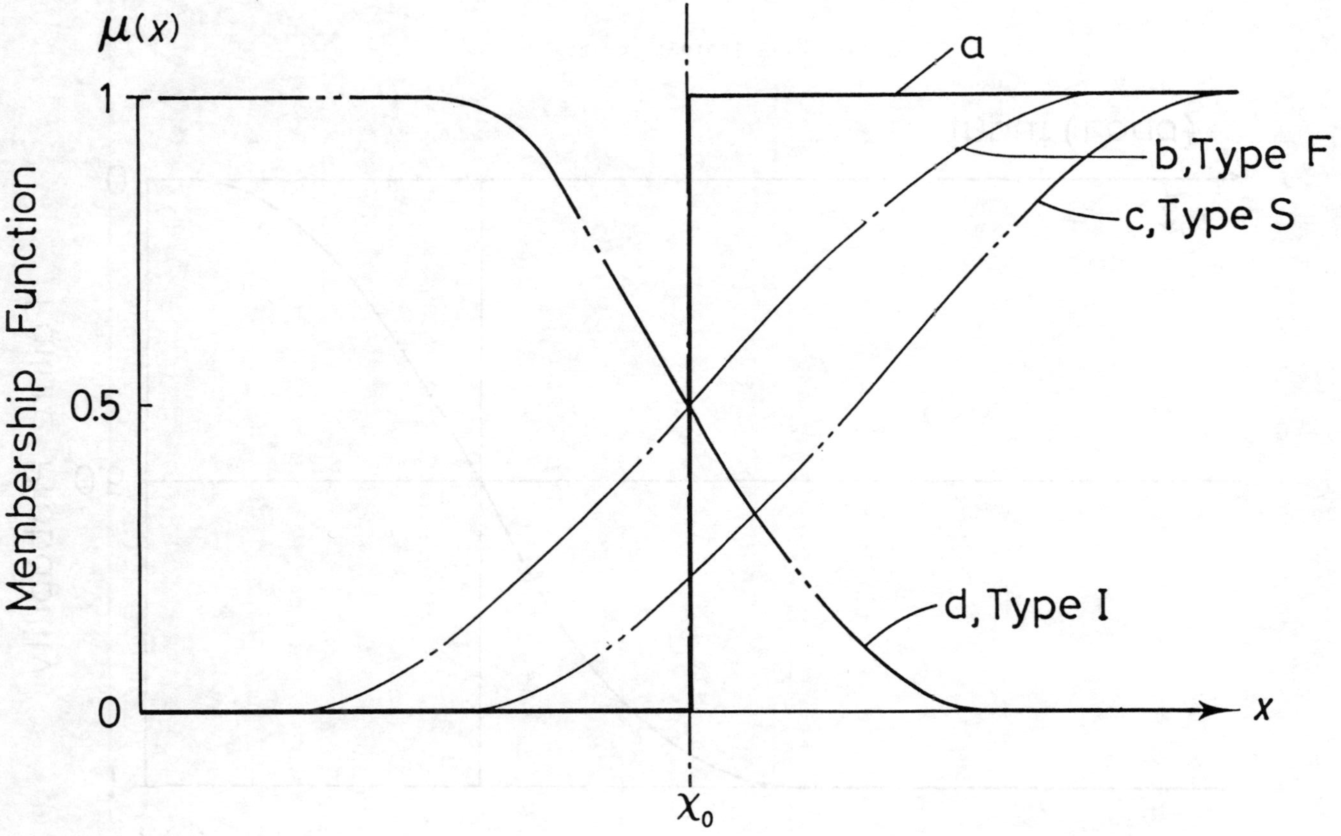

Fig. 2 Membership Function *v.s.* Input for Judgement
--- Types of Subjective Judgements

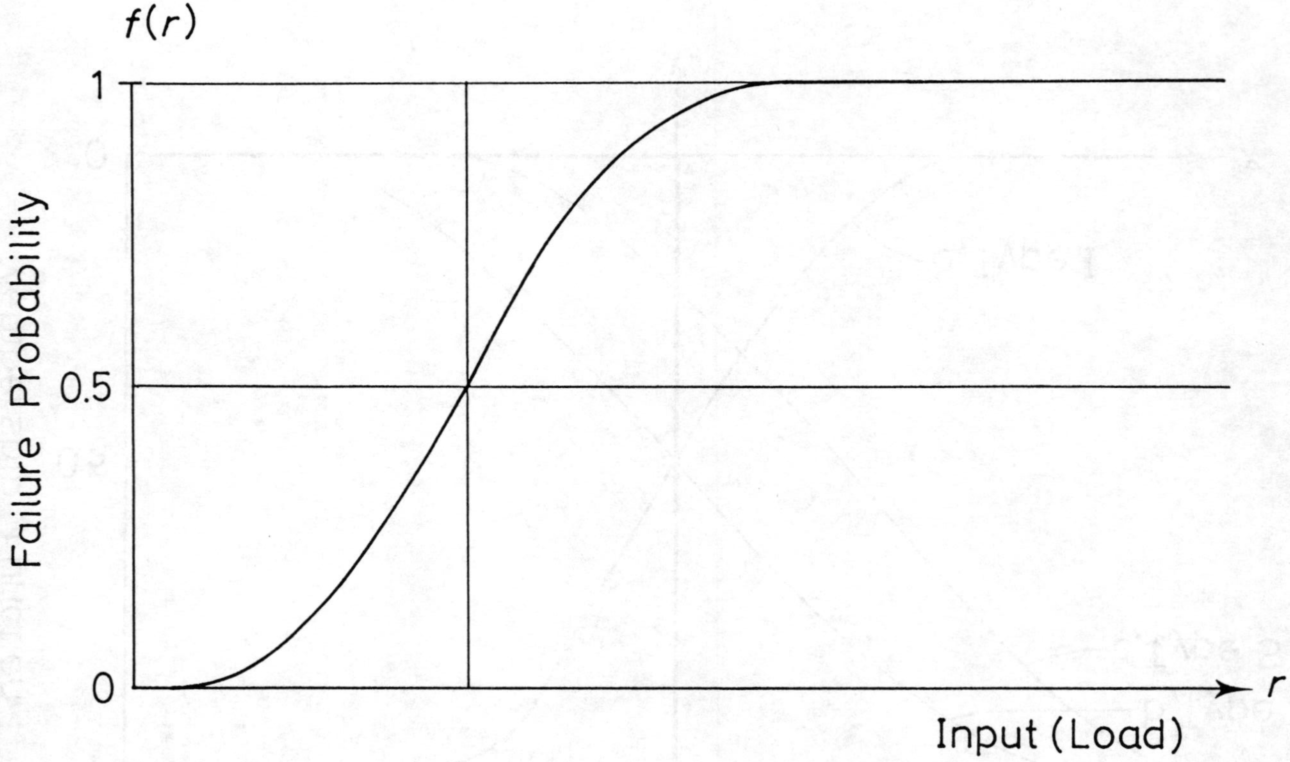

Fig. 3 Typical Fragility Curve

RELIABILITY AND COST–EFFECTIVENESS OF STRUCTURES WITH ACTIVE CONTROL

Y. K. WEN AND A. H–S. ANG
Department of Civil Engineering, University of Illinois, Urbana, IL USA and
Department of Civil Engineering, University of California, Irvine, CA USA

ABSTRACT

Reliability and cost–effectiveness of active control in structures are examined and a conceptual model for expected total life cycle cost analysis is presented. It is shown that full reliance on control to protect life safety is not a viable design alternative at present. Control can be used as an effective redundant component to reduce cost of damage and unserviceability and in such a design the cost–effectiveness does not require high level of reliability of the control system.

INTRODUCTION

As described by Yang and Soong [1], the essential components of a structure with active control system (closed–open–loop type) would be as shown in Fig. 1, consisting of the following:

- the structure
- the actuator system
- the sensor system
- the computer system, composed of the computer hardware and the operating software

Each of these components, of course, would have a limited reliability (i.e., < 1.0) over a specified duration. Moreover, the functional reliability of the control system (consisting of the computer processor, the sensors, and the actuators) will obviously affect the reliability of the structure, and the advantages of the controlled structure will be realized only if structural safety can be assured with high reliability –– e.g., relative to that of a conventional structure without control. However, even if all the components of the control system can be assured, it will remain difficult for the "owner" or occupants to accept complete reliance on the control system for the safety of a structure during an earthquake or other extreme hazard. Nevertheless, an active control system can contribute to structural safety in a cost–effective way.

DESIGN ALTERNATIVES

Obviously, the cost–effectiveness of an active–controlled structure would be enhanced only if the reliability of the entire system can be assured. The perceived reliability, or lack thereof, is an important consideration in the practical implementation of active control system in structural engineering.

For practical implementation, the foremost consideration is the safety of the structure, particularly against collapse, e.g., under the design earthquake. In order to insure this fail–safe structure, it would be prudent to design the structure under the assumption of a worst case scenario; i.e., that during a strong earthquake, the control system will fail to function properly and thus the uncontrolled structure must be able to withstand the seismic forces without collapse. However, as is implicit in current seismic codes, a tolerable degree of damage may be acceptable.

Several alternative scenarios may be examined with reference to earthquake safety, as follows:

Alternative A: Conventional structure plus active control system –– The control system serves as a redundant system to increase structural reliability.

Alternative B: Full reliance on control system –– Design the structure assuming the control system is fully reliable during an earthquake.

Alternative C: Control system to avoid damage only –– Design the structure with reduced seismic capacity to tolerate damage but without collapse.

These alternative scenarios are also summarized in the table below.

TABLE 1
Alternative Scenarios

Alternative	Structural Design Basis	Role of Active Control
A	As a conventional structure	To provide additional safety; redundant safety system
B	Controlled response	To reduce response motion necessary to insure structural safety
C	Reduced seismic capacity to tolerate damage without collapse	To reduce response motion necessary to prevent structural damage

In Alternative A, the active system is completely redundant as far as structural safety is concerned. The structure is fully capable of resisting the earthquake without the control system even though some damage may be expected.

In Alternative B, structural safety is dependent on the operational reliability of the control system; i.e., if the control system should fail during an earthquake the structure would have high probability of collapse.

In Alternative C, a structure would be designed with reduced capacity (from that of the conventional design) admitting a tolerable degree of structural damage under the design earthquake but will not collapse assuming the absence of control. That is, in this alternative the uncontrolled structure will likely sustain damage but is fail–safe against collapse even if the control system should fail during the design earthquake.

COST–EFFECTIVENESS OF ALTERNATIVES

In Alternative A, the control system is completely redundant as far as the safety of the structure is concerned and thus would involve additional cost. The benefit of the control system is that the occupants, in the case of a building, will feel lower intensity motions than would otherwise; this may be important for protection of sensitive equipment. Also, if the system is operational during the earthquake there will be no damage to the structure and, therefore, a saving is realized in terms of damage cost in the total life cycle cost of the structure.

Alternative B would be difficult to accept, especially for structures such as buildings and bridges where the loss of human lives may be involved in case of structural collapse should the control system fail to operate properly. Theoretically, this alternative would be most cost–effective if the control system is reliable. However, regardless of its reliability, complete reliance on its proper operation during all earthquakes over the life of the structure would be extremely difficult to "sell". For this reason, Alternative B cannot be viable from a practical standpoint.

Alternative C is a compromise between alternatives A and B. In this alternative, there is a saving in the cost of the structure derived from a lower capacity requirement in its design than that of the corresponding conventional structure. Safety against collapse is assured with high reliability without relying on the control system, even though the structure can be expected to sustain a tolerable degree of damage. The pay–off with this alternative is that damage cost can be avoided when the control system is operational, and the occupants would experience lower intensity of response motions. To implement this alternative requires a precise and quantitative method for assessing structural damage of an uncontrolled structure as well as for estimating its reliability against collapse.

EXPECTED LIFE CYCLE COST ANALYSIS

To demonstrate the actual benefit of the Alternative A or C, a life cycle cost–effectiveness analysis is necessary. It needs to take into consideration the randomness of the loading, consequence of "failure", exposure time, and reliability of the control system. Earthquakes, severe winds and waves and other man–made dynamic loadings are the major concerns for structural control. Obviously, the frequency and intensity of these loadings and the consequence of failure have a bearing on whether control devices and how much control are needed. Risk of failure increases with exposure time; also environmental loads such as earthquakes occur infrequently and consequence of distant future failure needs to be properly discounted. Finally, the reliability of the control unit obviously affects the overall reliability and cost effectiveness of the system.

Assumptions

To allow a closed form solution and an easy parametric study of the problem, the following assumptions are made:

i) The loading occurs in time according to a simple Poisson process with an occurrence rate ν and the intensities are random and independent from occurrence to occurrence.

ii) The structural resistance is time invariant and the probability of failure (limit state being exceeded) does not depend on time.

iii) The limit state can be adequately described in terms of the maximum response of the structure.

iv) To discount future cost in present value a constant discount rate can be used.

Formulation

The expected total cost of a structure for a given time period (0,t) can be formulated as follows:

$$E(C) = C_o + E[\sum_{i=1}^{\infty} C_f(t_i)\ P_f\ ] + \int_0^t C_m(t)\ dt \tag{1}$$

in which
- E = expected value
- C_o = initial cost of the structure with control
- C_f = cost of failure (limit state being exceeded), e.g., including damage cost and cost of loss of service
- t_i = time of i–th occurrence of loading
- P_f = probability of failure given the occurrence of loading at $t = t_i$; assumed to be time–invariant
- C_m = maintenance cost of the control system per year, significant for active systems, small for passive systems

Using a constant discount rate λ per year, $C_D(t_i)$ and $C_m(t_i)$ are given by

$$C_f(t_i) = C_f\ e^{-\lambda t_i}$$

$$C_m(t) = C_m\ e^{-\lambda t} \tag{2}$$

in which C_f and C_m are in present dollar value.

The limit state probability of the structure with control depends on whether the control device operates properly under loading. As shown in Fig. 1, the mechanical system and the associated reliability problems of the structure with control can be quite involved. Conceptually, however, the following simple model can be used

$$P_f = P(X > x_o|C)\ P(C) + P(X > x_o|\overline{C})\ P(\overline{C}) \tag{3}$$

in which X = maximum response under consideration, e.g., interstory drift or maximum displacement at a critical location of the structure

x_o = allowable maximum response

C = event that the control system operates properly

$\overline{C}$ = event that the control system does not operate properly

$P(X > x_o | C)$ = probability of failure given C

$P(X > x_o | \overline{C})$ = probability of failure given $\overline{C}$

Determination of the probabilities in Eq. 3 requires reliability analyses of the structural and the control systems. Several general observations, however, can be made:

i) $P(X > x_o | \overline{C})$, the probability of failure of the structural system without control is small for Alternative A since the control system acts as a redundant component. It would be higher for Alternative C because of the weaker structural design and high for Alternative B since it fully relies on the control system to resist the loading.

ii) $P(\overline{C})$ requires a detailed system reliability analysis of the control system; very little information is currently available regarding either the components or the system.

Substituting Eq. 2 into Eq. 1, one obtains.

$$E(C) = C_o[1 + rP_f \frac{\nu}{\lambda}\ (1 - e^{-\lambda t})] + \frac{C_m}{\lambda}\ (1 - e^{-\lambda t}) \tag{4}$$

in which $r = C_f/C_o$, the cost ratio; it can be interpreted as a nondimensional parameter of consequence of "failure". Some details of derivation are given in the Appendix.

The ratio of the expected total cost of a structural system with control to that of a conventional system can be obtained as

$$\frac{E(C_2)}{E(C_1)} = (\frac{C_2}{C_1}) \left[\frac{1 + \frac{1}{\lambda}(r_2\ \nu p_2 + r_m)(1 - e^{-\lambda t})}{1 + \frac{1}{\lambda}\ r_1\ \nu p_1\ (1 - e^{-\lambda t})} \right] \tag{5}$$

in which $E(C_2)$ = expected total life cycle cost of a system with control

$E(C_1)$ = expected total life cycle cost of a conventional system

C_2 = initial cost of a system with control

C_1 = Initial cost of a conventional system

t = lifetime or exposure time of the system

p_2 = probability of "failure" of the controlled system given the occurrence of the loading (Eq. 3)

p_1 = probability of "failure" of the conventional system given the occurrence of the loading

r_2 = ratio of cost of "failure" (in present value) to initial cost for the system with control

r_1 = ratio of cost of "failure" (in present value) to initial cost for the conventional system

r_m = C_m/C_2, maintenance cost of the control system per year as a fraction of the initial cost

It is obvious that in order for a system with control to be cost–effect compared with a conventional system, the expected total life cycle cost ratio given by Eq. 5 should be less than unity.

Parametric Study

Equations 3 and 5 allow one to perform a parametric study of the cost–effectiveness of a system with control. In this study, we will use Alternative A as an example and make comparisons with a conventional system which is simply the Alternative A design without the control system. The loading is that due to earthquakes in the structure's lifetime with a mean occurrence rate of 0.1 per year.

Recent research [1,2] indicates that active control can reduce the response level of a structure by 30% to 50% under realistic loading conditions. A reduction of 40% of the maximum response of the structure will be assumed in this study. Under seismic loading, the variability of the response is high largely due to the variability and uncertainty in the loading. Experience indicates that a coefficient of variation of 60% is a realistic number and that the maximum response can be approximately modeled by an extreme value type I distribution [3]. The above information allows one to calculate the conditional probabilities in Eq. 3 by assuming that the mean value of the maximum response for system with control is 60% of that of the conventional system and a coefficient of variation of 60%. Table 2 shows the calculated conditional limit state probabilities based on this information. The probability levels represent different limit states under consideration.

TABLE 2
Conditional Limit State Probability

$P(X>x_o\|\bar{C})$	10^{-1}	10^{-2}	10^{-3}
$P(X>x_o\|C)$	8.3×10^{-3}	1.7×10^{-4}	3.8×10^{-6}

In Alternative A, the cost increase is generally only a few percentage of the initial total cost [4]. The cost ratios r_1 and r_2 for the two systems would be similar and are assumed to be the same in this study, i.e., $r_1 = r_2 = r$. A wide range of the cost ratio is used reflecting limit states of different levels of consequence. Similarly, a wide range of $P(\bar{C})$ is used to examine the effect of reliability of the control systems.

The term νp_1, in Eq. 5 corresponds to the annual probability of the limit state being exceeded in current design. For example, in ATC–3–06 [5], the implied risk of "life–

threatening damage" in current design is of the order of 10^{-3} per year and risk of collapse is 10^{-4} to 10^{-5} per year. More specifically, Wen, et al. [4] found that the risk of 2% interstory drift ($\approx$4 in.) being exceeded in a Special–Moment–Resistant–Space–Frame (SMRSF) building designed according to 1988 UBC at a site in Imperial Valley, California, 5km from the Imperial Fault is of the order of 10^{-3} per year. The Imperial Valley is a high seismicity area, well known for the damaging earthquake ground motions recorded in the last 50 years. These numbers can be used as a reference in the parametric study. It is pointed out that in reality, for a particular design, these parametric values are related. For example, for an important structure whose proper performance under the loading is crucial, r would be large and νp would be small and so on.

Figure 2 shows the ratio of the expected total life cycle costs as a function of exposure time for different values of cost ratio r. The discount rate λ, the initial cost ratio C_2/C_1, the maintenance cost ratio $r_m = C_m/C_2$, and the annual risk of the limit state in the structure without control νp, are also indicated in the figure. For $\nu p_1 = 10^{-4}$, the level corresponding to collapse or life threatening damage, the expected life cycle cost ratio is always greater than one, even for large value of r. This indicates that the control system is not cost–effective simply because the risk is small enough in current design against such limit states that there is not much to improve upon. For $\nu p_1 = 10^{-3}$, the level corresponding to serious damage, the control system would be cost–effective if the consequence is large, say $r > 5$.

Figure 3 shows the results for $\nu p_1 = 10^{-2}$, which may be interpreted as the level corresponding to moderate damage and serious serviceability problem. Only small values of r are considered. It is clear that control can be cost–effective against this type of limit states. The effect of the reliability of the control system is shown in Fig. 4. It is seen that in Alternative A, the expected life cycle cost ratio is not particularly sensitive to the reliability of the control system as long as the failure probability is kept below a reasonable level (say < 0.1).

In the above discussion, a constant initial cost ratio $C_2/C_1 = 1.06$ is used. From Eq. 5 one can also see that in order for the control system to be cost–effective, C_2/C_1 cannot be larger than the reciprocal of the expression in the square bracket. The results are shown in Fig. 5. For example, for control against a limit state of serious damage ($\nu p_1 = 10^{-3}$), the cost should not be more than a few percentage of the initial total cost of the conventional design unless the consequence of the limit state is large.

The annual maintenance cost has been assumed to be only about 1% of the cost of the control in the above analysis. It is obvious from Eq. 5 that it may become significant and have an impact on the cost effectiveness if the maintenance cost ratio r_m approaches the value of $r_2 \nu p_2$ or $r_1 \nu p_1$. For example, for the case $\nu p_1 = 10^{-3}$ ($\nu = 0.1$, $p_1 = 10^{-2}$) from Table 2, $\nu p_2 = 1.7 \times 10^{-5}$. The annual maintenance cost (with $r_m = 0.0005$) would be actually higher than the expected annual failure cost for $r_2 < 30$. More importantly, if $r_1 < 0.5$, it would be higher than $r_1 \nu p_1$, the annual expected cost of the uncontrolled structure, hence, making the control not cost effective.

SUMMARY AND CONCLUSIONS

The reliability and cost–effectiveness of a structure with active control are examined in the context of three design alternatives: A – conventional structure plus active control system; B – full reliance on control system; and C – control system to avoid damage only. A conceptual model for the cost–effectiveness analysis is also presented considering the loading uncertainty, consequence of failure and effect of reliability of the control system. It is shown that Alternative B is presently not a viable design because of the requirement of high reliability of the control system which cannot be assured since very little information is available at present. Parametric study based on this model shows that in an Alternative A design.

i) It is not cost–effective to protect life safety;

ii) It is cost–effective against severe to moderate damage and unserviceability if the consequence is large;

iii) High level of reliability of the control system is not needed to ensure the cost–effectiveness of the system; and

iv) Maintenance cost could become an important factor.

ACKNOWLEDGMENTS

This study is supported by the National Science Foundation under Grants NSF CES–88–22690 and NSF BCS–89–03080. This support is gratefully acknowledged.

REFERENCES

1. Yang, J. N. and T. T. Soong. "Recent Advances in Active Control of Civil Engineering Structures," Probabilistic Engineering Mechanics, Vol. 3, No. 4, 1988, pp. 179–188.

2. Soong, T. T. and A. Reinhorn. "Full Scale Structural Applications of Active Control Systems," NCEER Bulletin, Vol. 5, No. 2, April 1991.

3. Wen, Y. K.; D. A. Foutch; D. Eliopoulos; and C. Y. Yu. "Reliability Evaluation of Current Design Procedures for Steel Buildings Under Seismic Loads," Proceedings, 6th International Conference on Applications of Statistics and Probability in Civil Engineering (ICASP6), Mexico City, Mexico, June 1991.

4. "Computerized Counterweight System Actively Controls Building Vibrations," Perspectives, Vol. 9, No. 2, Kajima Corporation, 1989.

5. Applied Technology Council (ATC). "Tentative Provisions for the Development of Seismic Regulations for Buildings," ATC Report, ATC–3–06, Palo Alto, California, 1978.

APPENDIX: Derivation of Expected Failure Cost

In Eq. 1, the expected value of summation of cost of future failure needs to be evaluated. Since the occurrence of the loading is modeled by a Poisson process, the number of occurrences N is a random variable. Conditional on N = n in (0,t), the occurrence time $T_1, T_2, ..., T_n$ are independent and uniformly distributed in (0,t). Therefore, in Eq. 1 one obtains,

$$E[\sum_{i=1}^{n} e^{-\lambda t_i}] = n \int_{o}^{t} \frac{e^{-\lambda \tau} d\tau}{t} = \frac{n}{t\lambda} [1 - e^{-\lambda t}] \qquad \text{(A-1)}$$

The unconditional expected value considering the random number of occurrences is given by

$$\sum_{i=1}^{\infty} \frac{n}{t\lambda} (1 - e^{-\lambda t}) \frac{(\nu t)^n}{n!} e^{-\nu t} = \frac{\nu}{\lambda} [1 - e^{-\lambda t}] \qquad \text{(A-2)}$$

which can be substituted back into Eq. 1 to obtain Eq. 4.

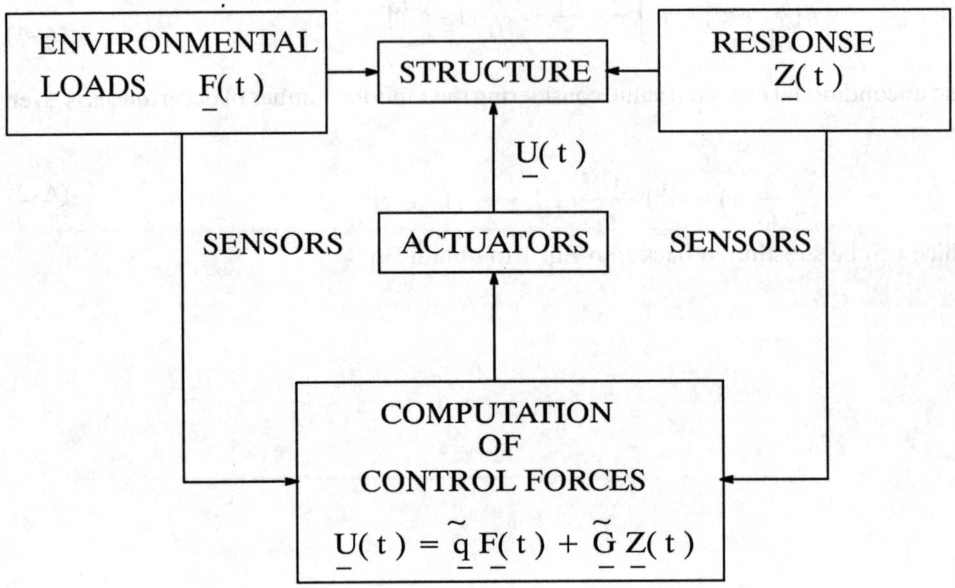

Fig. 1 Closed–Open–Loop Control

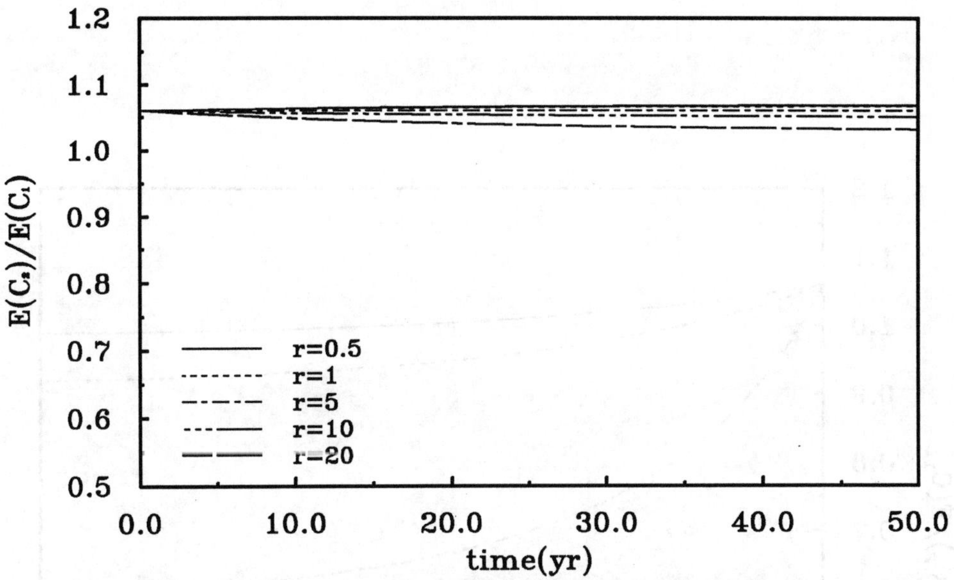

λ =0.05 C_2/C_1=1.06 C_m/C_2=0.0005 νP_1=0.0001 $P(\bar{C})$=0

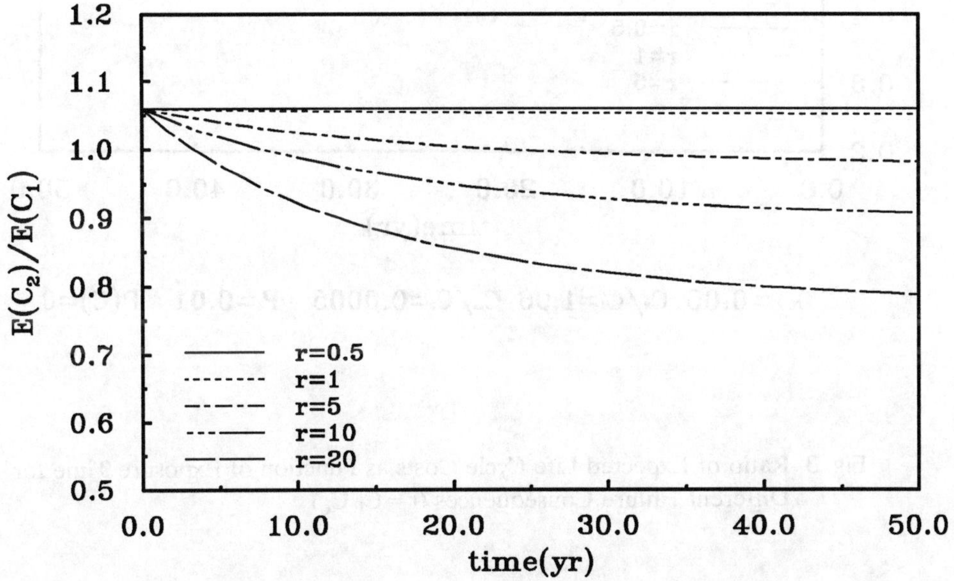

λ =0.05 C_2/C_1=1.06 C_m/C_2=0.0005 νP_1=0.001 $P(\bar{C})$=0

Fig. 2 Ratio of Expected Life Cycle Costs as Function of Exposure Time for Different Failure Consequences ($r = C_f/C_O$)

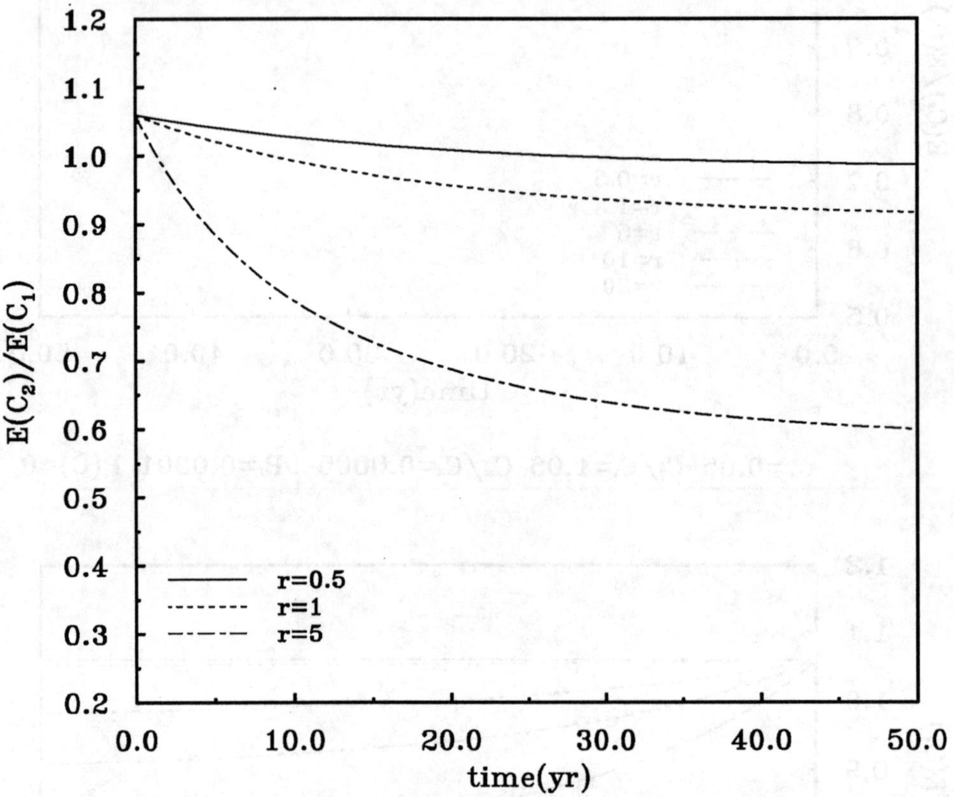

λ =0.05 C_s/C_1=1.06 C_m/C_s=0.0005 νP_1=0.01 $P(\bar{C})$=0

Fig. 3 Ratio of Expected Life Cycle Costs as Function of Exposure Time for Different Failure Consequences ($r = C_f/C_o$)

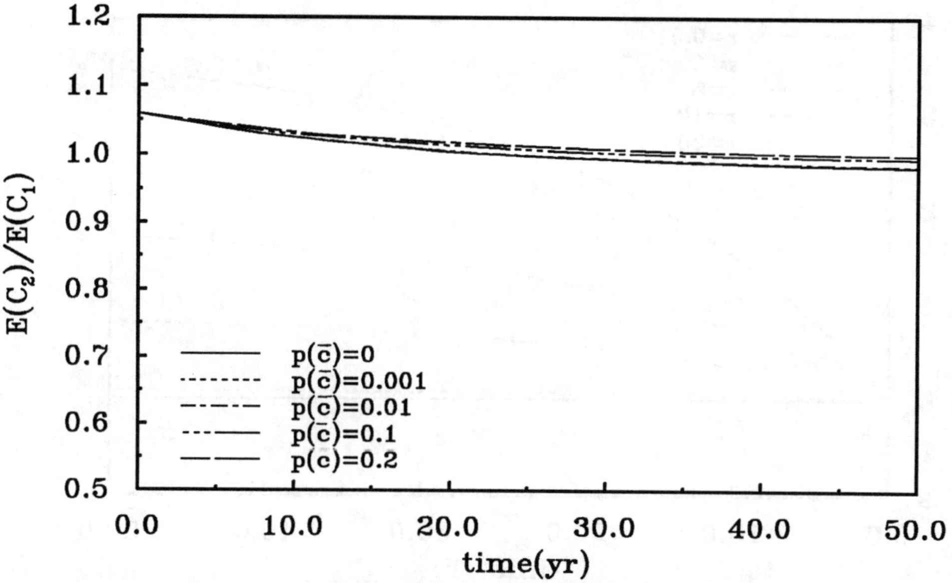

$\lambda=0.05$ $C_2/C_1=1.06$ $C_m/C_2=0.0005$ $\nu P_1=0.001$ $r=5$

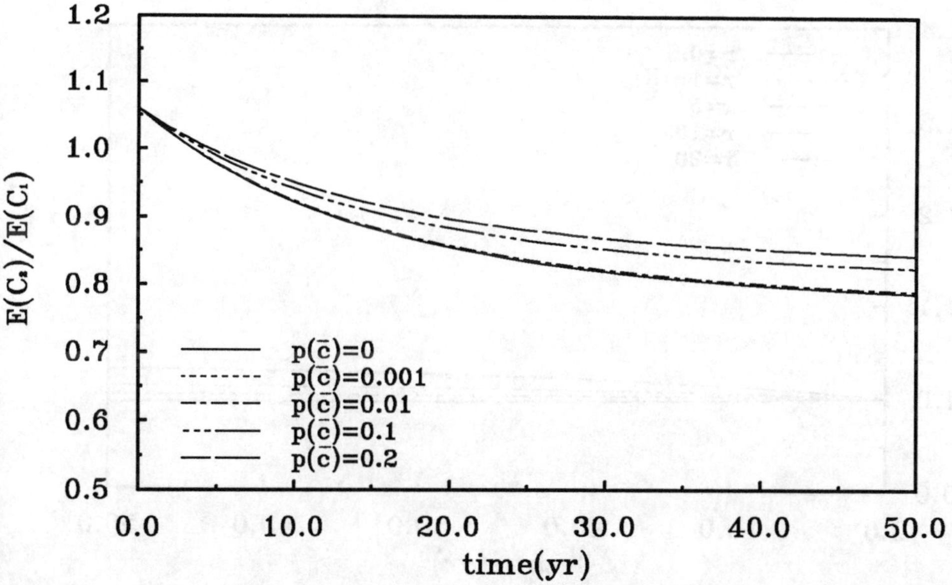

$\lambda=0.05$ $C_2/C_1=1.06$ $C_m/C_2=0.0005$ $\nu P_1=0.001$ $r=20$

Fig. 4 Ratio of Expected Life Cycle Costs a Function of Exposure Time for Different Reliability Levels (or Failure Probabilities) of the Control System

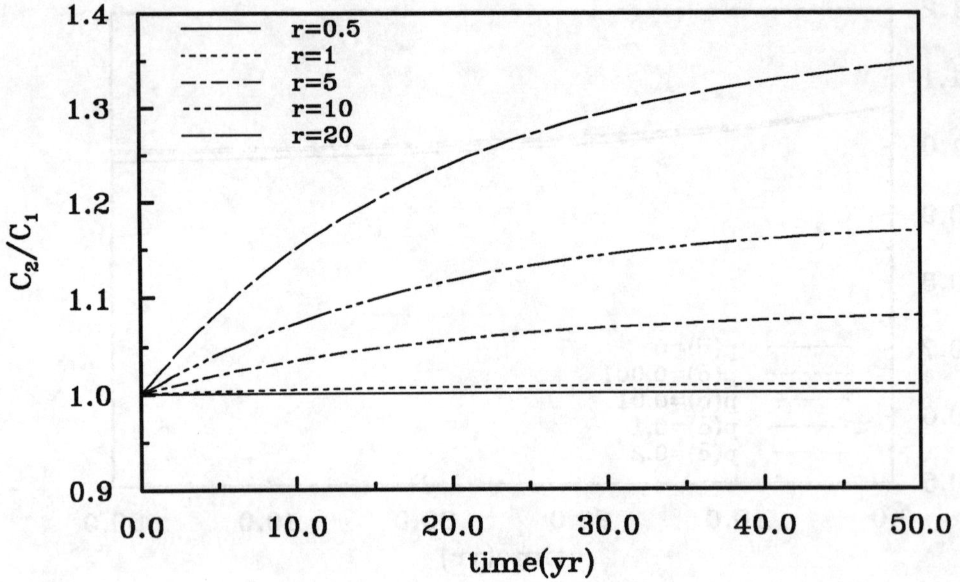

λ =0.05 C_m/C_1=0.0005 νP_1=0.001 $P(\bar{C})$=0

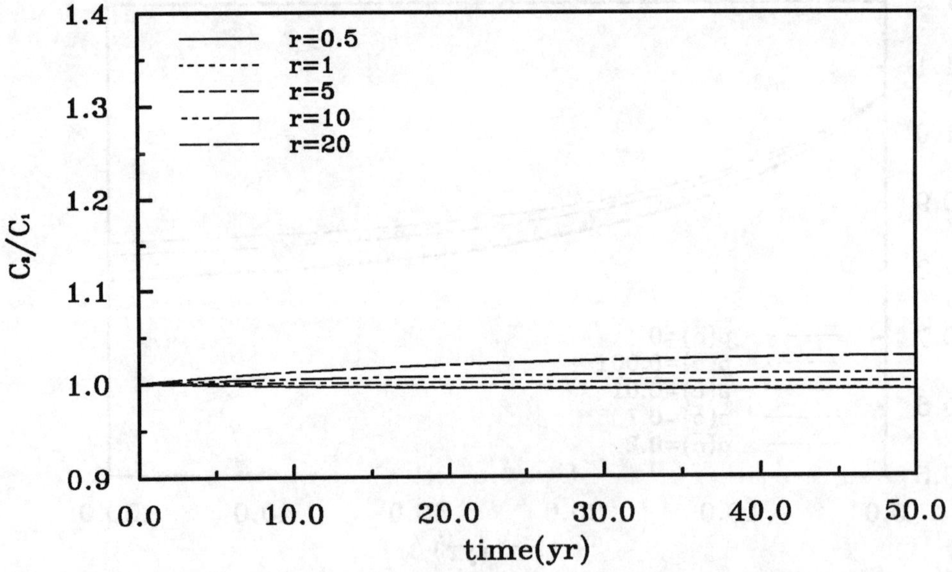

λ =0.05 C_m/C_1=0.0005 νP_1=0.0001 $P(\bar{C})$=0

Fig. 5 Maximum Initial Cost Ratio Allowable for Cost–Effective Design as Function of Exposure Time

KBES and Neural Networks

UNCERTAINTY TREATMENT FOR INTELLIGENT SYSTEMS - A NEUROCOMPUTING ALTERNATIVE

Fabio Casciati

Professor , Dept. of Structural Mechanics
University of Pavia
via Abbiategrasso 211, I27100 Pavia, ITALY

ABSTRACT

The state of health of a structure, with control devices, can be assessed from sensor measurements, but its estimate is very sensitive to any sort of uncertainty. This paper investigates the potentialities offered by neural computing for building a different way of manipulating sensor measurements. The main goals are: 1) to avoid mathematical models for the definition , in the state variable space, of the region where the actuators may remain inactive; 2) to give the system control 'brain' the possibility of learning from the experience of service.

INTRODUCTION

Consider an intelligent system. The assessment of the state of any single component of it is the link by which any piece of information collected by the sensors is converted into a decision: to activate or not to activate the actuators. A common approach to this problem requires the availability of a first mathematical model which transforms the sensor signals into state variables and of a second mathematical model which separates the safe region from the unsafe one in the space of these state variables. Due to their dominant role, uncertainties must be included in the models and this often leads to the introduction of subjective aspects. As an alternative to the previous approach, this paper investigates the potentialities which neural computing offers in this field in order to simplify the whole decisional process. Two additional goals are also pursued: to make more objective the process and to give the control device the possibility of learning from experience.

UNCERTAINTY TREATMENT

Let **q** be the vectors of measurements collected from the sensors distributed along the intelligent system [1]. The state variables **x** represent the mechanical state of the system. They can be stresses, strains, internal forces or kinematic parameters; they are related to the variables **q** by a mathematical model, say **F**:

$$\mathbf{x} = \mathbf{F}(\mathbf{q}) \tag{1}$$

A second mathematical model, say R, converts the vector $\mathbf{x}$ into the scalar y

$$y = R(\mathbf{x}) \tag{2}$$

When the response variable y is lower than or equal to a threshold value y_u , the intelligent system does not activate any actuator (Z=0, with Z a Boolean auxiliary variable); otherwise, for $y > y_u$, the control device is turned on (Z=1):

$$Z = f(y) \tag{3}$$

where $f(\ )$ is called the activation function. $f(\ )$ can be a Heaviside function ('hard limiters'), a linear function till a given threshold ('threshold logic') or a sigmoidal nonlinearity.

Uncertainty in the previous scheme comes out from several aspects:

- the measurements $\mathbf{q}$ are affected by uncertainties coming from instrument accuracy, signal transmission and its digitalization;
- the mathematical model $\mathbf{F}$ is an idealization of the actual behaviour of the system; its randomness will be more and more significant as deeper mechanical aspects, like hysteresis, P-Δ effects and so on, are included in the analysis;
- the mathematical model R is an average of scattered experimental results whose coefficient of variation is generally significant;
- the value of the parameter y_u results from the same regression scheme which leads to R and hence also this threshold cannot be regarded as deterministic

Assume that the 'brain' of the intelligent system is conceived as an expert- system. As already discussed in Ref. [2], the basic role of uncertainty treatment in the design of an expert system turns out from the subjective, incomplete and imprecise nature of the information. Different techniques from different disciplines are available for handling uncertainty. The 'logical model' makes use of 'endorsements' for believing or disbelieving hypotheses. To quantify the extent to which the imprecise language statements match the proposition to be expressed is the basis of fuzzy reasoning in the 'linguistic model'. The 'legal model' works whitin evidence theory and the 'statistical engineering model' uses probability calculus. The last one is generally a form of Bayesian revision [3], i.e. an updating scheme that incorporates new information in the form of observations of evidence with prior information. The quantitative nature of the sensor indications suggests the adoption of the probabilistic scheme, but this essentially collides with two obstacles: 1) the ambiguity of data on the mathematical models $\mathbf{F}$ and R; 2) the computational effort that any uncertainty propagation algorithm generally requires.

NEURAL COMPUTING

A possible alternative to the previous classical approach introduce a 'black box' operator Θ such that

$$Z = \Theta(\mathbf{q}) \tag{4}$$

Here the auxiliar Boolean (0=do not activate ; 1=activate) variable Z is just the output of the aplication of the operator Θ to the vector $\mathbf{q}$. Θ is regarded as a 'black box' item, rather than as a mathematical model. Therefore, the operator Θ can also be built by just training an 'intelligent device', i.e. a neurocomputing scheme [4],[5]. A neural network consists of a large number of simple neural-like processing elements (units) connected to each other by weighted links. The learning strategies are based on changing the weights of the links connecting the nodes. The neural structure can also be changed during the learning stage. According to Ref. [6], five primary components can be distinguished:

1. the 'learning domain', which includes the input and output patterns for each training instance. In the case under investigation one has a vector of inputs (continuous, discrete and/or discretized) and one couple of values for the output (0,1);
2. the 'neural net': it is defined by a combination of input, hidden and output layers;
3. the 'learning strategy': in particular a back-propagation learning for multilayer neural networks has been implemented [7] and used in this paper;
4. the 'learning process': the weights assigned to the links connecting the nodes are changed in reducing the system error to a certain prescribed value;
5. the 'analysis process'. After the final weights are obtained, one verifies the learning process in 4. and starts the planned use of the built black box. Each further iteration is also used to improve the learning performance.

Despite several commercial programs are already available [8], the development of an own code was suggested by the following motivations:

- to have the possibility of introducing an algorithm for the selection of the number of hidden layers;
- to preserve the possibility of an automatic selection of the number of nodes for each hidden layer;
- to build several options for the connectivity of the neural network;
- to understand the meaning of the final results and to implement this experience into an algorithm of automatic selection of the network

The resulting algorithm presents two main advantages over the classical approach which makes use of mathematical models:

- it has a very high computational speed;
- its robustness is excellent, in the sense that the whole procedure is fault tolerant.

THE IMPLEMENTED BACK-PROPAGATION CODE

The code made available in Ref. [7] runs under the operating system MS-DOS on personal computers. The user interface is based on the classical windows showing the options at that time possible. During the different stages, the screen shows a scheme of the neural network where the activity levels of the single neurons and the strength of their links are illustrated by an appropriate scale of colours. The definition of the neural network requires:

- total number of neurons, number of input neurons, number of hidden layers and corresponding neurons, number of output neurons;

- the kind of connection, which can be total (each neuron is linked with all the neurons of the successive layers), total between layers (each neuron of the i-th layer is linked with all the neuron of the (i+1)-th layer), free according to the user selection;
- the way of introducing the data: by keyboard or by files

The flow-chart of the code is illustrated in Figure 1. In it, two main branches can be distinguished: on the left the training branch and on the rigth the analysis branch. Three successive sections from top to bottom can be identified: section 1 governs the definition of the neural network; section 2 performs the numerical elaborations and shows the neural network; section 3, eventually, provides the final results and builds a report of the developed activity.

The back-propagation algorithm is a generalization of the least-mean squares one. The objective is the minimization of the mean quadratic error between the output of the multi-layer perceptron and the actual response. The algorithm requires the activation function [4] in Eq. (4) to be continuous and non linear; the following sigmoid function is used:

$$f(y) = \frac{1}{1 + exp(-y))} \tag{5}$$

In the problem under discussion one has a single output unit which receives y and produces the output:

$$Z = f(y) \tag{6}$$

The input y is a linear combination of the outputs X_{lK} produced by the hidden units of the K-th layer, i.e. the last of the hidden K layers:

$$y = \sum_l W_l X_{lK} \tag{7}$$

The weights of the links beetween the final unit and the hidden units of the last layer are W_l. The ones of the links between intermediate layers are denoted by ω_{lIk}, l being the index of the unit in layer k (k=1,....,$K-1$) and I the one of layer $k+1$. Of course, for connectivity among units of two successive layers one has:

$$X_{I(k+1)} = f(\sum_l \omega_{lIk} X_{lk}) \tag{8}$$

The first layer units, finally, receive the net input $\sum_i w_{iI} q_i$ and produce output (k=0):

$$X_{I1} = f(\sum_i w_{iI} q_i) \tag{9}$$

w_{iI} being the corresponding weights. For sake of simplicity, one assumes in the following K=1, i.e. a single layer of hidden units is considered. The weights to be determined in this case are W_l and w_{iI}. The algorithm steps can then be listed as follows:

- initial random values are given to the weights w_{iI} and W_l; the 'offset' or 'bias' θ_I which characterizes each hidden output neuron is also selected: its value is detracted from the neuron input weighted sum before the output is computed;
- initial patterns are provided: a set of n vectors q* and the corresponding values of the variable Z^*;

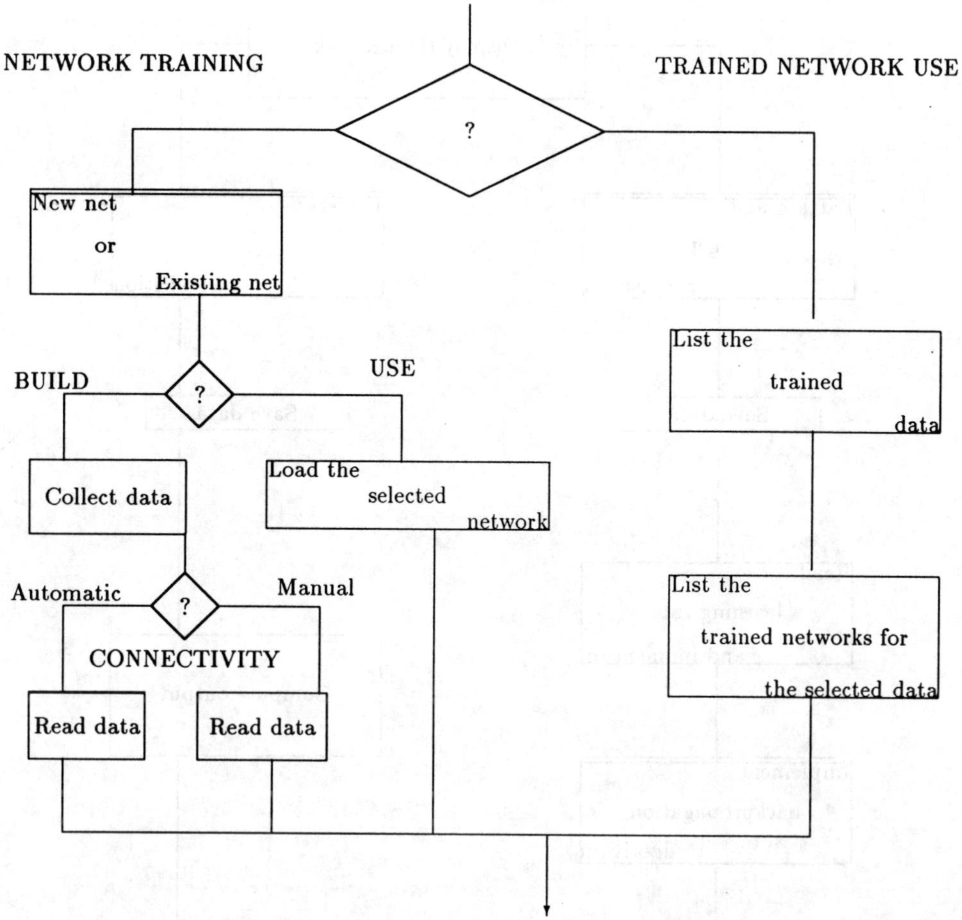
NETWORK TRAINING
TRAINED NETWORK USE
?
New net
or
Existing net
BUILD
?
USE
Collect data
Load the
selected
network
Automatic
?
Manual
CONNECTIVITY
Read data
Read data
List the
trained
data
List the
trained networks for
the selected data

Figure 1

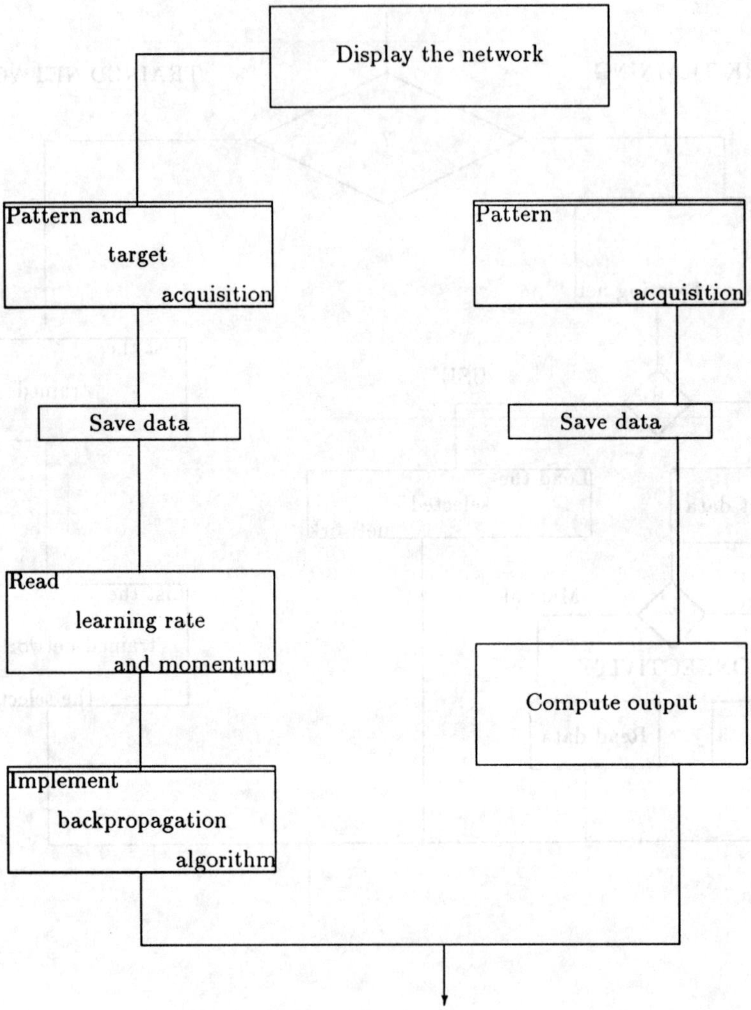
Display the network
Pattern and
target
acquisition
Pattern
acquisition
Save data
Save data
Read
learning rate
and momentum
Compute output
Implement
backpropagation
algorithm

Figure 1

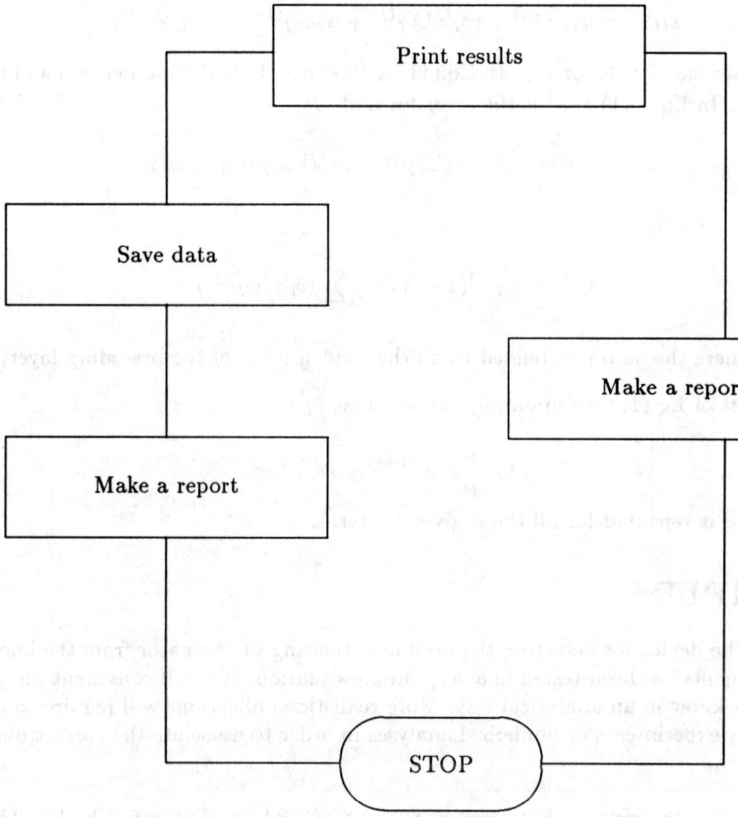

Figure 1: Flow-chart of the code implemented in Ref. [7] .

- given pattern j, the output of each neuron is captured; the final one, for istance, is obtained by Eq. (6):

$$Z^{(j)} = \frac{1}{1 + exp(-(\sum_I W_I{}^{(j-1)} X_I{}^{(j)}))} \tag{10}$$

with $X_I = X_{I1} = X_{IK}$, $I = l$. Their values are computed from $\mathbf{q}^{*(j)}$ by Eq.(9);

- the single weights of the links between neuron i and I, are updated by the formula:

$$w_{iI}{}^{(j)} = w_{iI}{}^{(j-1)} + r\delta_I{}^{(j)} X_I{}^{(j)} + \alpha(w_{iI}{}^{(j-1)} - w_{iI}{}^{(j-2)}) \tag{11}$$

A similar espression holds for W_I. In Eq.(11) α $(0 < \alpha < 1)$ is the momentum and r is the learning rate. In Eq. (11) $\delta_I{}^{(j)}$ is the error for node I:

$$\delta^{(j)} = Z^{(j)}(1 - Z^{(j)})(Z^{*(j)} - Z^{(j)}) \tag{12}$$

for the output node;

$$\delta_I{}^{(j)} = X_I{}^{(j)}(1 - X_I{}^{(j)}) \sum_l (\delta_l{}^{(j)} w_{Il}{}^{(j)}) \tag{13}$$

otherwise, where the sum is extended to all the hidden units of the preceding layer;

- the equivalent of Eq.(11) for updating the unit bias θ_I is

$$\theta_I{}^{(j)} = \theta_I{}^{(j-1)} + r\delta_I{}^{(j)} \tag{14}$$

- the procedure is repeated for all the n given patterns.

AN EXAMPLE

The feasibility of the device for detecting the need of activating the actuator from the knowledge of sensor measurements has been tested in a very simple situation. It is self consistent but assume that the safe set is know in an analytical way. More realistic applications will require to develop 'ad hoc' laboratory experiments or numerical analyses in order to associate the correct output to any training pattern.

Let σ_I and σ_{II} be the principal stresses in a point of a plane continuous body. The simple Tresca yielding criterion regards as elastic (safe) the points for which:

$$({\sigma_I}^2 - 4C^2) + ({\sigma_{II}}^2 - 4C^2) + [(\sigma_{II} - \sigma_I)^2 - 4C^2] \leq 0 \tag{15}$$

with

$$C = \sigma_y/2 \tag{16}$$

and σ_y = yield stress. Let σ_y be 240 N/mm^2. Each of the variables σ_I and σ_{II} is initially discretized in 5 classes, identified by 4 Boolean variables:

0000 = either (-∞,-240) or (240, ∞)

1000 = (-240, -120)

0100 = (-120, 0)

0010 = (0, 120)

0001 = (120,240)

One has therefore 8 input Boolean units defining a couple (σ_I, σ_{II}). Moreover, 25 different outcomes are possible for each couple (σ_I, σ_{II}). The response variable for each couple is selected 0 if all the points of the class are safe; 1 otherwise. This substitute Eq.(15) with the safety domain of Figure 2. A single layer of 4 hidden nodes was considered. It is immediate to see that using 25 patterns (one for each class) for training the neural network, all the possible contributions can be investigated and the neural network is perfectly trained.

The problem is now complicated by a new discretization of each σ axis into 11 classes and, hence, of the (σ_I, σ_{II}) plane in 121 cells (Figure 3). In order to maintain low the number of input units, the class number of each principal stress is given by four Boolean variables in binary form:

1 = 0001 = (-∞, -240)

2 = 0010 = (-240, -180)

3 = 0011 = (-180, -120)

4 = 0100 = (-120, -60)

5 = 0101 = (-60,0)

6 = 0110 = (0)

7 = 0111 = (0, 60)

8 = 1000 = (60, 120)

9 = 1001 = (120,180)

10 = 1010 = (180,240)

11 = 1011 = (240,∞)

The full training developed for the previous discretization becomes cumbersome in this case. The training has therefore been conducted by using: i) 3 or more central composite designs [9] [2] of different width; ii) a mesh of cells covering the half-plane $\sigma_I \leq 0$ as a chessboard. The first training is repeated by considering different numbers of hidden nodes and layers:

a) 1 hidden layer of 4 nodes;

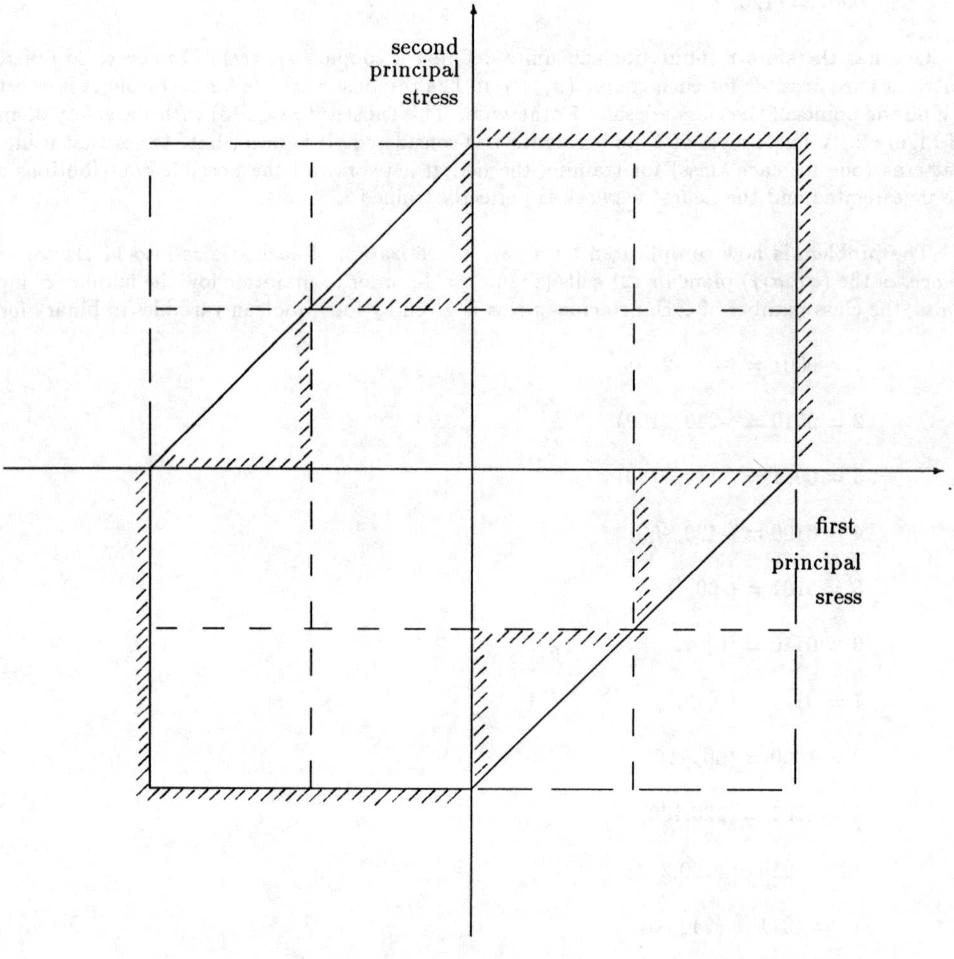

Figure 2: Tresca's safety domain discretization in 25 safe/unsafe cells.

Table 1: Check of the neural networks trained by the policy i) (repeated central composite designs).

network type	number of cells with output: wrong	too approximate	correct
a)	21	24	76
b)	24	26	71
c)	39	12	70

b) 2 hidden layers of 8 and 4 nodes, respectively;

c) 3 hidden layers of 4, 4 and 2 nodes, respectively (Figure 4).

The validation is obtained by checking the trained network in all the 121 classes. The number of unaccurate results gives an estimate of the goodness of the training process. This check was quite unsatisfactory for the neural networks trained by policy i). A first central composite design (9 experiments with 1 central point) of radius 120 N/mm^2 was considered: this means $Z(0,0) = 0$ (central point); $Z(120,0) = 0$; $Z(120\sqrt{2}, 120\sqrt{2}) = 0$ and so on. The output Z is zero in these points of the plane (σ_I, σ_{II}) since they lie inside the Tresca's region of Figure 3. Otherwise the corresponding value of Z is put equal to 1. The experiment plan was completed by two further central composite designs, of radius 240 and 260 N/mm^2 respectively, without central points. They were replaced by the assignment: $Z(260, 260) = 1$ and $Z(-260, -260) = 1$. The results are summarized in Table 1.

A graphical inspection of the results, which is not provided here, shows that the lack of accuracy occurs in the same cells for the three different networks. These cells, of course, are the ones which were not covered by the training experiments. Thus, the conclusion is that a denser coverage of the state variable space should be pursued. This is the aim of policy ii): the half-plane $\sigma_I \leq 0$ is covered as a chessboard by 27 experiments. The validation process, in this case, only detects 5 wrong results over a total of 51 cells. Each of these wrong results is located along the boundary of the Tresca's region (see in Figure 3 the cells denoted by 'o').

The teaching of this example is that, even in a simple case, the learning process must be as much accurate as possible. But realistic problems [10] require a finite-element analysis of complex structural systems in order to estimate the value of Z^* for a given pattern $\mathbf{q}^*$. Therefore, to cover the whole space of the state variables can result too expensive. The experience of this paper suggests the adoption of the following procedure:

- to conduct finite-element analysis experiments ($Z^*_{1i} = Z^*(\mathbf{q}^*_{1i})$) according to the densest scheme compatible with the computational effort one can afford;
- to build on these results appropriate local interpolations (localized response surfaces [9]), in order to estimate the output variable Z^*, say Z^*_{2i}, for additional patterns $\mathbf{q}^*_{2i}$;
- to train the neural network for both the sets of patterns $(\mathbf{q}^*_{1i}, Z^*_{1i})$ and $(\mathbf{q}^*_{2i}, Z^*_{2i})$, in order to ensure the largest coverage of the space of the input variables $\mathbf{q}$.

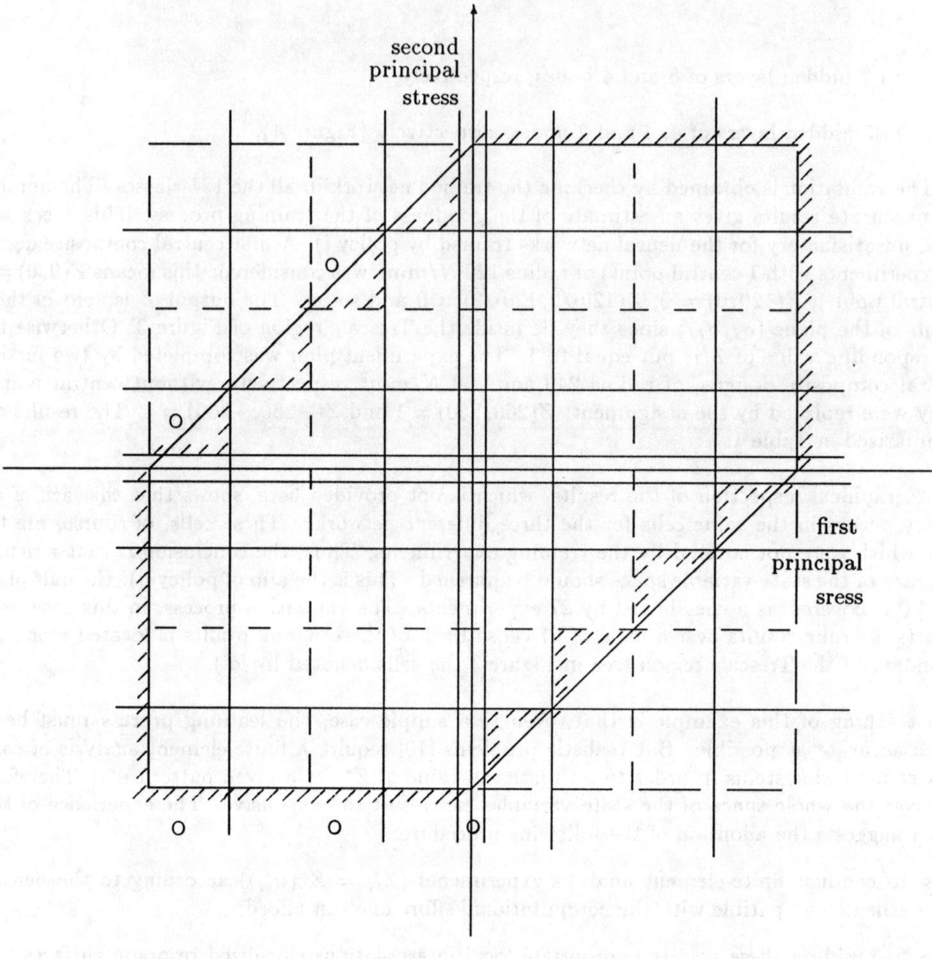

Figure 3: Tresca's safety domain discretization in 121 safe/unsafe cells.

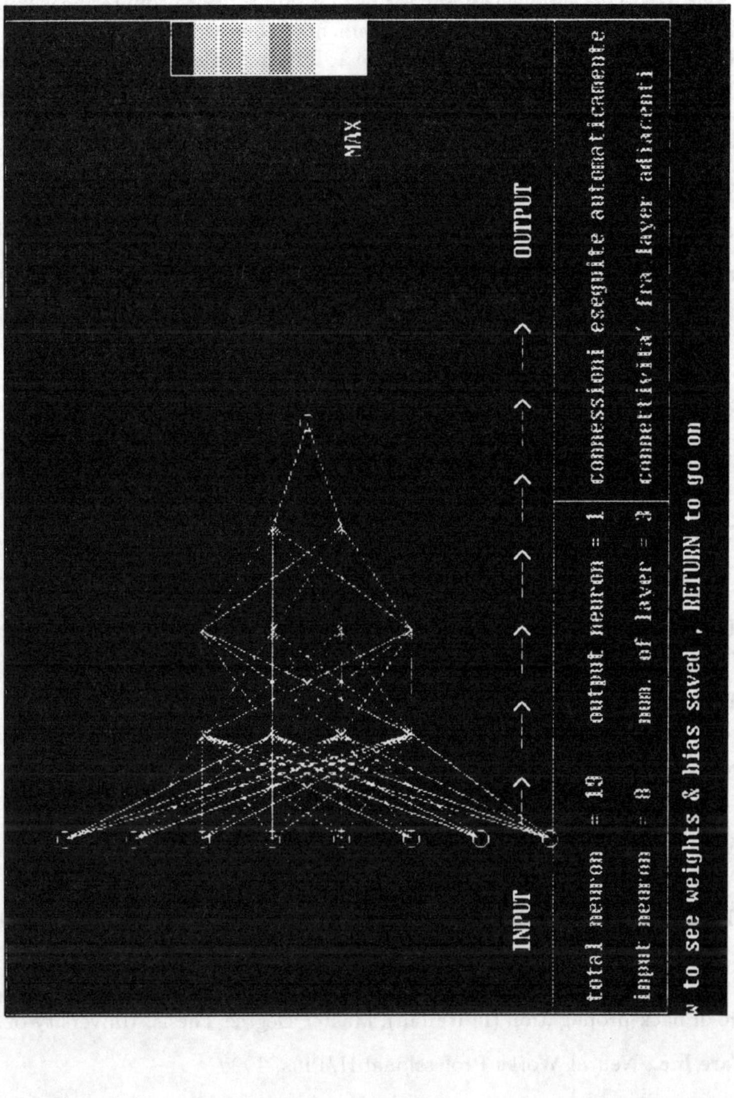

Figure 4: Neural network with three layers of hidden nodes as it appears on the computer screen during the training.

CONCLUSIONS

This paper discusses the possibility of training a 'black box' (a multilayered neural network) to convert a set of monitoring measurements, coming from an intelligent system, into (0,1) response about state of alert, damage seriousness and/or need of appropriate countermeasures (activation of actuators, ...) The present version of the implemented code works on discrete (or discretized) sensor measurements and associate state 0 or state 1, in terms of alert, with each analysed vector of measurements. High speed and fault tolerance are the two main features of the procedure. Further investigations, which are presently in progress, seem to be necessary before the studied neural network can be proposed as 'brain' for an active control device. They cover:

1. the extension to real valued input variables;
2. the identification of the more appropriate experiment design plan for the learning process;
3. a rational criterion for the selection of the number of layer and of the number of hidden nodes per layer.

ACKNOWLEDGEMENT

This research was supported by a grant from the Italian Ministry of University and Research (MURST).

References

[1] Chong K. P., Liu S.C. and Li J. C., Intelligent Structures, Elsevier, London, 1990

[2] Casciati F. and Faravelli L., Fragility Analysis of Complex Structural Systems, Research Studies Press, Taunton, UK, 1991

[3] Casciati F. and Faravelli L., Expert Systems and Seismic Vulnerability, Proc. 9th ECEE, Moscow, 1990, Vol I, pp. 23-32

[4] DARPA Neural Network Study, AFCEA Internatioanl Press, 1988

[5] Hertz J., Krogh A. and Palmer R.G., Introduction to the Theory of Neural Computation, Addison-Wesley Pubbl. Co, Redwood City, USA, 1991

[6] Hung S.L. and Adeli H., A Neural Network Environment for Intelligent CAD, in Adeli H. and Sierakowski R. L. (eds.), Mechanics Computing in 1990's and Beyond, ASCE, 1991, pp. 93-97

[7] Benzi R., PROSIT: un programma per l'implementazione del multy-layer perceptron secondo l'algoritmo di back-propagation (in Italian), Master Degree Thesis, University of Pavia, 1991

[8] Neural Ware Inc., Neural Works Professional II/Plus, 1990

[9] Faravelli L., Finite Element Analysis of Stochastic Nonlinear Continua, in Liu W. K. and Belitschko T. (eds.) Computational Mechanics and Probabilistic and Reliability Analysis, Elmepress Int., Lausanne, 1989, pp. 263-280

[10] Shinozuka M., Wen Y.K. and Casciati F., Seismic Damage and Damage Control Design for RC Frames, in Chong K.P., Liu S.C. and Li J.C. (eds.), Intelligent Structures, Elsevier, 1990, pp. 315-334

EXPERIENCES IN THE DEVELOPMENT OF EXPERT SYSTEMS FOR STRUCTURE AND SYSTEM MONITORING

GIANALBERTO RIGHETTI
D'Appolonia S.p.a.
Via Siena, 20 - 16146 Genoa, Italy

ABSTRACT

This paper focuses on engineering frameworks based on a monitoring system and on expert modules for interpretation of data and evaluation of system behavior. Aspects related to the implementation of products of this kind, which are also required of being operative in an industrial context, are first presented. In the second part of the paper, five related projects are described in more detail. Structural, geotechnical, environmental, hydrological, geological and transportation systems are covered by the mentioned applications. In the conclusions, trends for future developments are envisaged.

INTRODUCTION

Sensors and instrumentation systems are widely used in civil engineering practice for monitoring structure and foundation behavior under field conditions [1]. Complex mechanical systems, such as engines of ships, airplanes and helicopters, plants for oil and gas production and treatment, machines of manufacturing industries, are often controlled by use of a monitoring system. Actually, remote monitoring is increasingly applied to larger systems such as transportation, hydrological and environmental ones.

A monitoring system can provide a complete picture of the "status of system" in relation to specific physical phenomena. However, the evaluation of the score of data involved, which is often required on a real-time basis, represents the crucial task of the entire framework and should be cause of the failure of the monitoring-based project. A beneficial solution has been found by including, within the framework, computer-based expert modules able to perform a systematic interpretation of large amount of data and the subsequent engineering decision-making process.

This work presents an overview of projects of this kind where the System Engineering Group of D'Appolonia has been involved. A common point was always the need to solve a specific problem of an industrial client by use of the most appropriate engineering solutions. Therefore, the greater effort has been dedicated, in particular, in defining the best approach to include knowledge-based criteria and AI skills within the operative framework to be installed in the industrial context and to be used in routine work.

Generic aspects related to the above problem are presented in the next chapter. The rest of the paper describes five projects where monitoring frameworks have been applied.

EXPERT MONITORING OF INDUSTRIAL SYSTEMS

According to the definition proposed by Hayes-Roth [2], KBES for monitoring of systems mainly compare observations of system behavior to features that are essential to the successful outcome of an event. Such crucial features correspond to potential flaws in the assumed behavioral pattern of the system and are considered "vulnerabilities" of the system.

As a matter of fact, several engineering frameworks which are based on a monitoring system have been designed not only to detect flaws, but also to perform diagnosis of the causes and prognosis of the consequences, to select counter-measures, or to optimize a given operation on the system according to the real-time detected status of relevant parameters. Thereby, to such frameworks it is normally required to perform multi-objective tasks, in particular to interpret the data, to understand the actual behavior, and to support the decision process of the user by formulating operative plans.

Regardless to the specific engineering application context, a framework of the aforementioned type normally relies on the following basic elements:

- monitoring devices (sensors, instruments, etc.);
- a data collector and storage system (data bases and relevant management modules);
- several "expert modules" dedicated to perform specific tasks (interpretation of data, definition of scenarios, diagnosis, prognosis, optimization of plans, decision support, etc.);
- internal module-interface subsystems;
- external and user-interface subsystems.

Operative frameworks are usually designed to be installed on the field (i.e. inside the monitored environment), and to be used on a real-time basis. Real-time is normally related to a frequent use (when it is required or/and at constant intervals in the time). As a consequence, specific requirements are therefore the capability of installing the software on commercially available hardware and of easy interfacing of the expert modules software with the instrumentation and the database systems. To release these constrains, programming languages and tools which are not specific to artificial intelligence are often adopted [3].

In the last few years, D'Appolonia S.p.A. has been involved in several projects related to the design and the development of expert monitoring frameworks applied to various kind of engineering systems. Structural, geotechnical, hydrogeological, mechanical, environmental and transportation systems have been considered so far. Table 1 presents a summary of the relevant projects; in the table it is indicated whether the system has been operatively designed and implemented, designed at a feasibility level, or only studied at a conceptual level.

NPP FOUNDATION SETTLEMENTS MONITORING

Instrumentation systems are extensively employed to monitor the performance of soil and building foundations in Nuclear Power Plant (NPP), where both alignment of equipment and differential settlements between adjacent foundations may be critical to safety. A knowledge based expert system has been successfully employed at Caorso NPP in Northern Italy in performing real-time evaluation of measured data allowing also for immediate detection of any vulnerability which may affect the behavior of the soil-structure system under control.

TABLE 1
Relevant Projects Related to Monitoring Frameworks

PROJECT	CLIENT	AREA	YEAR	STATUS
1 - Monitoring of Geotechnical Settlements of NPP Structures	ENEL	Geotechnical	86	Operative
2 - Monitoring of Ship Engine	Internal Research	Mechanical	87	Feasibility
3 - Production Process Monitoring of Unmanned Offshore Platforms	Agip	Mechanical	88	Under Develop.
4 - Torre Melissa Landslide Monitoring for Safety Control	FS	Geotechnical	88	Conceptual Study
5 - Scrivia River Water Quality Monitoring	ENEA	Environmental	88	Conceptual Study
6 - Traffic Bologna Belt Monitoring	Sotecni	Transportation	89	Feasibility
7 - Alta Valtellina Hydric Resources Management	AEM	Hydrological	89	Conceptual Study
8 - Monitoring of Highway Civil Structures and Pavement	SAT	Structural	90	Feasibility
9 - Mincio River - Mantova Lakes Hydric System Monitoring and Management	TEMA	Hydrological	90	Operative
10 - Monitoring of Waste Disposal	Internal Research	Environmental	91	Feasibility
11 - PIUS Vessel Structural Monitoring	ENEL	Structural	91	Under Develop.

The Caorso NPP site subsoil comprises two separated alluvial aquifers, whose piezometric regime is controlled by the nearby Po River. The heavy buildings of the plant and the massive soil embankment have produced large settlements, which require continuous monitoring: the average settlements of the buildings vary in the range 20-35 centimeters. An additional few centimeters of settlement are predicted for the next 30 years, based on the long-term settlements statistical forecast.

To monitor in a continuous and automatic way the settlements of the foundations, a system of level indicators has been installed on the Main Buildings of the plant. Groundwater monitoring is performed by a Casagrande and Wellpoint piezometers, installed within both aquifers inside and outside the diaphragm, and by sensors (water level/pumping rate) installed in the dewatering wells.

The periodic evaluation of the instrumentation data have been automatized by GEOMAN (GEOtechnical Monitoring ANalyzer) which performs all steps previously described simulating the reasoning process of the geotechnical experts [4].

The knowledge base is parted into six major groups of rules (Figure 1), managed by six expert modules: a) the GEOMAN module, which controls the system reasoning process and includes the data definition; b) the GENERAL module, which manages the other modules, contains the user interface, and interacts with the site information data base; c) the module PIEZOM, which performs the piezometric data interpretation; d) the module PUMP, which evaluates the dewatering system data; e) the module SETTL, which performs the settlement data interpretation; and f) the module REPORT, which plays as the interpreter of the anomalies found combining the results obtained by c), d) and e).

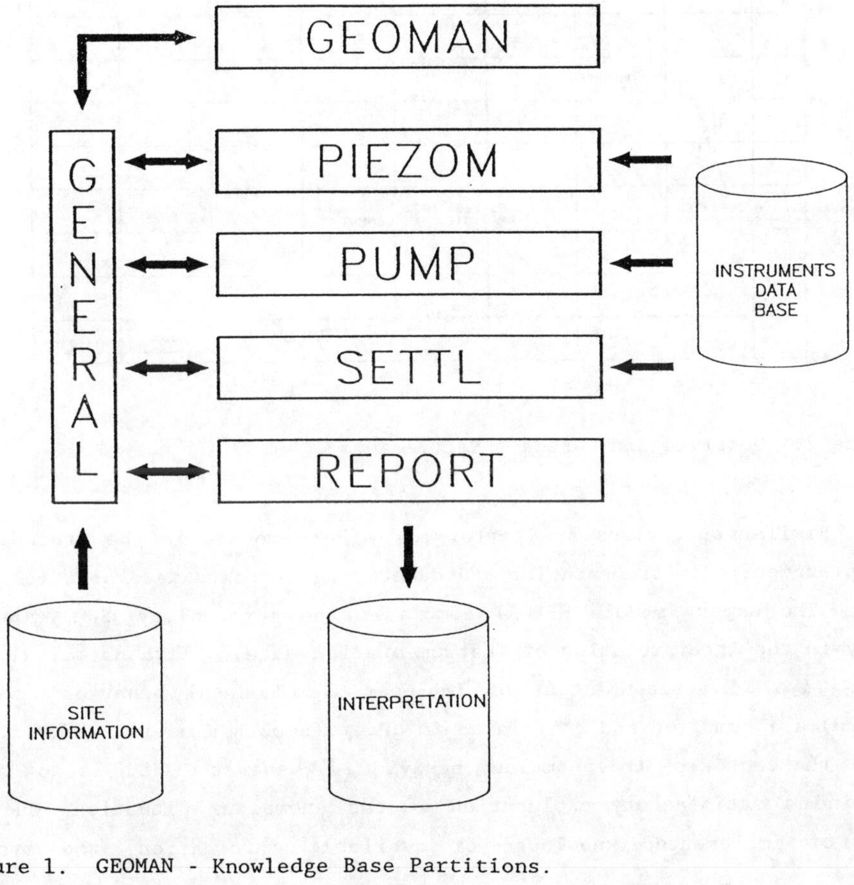

Figure 1. GEOMAN - Knowledge Base Partitions.

In the process of settlement data evaluation, SETTL first analyzes each instrument a) searching for instruments whose readings result anomalous with reference to the average settlement behavior of the foundation and b) searching for instruments whose readings disagree with the forecast values (Figure 2); then, the system analyzes groups of instruments and evaluates the settlement measurement by a) evaluating the foundation average settlement, tilt and tilting direction with reference to the measured vs. the predicted values and b) comparing the amount of differential settlement in correspondence of the rigid connection between adjacent buildings to the design limits.

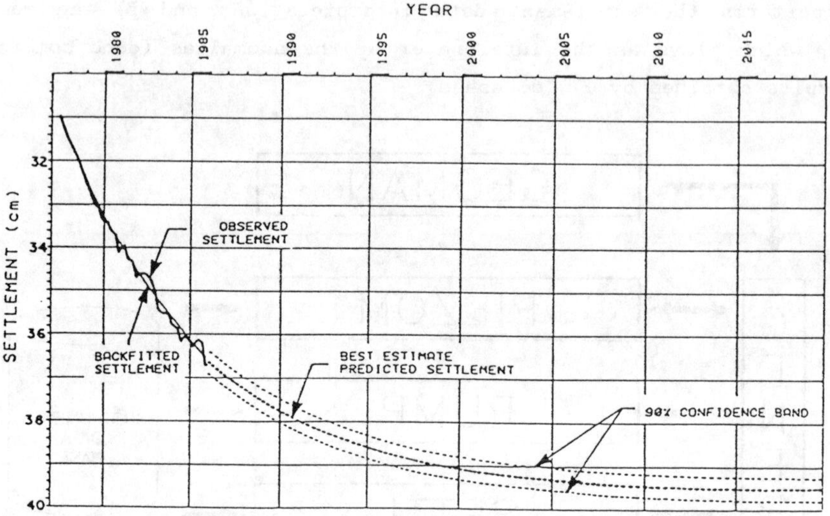

Figure 2. Observed and Forecast Settlements.

Similar operations are performed by PUMP module in the process of the piezometric instrumentation and dewatering system data evaluation.

The expert module REPORT simulates the methodology that experts apply in the interpretation of instrumentation data. All abnormal situations (i.e. any inconsistent or improper instrumental behavior in the period) are compared and correlated in order to obtain useful information about the causes of the anomalous behavior. Whenever GEOMAN is not able to find a satisfactory explanation of the anomalous situation, due to lack of implemented knowledge or available information, the system signals the anomalous situation. Any conclusion is provided with a certainty factor which represents a relative measure of the confidence of the logical reasoning process.

GEOMAN states also about the absolute and differential settlement trend, making hypotheses about possible causes (i.e. variations of live loads during the refueling activities, elastic effects due to changes in the piezometric level in the foundation subsoil, etc.), and interprets the piezometric regime in the aquifers with respect both to known seasonal variations of the average piezometric levels and design limit situations. The effects related to the dewatering activities are also taken into account.

GEOMAN has been developed within the DAISY programming environment [5], a production system developing tool featuring in particular portability on commercial available hardware and software and direct interfacing with external programs and data collection systems.

System performance is very effective allowing for a continuous and real-time surveillance of the site. However, the processing time is of over 24 hours when the instrumentation data set of an entire year is processed, using a HP-9000 for the expert system. The session related to only three months of data requires about six hours.

The KBES interpretation resulted as close to the expert's conclusions within a sample period of two years (1986-1987), where "traditional" reports from geotechnical experts were available. All available data were processed, regardless the typical selective approach used by experts in the analysis of large data-bases.

PROCESS MONITORING OF UNMANNED OFFSHORE PLATFORM

Within a research project under development by Agip and ELF, D'Appolonia has given a contribution to the feasibility design of an expert system for control of production processes of an unmanned offshore platform.

The principal aim of the system is to control the gas production processes, performed on the platform and managed by an automation system, in order to detect possible malfunctions, diagnose the causes, and support the supervisor decisions about type and priority of the actions to be taken in order to restore the normal operation. It is noted that the final expert system will be located on-shore and is supposed to service more than one platform at the same time.

The general structure of the proposed system is represented in Figure 3. The framework relies on several independent expert modules (MECS), each one dedicated to control the behavior of a given process subsystem. Each MECS obtains data about the physical parameters of the relevant process both from sensors (s), located at selected nodes of the relevant plants, and from the automation system (which is responsible for the normal management of the production process). Data are collected in real-time on the MECS local working memory and analyzed according to the knowledge relevant to the specific process under consideration.

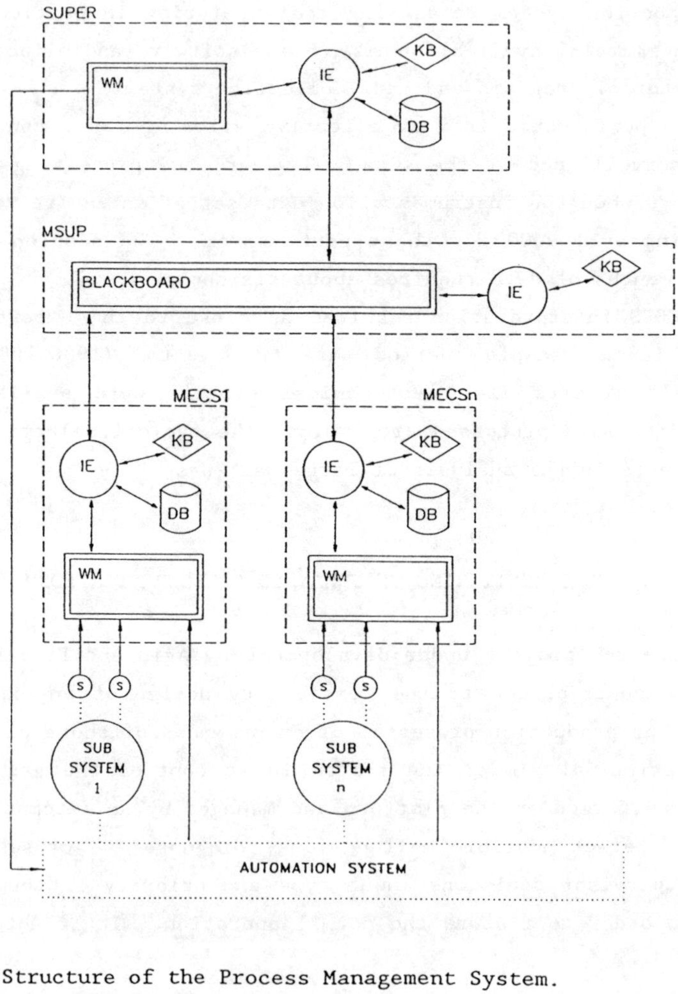

Figure 3. Structure of the Process Management System.

When the local knowledge is insufficient, additional data are asked to the other MECS by a codified request written on the MSUP blackboard. Thus, MSUP provides to coordinate data transfer between the MECS. Moreover, the results of MECS operations are transferred, via blackboard, to the SUPER module, which is able to coordinate the entire process and to derive final conclusions on the basis of a global view of the entire set of MECS results. SUPER can communicate its final conclusions to the user or directly to the automation system.

D'Appolonia has suggested the use of the DAISY environment to develop such a framework based on several expert modules required to work simultaneously. In fact the DAISY environment is well suited to be installed on parallel machines and to handle partitioned knowledge bases and working memories [6].

HIGHWAY MAINTENANCE BASED ON MONITORING

Within a multi-objective project for the management of the new Livorno-Civitavecchia Highway (actually under design), monitoring facilities and an integrated expert system are provided.

The monitoring system will be applied to some of the civil structures of the highway in order to control, during lifetime, the most critical parameters related to deterioration and the actual structural behavior. In principle, the following data will be monitored:

- traffic flows and loads;
- meteorological data;
- displacements of some bridge decks;
- dynamic accelerations at selected points of few bridge decks;
- displacements of other civil structures.

The expert system has been designed in order to select optimized maintenance programs for the civil structures and the pavement based on their present conditions, defined on the basis of the measured data or forecast by statistical analysis of the damage evolution (risk maps).

The expert system will be able to take into account all aspects related to inspection, monitoring, maintenance and rehabilitation, and also to accept real-time modifications to the strategy followed when programming priorities and constraints of maintenance operations.

MANAGEMENT OF MINCIO HYDRAULIC SYSTEM

The Mincio river and Mantova lakes hydraulic model is represented in Figure 4.

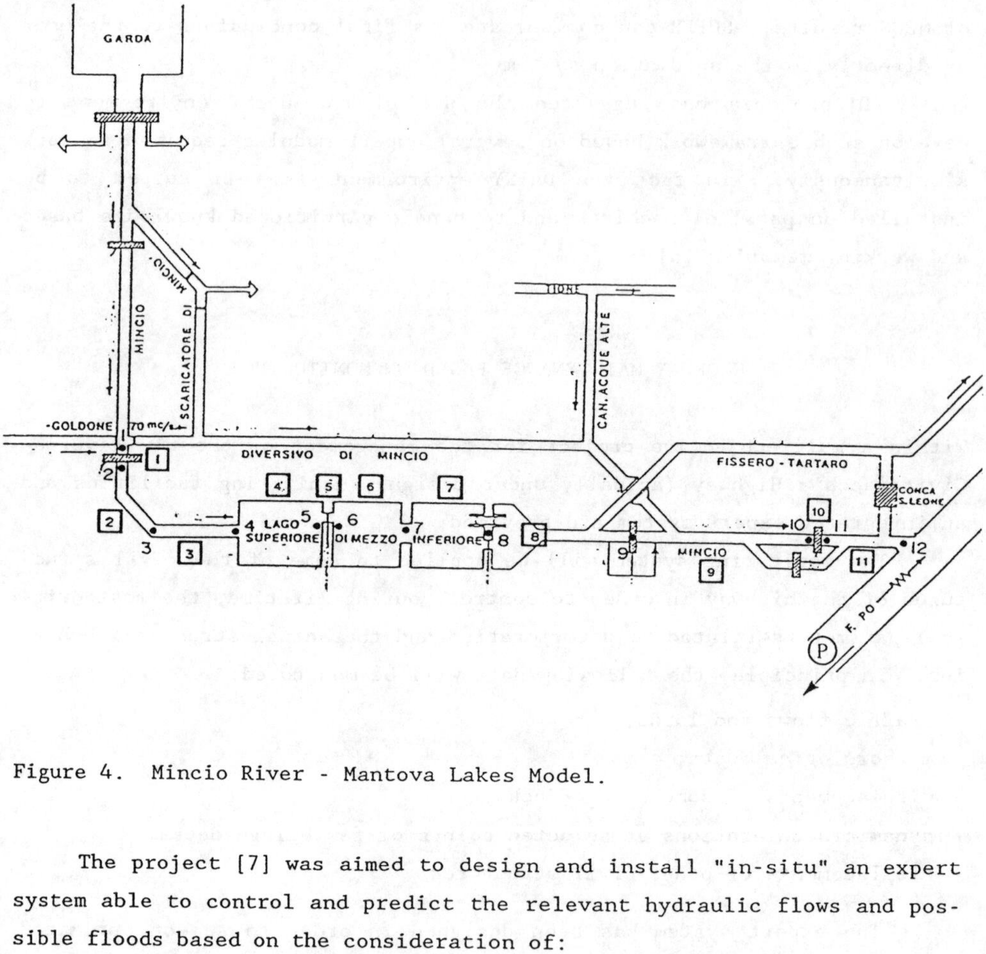

Figure 4. Mincio River - Mantova Lakes Model.

The project [7] was aimed to design and install "in-situ" an expert system able to control and predict the relevant hydraulic flows and possible floods based on the consideration of:

- a set of known actual hydraulic levels, measured by monitoring devices at few selected locations;
- a set of data about the sluice gate actual status;
- a set of approximate flow models;
- historical experience about manual management of the hydraulic system.

Basically, the expert system is structured into five modules, as represented in Figure 5.

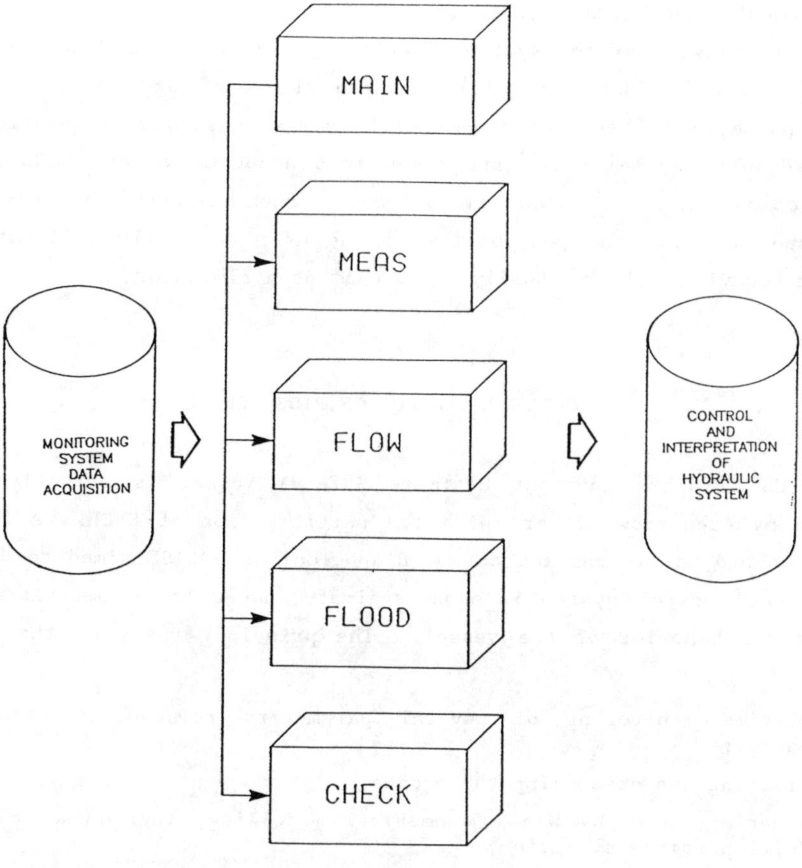

Figure 5. Mincio Management System Expert Modules.

The MAIN module manages the other modules and the data base where the hydraulic model is memorized; the MEAS module visualizes the actual status of the hydraulic system according to monitored data; the CHECK module performs comparisons between predicted flows and floods and recorded values related to optimum operation and safety.

FLOW and FLOOD modules represent the expert part of the framework. Based on the aforementioned data, and in particular on heuristic knowledge about the sluice gates and their operations consequences on the hydraulic system, combined to the actual measured data, the first evaluates the expected hydraulic flows, the second, by integrating also monitored data about the Po river hydrological status, predicts possible flood conditions for the system.

The software of the system is written in FORTRAN77 and installed on IBM-PS. Special data base files encode the knowledge about flow and flood patterns of behavior and several typical hazard configurations of the hydraulic system which are taken into account by the CHECK module when comparing the actual situations. Such a structure has been developed to give the possibility to the user to easily calibrate the system behavior and, eventually, to use it as a simulator.

STRUCTURAL CONTROL OF PIUS VESSEL

The PIUS (Process Inherent Ultimate Safety) Vessel is actually under design by Asea-Brown Boveri with the participation of ENEL. A related research project is carried out by D'Appolonia and CISE aimed to develop an expert control system able to evaluate, on a real-time basis, the structural behavior of the vessel. The principal tasks of the system are:

- real-time monitoring of several parameters related to structural behavior;
- validating and organizing these data;
- evaluating, also by use of numerical modelling, the actual behavior and its possible evolution;
- evaluating, by an expert module, possible critical conditions and, in this case, detecting causes and producing alarms at selected levels.

Figure 6 gives an idea of the entire framework.

Uptodate, the feasibility study phase has been completed. This study concerned identification of parameters, availability of instruments for monitoring system, design of basic structure of data base, definition of numerical models to simulate physical behavior, and definition of the basic logic of the expert module and related knowledge base. The operative phase will start in the next future.

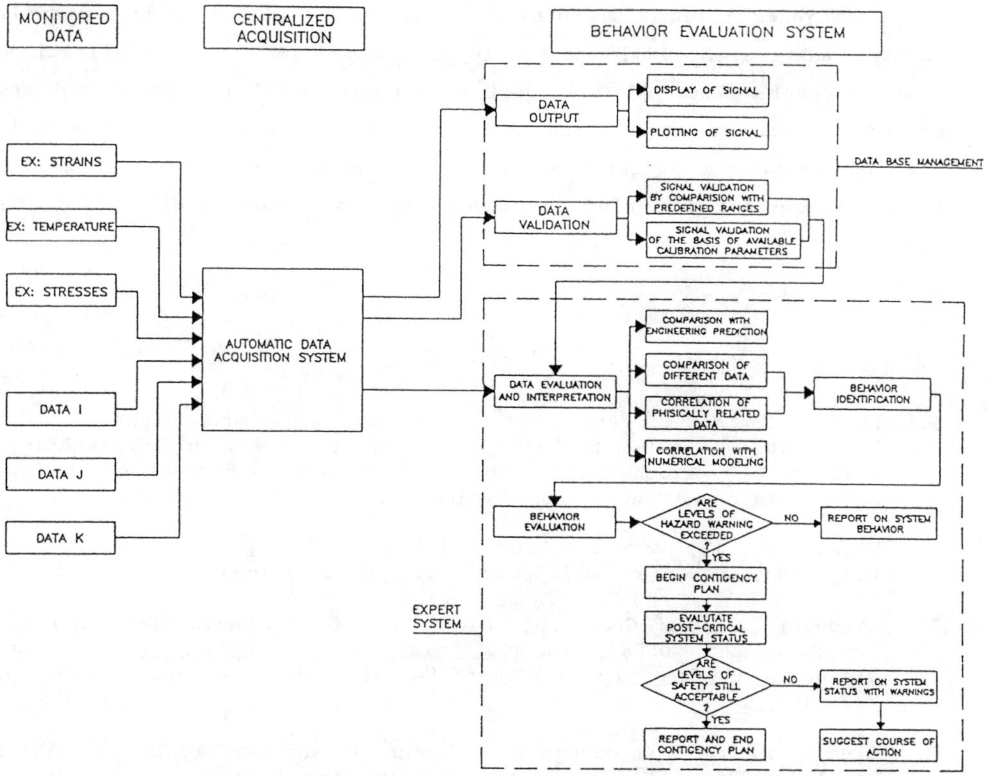

Figure 6. Framework for PIUS Vessel Structural Control.

CONCLUSIONS

Aspects related to the development of engineering frameworks encompassing, monitoring of a system and expert modules, have been presented. Some of these frameworks are actually operative and their performance is kept under control in order to derive indications for new releases and future projects.

From the experience gained so far, it can be envisaged a promising future in the application of "expert monitoring" also to large and complex systems outside the typical fields of application; as an example, environmental, transportation, economical and biological systems can be mentioned.

Projects related to intelligent monitoring of such systems will involve particular skill in knowledge acquisition and formalization, since several unknown domain-specific aspects shall be elicitated and merged with modern engineering methods. For the knowledge engineer is thereby envisaged a position of increasing importance and responsibility in organizing and designing the development of future projects in this area.

REFERENCES

1. Cremonini, M.G., Ferro, G., Jenkins, R.D. and Righetti, G., "Applications of Monitoring Systems in Structural and Geotechnical Engineering", Proceedings, 2nd International Conference on On-Line Surveillance and Monitoring, Venice (Italy), 1986, pp.51-60.

2. Hayes-Roth, F., Waterman D.A. and Lenat D.B., Building Expert Systems, Addison-Wesley Publ. Comp. Inc., London, 1983.

3. Righetti, G. and Cremonini M.G., "The DAISY Environment and the Expert System GUESS", in Artificial Intelligence in Engineering: Diagnosis and Learning, J.S. Gero Ed., Elsevier, New York, 1988, pp. 227-244.

4. Cremonini, M.G., Vardanega C., Mirone, M. and Saveri, E, "GEOMAN: A Knowledge Based Expert System for On-Line Monitoring of Nuclear Structures", Proceedings, 10th International Conference on Structural Mechanics in Reactor Technology (SMIRT), Anaheim, CA, 1989.

5. Righetti, G., DAISY: A Production System Environment for Expert System Implementation, User's Manual, D'Appolonia Report, Vers. 1.0, Genoa, 1986.

6. Del Grosso, A. and Righetti G., "Finite Element Techniques and Artificial Intelligence on Parallel Machines", Proceedings, International Conference on Computational Engineering Mechanics, Beijing, China, 1987.

7. Cremonini, M.G. and Manfredini G.M., Impianti di Controllo e Telemisura, Sistema Idraulico Mincio - Laghi di Mantova, Manuale D'Uso del Software Applicativo, D'Appolonia Report No. 87-502, Genoa, 1990.

QUALITATIVE PHYSICS AND ITS APPLICATIONS TO STRUCTURAL ENGINEERING SYSTEMS

ANDREA DEL GROSSO and ANDREA ZUCCHINI
Istituto di Scienza delle Costruzioni - Facoltà di Ingegneria
Via Montallegro, 1 - Genoa - Italy

ABSTRACT

The paper discusses application potentials of the so-called qualitative physics theories in the field of structural mechanics. These theories are intended to provide mathematical basis to the development of computerized procedures able to simulate expert reasoning based on qualitative models of the behavior of physical systems. The developments presented in the paper are concerned with the study of plane elastic frames and nonlinear shear-type frames. Qualitative analyses of such structures are performed in terms of informations only characterized by the sign of their variation. Extensions to the handling of more precise informations are also discussed in view of applications to the field of seismic vulnerability analysis of structures.

INTRODUCTION

In the study of real world phoenomena, many meaningful conclusions can be drawn out of a reasoning exclusively or mainly based upon qualitative informations concerning those phoenomena. The term qualitative, in the present context, refers to informations that allow the description of the dynamics of phoenomena by means of parameters and variables not possessing precise numerical values.

For example, in many cases it is possible to infer important informations on the behavior of physical systems, simply on the basis of their description. Figure 1 shows, with reference to structural systems, some situations where a <u>qualitative</u> response of the systems can be easily determined by people aware of the fundamentals of structural mechanics.

The scope of <u>qualitative physics</u> is to develop logical and mathematical fundamentals, together with information processing techniques, able to construct models of the qualitative reasoning on the behavior of physical systems.

The interest toward construction of qualitative models is obviously greater in those cases where the representation of phoenomena by means of quantitative models is impossible or, although generally possible, involves parameters the numerical values of which can be only roughly estimated. Studies devolved to the mathematical formalization of qualitative models are indeed known since a relatively long time in economical sciences [1,2], but major developments are to be ascribed to the present decade, and mainly belong to applications of artificial intelligence techniques in innovative fields of biophysical sciences and engineering.

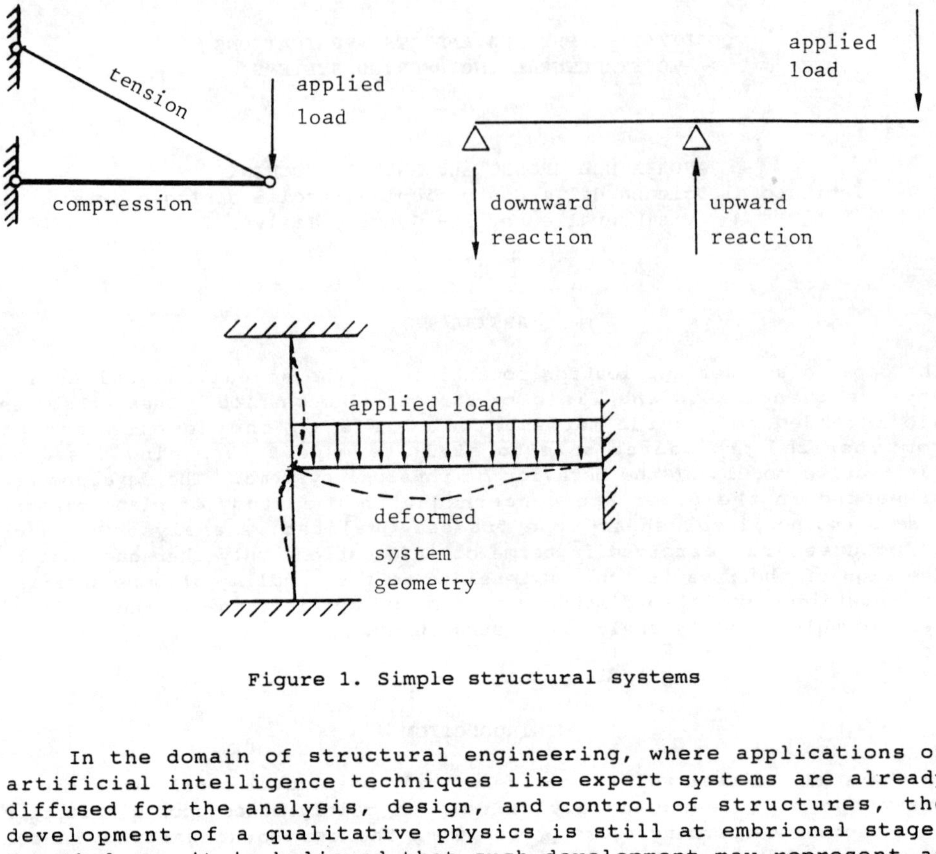

Figure 1. Simple structural systems

In the domain of structural engineering, where applications of artificial intelligence techniques like expert systems are already diffused for the analysis, design, and control of structures, the development of a qualitative physics is still at embrional stage. Nonetheless, it is believed that such development may represent an interesting opportunity for future applications, especially in monitoring and diagnosis of structural performance.

In these problems, extremely large amounts of data as well as very few incomplete informations may need to be processed. Conventional rule based reasoning systems built on shallow, domain specific knowledge may often fail in providing causal explanation of observed facts, where model based reasoning schemes based on fundamental, deep knowledge would be able to provide autonomous interpretation.

The present paper, based on previous work performed by the Authors [3-6], is aimed to discuss some of the peculiar aspects of qualitative modelling through qualitative physics approaches, and the potential for application in the field of intelligent structural systems.

GENERAL ASPECTS OF QUALITATIVE PHYSICS

In the context of AI research, qualitative physics has the aim of constructing models of the qualitative knowledge of physical world, endowed with generality comparable to the one of the models of quantitative physics.

General reference frameworks for the definition of qualitative models in the physical domain can be derived from the fundamental works by Hayes [7,8] and by the theories due to de Kleer and Brown, Kuipers, and Forbus published in [9].

As any model of the physical reality, a qualitative physics theory requires that the three following fundamental steps are adequately formalized:

1. definition of the physical primitives, i.e. of the objects used in the description of the physical situation or device;
2. translation of the physical description into a mathematical model;
3. model resolution and interpretation of results.

The three theories mentioned above mainly differ because of the selection of the physical primitives, while they share fundamental mathematical aspects. These aspects will be briefly addressed in this paragraph, in order to provide a general understanding of what should be expected from a qualitative physics approach. More comprehensive discussions can be found in the paper by Struss [10].

In the cases considered in the present context, a qualitative mathematical model is generally obtained through qualitative translation from a quantitative model. Due to that, the characteristics of the qualitative model can be studied with reference to the characteristics of the parent quantitative model.

For example, we shall say that a quantitative mathematical model is defined by a space of descriptions in terms of n real-valued variables (e. g.: a system of linear equations), and that a solution process maps a description into the corresponding set of solutions:

$$sol_{quant} : D_{quant} \rightarrow S\,(R^n) \qquad (1)$$

Analogously, the construction of a qualitative model will require:

1. a set of n spaces of qualitative values Q_i, appropriate for the different variables to be used in the description of the problem (quantity space);
2. a set of descriptions in terms of the corresponding qualitative variables;
3. a solution process, able to map a qualitative description into the corresponding set of solutions, i. e.:

$$sol_{qual} : D_{qual} \rightarrow S\,(Q_1 \times Q_2 \times \ldots \times Q_n) \qquad (2)$$

Moreover, the use of qualitative modelling procedures will be considered suitable for the simulation of reasoning about the behavior of a physical object provided all the important distinctions in the behavior of the object are retained, and the same essential descriptions, derivable from interpretations of the solutions of a conventional quantitative model, can be obtained.

From the above, conditions, properties, and requirements concerning the most relevant aspects of qualitative modelling can be derived.

A first consideration is concerning the space Q of the qualitative values. Such space must be composed by a finite set of values, possibly orderable, interpretable and exclusive (i.e.: not giving rise to superpositions in the interpretation of values), covering the entire behavioral field under study, and allowing adequate description of the different behavioral fields (granularity). As a matter of fact, the quantity space Q most commonly used is represented by the elementary sign space: {-,0,+}, where variables can only take an albebraic sign. Other choices are however possible, as for example sets of intervals defined on the real axis, orders of magnitude [11] etc.

Another key aspect is that, in order to formulate a qualitative description and a corresponding solution method, a qualitative algebra (Q-algebra) able to operate upon qualitative variables should be defined. Comprehensive treatments and discussions over the characteristics of Q-algebras are presented in the papers already referenced, and in more recent works [12,13].

For example, if the quantity space is represented by the sign space, addition and multiplication between two qualitative variables X and Y are respectively defined by Tables 1 and 2.

TABLE 1
[X] + [Y]

[Y] \ [X]	-	0	+
-	-	-	?
0	-	0	+
+	?	+	+

TABLE 2
[X] x [Y]

[Y] \ [X]	-	0	+
-	+	0	-
0	0	0	0
+	-	0	+

The above example let us point out that the addition between two qualitative variables of different sign gives an ambiguous (undefined) result. This fact is represented by the presence of question marks in Table 1. Ambiguity is an intrinsic character of qualitative analyses and is depending on the fact that Q-algebras do not completely verify all the characteristic properties of the quantitative algebra, i. e.: commutative, associative, distributive, existence of the neutral, opposite and reciprocal elements. Sticking to the example, in the sign space the opposite for the addition does not exist.

Ambiguity correspond to a loss of information, and this can result in different types of consequences: while in general it will produce, correctly, the entire set of equally possible solutions of the problem, under certain circumstances it can give rise to spurious (i.e. physically impossible) solutions in addition to the possible ones, without the possibility of discerning among them.

As a consequence, the qualitative description of the problem and the corresponding solution process must take into account all the characteristics of the qualitative algebra in use and particularly the aspects related to ambiguity. Description and solution process shall lead to the individuation of : a) all the possible states characterizing the behavior of the physical system, compatible with the selected variables, b) all the transition mechanisms between states, and c) all the possible sequences of transitions between states (i.e.: the possible behaviors). As a matter of fact, in linear steady-state systems the analysis may be limited to the reference state (intra-state behavior), while inter-state behavior is of a paramount importance in dynamical systems or in evolutionary processes.

It can be demonstrated [10] that qualitative models translated from quantitative ones will satisfy the above requirements and will therefore give correct and useful answers if, with respect to the parent quantitative approach, the following conditions are met:

1. every quantitative solution has some correspondent in the set of qualitative solutions (completeness);
2. all the qualitative solutions do correspond to some quantitative solution (soundness);
3. qualitative solutions are invariant with respect to transformations of the parent quantitative model into an equivalent one (stability).

Completeness can be obtained with sufficient generality by all the approaches considered herein, but the other properties may be obtained only in a more or less strong sense, depending on the choice of the qualitative variables, qualitative description and solution process.

FORMULATION OF QUALITATIVE MODELS

As already mentioned, qualitative physics theories of sufficient generality have been originally established by Forbus, Kuipers, and by de Kleer and Brown.

The qualitative process theory, due to Forbus, is based upon the fundamental hypothesis that variations in physical systems are directly or indirectly caused by processes; consequently, the physics of a domain should include descriptions of the different characteristic situations and a library of the processes that can be active in that domain. Physical situations are described by so-called individual views, containing objects, their properties, and relationships among them; processes are represented by something developing in time, able to modify object parameters, create new objects or new situations. Qualitative variables associated to the parameters are defined in qualitative numeric spaces, where limit points exist denoting discontinuous variations in active processes.

According to this theory, the analysis of a problem consists in the determination of a sequence of situations generated by the interaction among the different processes that can be activated during a specified time window.

Kuipers' theory of qualitative simulation proposes a representation of the physical system similar to a differential equation and produces a behavioral description corresponding to a continuous function, solution of a differential equation, although more weak. Conceptually, the description of the system is obtained as a qualitative translation of an ordinary differential equation, with symbols representing the physical parameters of the systems and a set of constraints expressing inter-relations among the parameters. Qualitative variables associated with the parameters are defined in interval or sign spaces; every parameter is a function of a single independent variable (usually time).

The search for the solution consists in the determination of the tree containing all the states compatible with the given constraints, that the system can enter increasing the value of the independent variable. Paths through the tree represent possible behaviors.

The approach which has been recognized to be potentially the most suited for application in the domain of structural mechanics is the one due to de Kleer and Brown. This theory, called qualitative physics based on confluences, postulates that the behavior of a complex system can be determined only by the laws governing the components of the systems and by the interconnections among them. In practice, the theory leads to the formulation of a qualitative mathematical model of a system just from a description of the physical arrangement of the system itself, more or less in the same way in which quantitative numerical models are constructed.

Models are composed by physical constituents, originally classified into: materials, i.e. the entities upon which the physical device is intended to act; components, i.e. the parts of the device which can effectively act on the materials transforming their characteristics; conduits, used to transfer materials from one component to another (the terminology reveals that the applications for which the theory has been thought were related to electrical or hydraulic circuits).

The description of the model takes place over three different levels: a) the topology of the constituents, that only defines the structure of the system; b) the physical laws characterizing the behavior of the single constituents, and c) class wide assumptions, valid for the generality of the physical systems analyzed. Model parameters and their variations are represented by qualitative variables usually defined in the sign space. Laws are expressed in form of equations, called confluences, involving qualitative variables. Different types of confluences can be defined, globally representing a system of qualitative differential equations (algebraic with respect to parameter variations). Equations are considered as constraints to the behavior of the entire system. The solution process consists in satisfying the constraints by propagating the assignement of values to variables; of course propagation accounts for active constraints.

To some extent, the solution process is a qualitative analogous of the solution methods for algebraic systems. However, the qualitative nature of the problem introduces two major differences. The first one is related to the ambiguity which arises during algebraic operations, preventing direct prosecution of the process. To overcome this difficulty and resolve ambiguity, heuristic rules on local behavior able to generate interpretations of the system response are added to the solution process. The second one is related to the structure of the qualitative system which renders the sequence of the operations necessary to reach the solution undetermined. To improve efficiency and exaustiveness of the search, special logical procedures, called ATMS [14], can be introduced.

Confluences alone are not however sufficient to adequately model the generality of the physical systems. For this reason, the concept of qualitative state is introduced. A qualitative state is a well defined region in the behavior of a component characterized by a unique set of confluences. A qualitative state is defined in terms of inequalities involving qualitative variables.

A qualitative analysis of a system performed in accordance with this theory, produces the set of all the states that are possible for the system as a result of the application of an external disturbance.

A feature of the approach that needs to be pointed out for the sake of the successive discussions is that passage from the physical description to the qualitative mathematical model takes place automatically. Indeed, starting from the definition of the components and of their connections, the qualitative equations can be directly generated in two distinct types : component confluences, extracted from a predefined component library, and connection confluences, expressing equilibrium conditions at the nodes and compatibility conditions along the meshes, as in any network analysis problem.

If the method looks to be conceptually simple, actual implementations may hyde several difficulties, due to the nature of qualitative equations. To present a few examples, in the most simple space {+,0,-} a typical confluence looks like:

$$\sum_{i=1}^{n} T_i = C \qquad (3)$$

where C is a qualitative constant, and the terms T_i can be qualitative variables as well as multiplications of constants times variables. Addition and multiplication are defined according to Tables 1 and 2.

A set of values satisfies a confluence when one of the following conditions is verified:

- the qualitative equality holds according to the rules of qualitative algebra;
- the left-hand side of the confluence cannot be evaluated.

A set of values contradict a confluence when all the following condition are verified:

- every variable has a value;
- the left-hand side can be evaluated;
- the confluence is not satisfied.

A confluence in neither satisfied nor contradicted in all the other cases.

A <u>pure</u> confluence only involves addition of variables, whilst a <u>mixed</u> confluence is composed by additions of products. Model descriptions containing mixed confluences can be transformed into descriptions containing only pure confluences by separating the behavior into qualitative states.

Let consider for example a system described by the confluence :

$$\partial X_1 + [X_1]\, \partial X_2 = 0 \qquad (4)$$

where [X] indicates the qualitative value of X, and ∂X indicates the qualitative value of the derivative of X with respect to an independent variable (e.g. time). The confluence (3) may then be resolved into pure confluences by introducing three different qualitative states:

state 1 :	$X_1 > 0$	$\partial X_1 + \partial X_2 = 0$	$[X_1] = +$
state 2 :	$X_1 = 0$	$\partial X_1 = 0$	$[X_1] = 0$
state 3 :	$X_1 < 0$	$\partial X_1 - \partial X_2 = 0$	$[X_1] = -$

Although a confluence model could be written directly from the knowledge of the physical component, qualitative models are usually obtained by quantitative-to-qualitative translation rules similar to the ones indicated in Table 3.

TABLE 3
Examples of quantitative/qualitative translation rules

$[e_1 + e_2] \rightarrow [e_1] + [e_2]$
$[e_1 \, e_2] \rightarrow [e_1] \, [e_2]$
$[0] + [e] \rightarrow [e]$
$[0] \, [e] \rightarrow [0]$
$[+] \, [e] \rightarrow [e]$
$[-] \, [e] \rightarrow - [e]$

It should be noted that a linear system represented by a quantitative model of the type $\sum_{i=1}^{n} C_i X_i = C_0$, where C_is are constants, is directly transformed into the confluence $\sum_{i=1}^{n} [C_i][X_i] = [C_0]$. Linear systems are defined into a unique qualitative state.

QUALITATIVE ANALYSIS OF PLANE ELASTIC FRAMES

The qualitative physics theory of de Kleer and Brown has been applied in [3] to develop a qualitative model for plane elastic frames. In order to formulate a quantitative mathematical model suitable for the transformation into a confluence model, a matrix formulation of the displacement method in terms of network theory has been used. Basically the formulation has been derived from the work by Spillers [15].

The transformation of a structural problem in terms of an equivalent network problem do however require some interpretations leading to various consequences with respect to the application of de Kleer and Brown theory.

The major difference between problems in structural mechanics and a generic network problem is that forces and displacements, that are in the structural case the main behavioral variables, are typically of a vectorial nature. Besides producing a greater complexity in the algebra, this introduces the need of handling more informations concerning the geometry of the system, because the orientation of each member should be known in order to refer all the vectors to a unique coordinate system.

Another difference is represented by the presence of multiple support nodes, and of the rigid-body modes inside elementary stiffness matrices. For these two latter aspects, it can be directly accounted in the topological formulation of the matrix displacement method.

According to this approach, the topology of the structure is represented through a directed graph: the nodes of the graph correspond to the nodes of the structure and the branches to the structural members; branches are directed according to member orientation. External restraints are assumed infinitely rigid and are tied to a unique reference node, being all restrained nodes topologically equivalent. Figure 2 illustrates the process of selecting a graph (Figure 2b) to represent a typical frame (Figure 2a). The graph in itself is of course representing merely the topological aspect of the problem. Other informations, like geometrical and mechanical data, can however be attached to the elements of the graph.

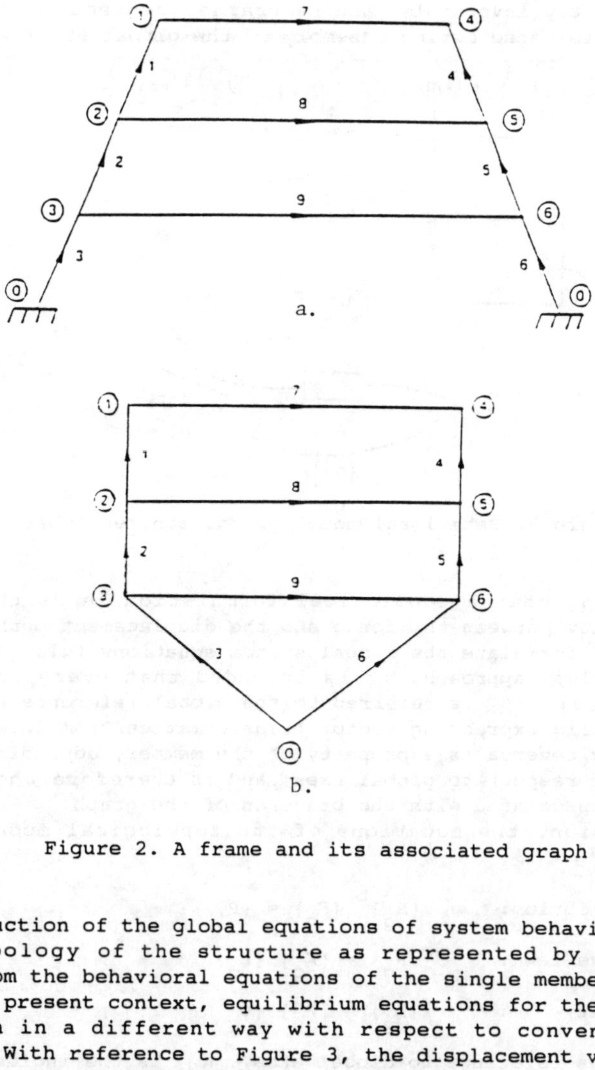

Figure 2. A frame and its associated graph

Construction of the global equations of system behavior is dictated by the topology of the structure as represented by the graph, and proceeds from the behavioral equations of the single members.

In the present context, equilibrium equations for the single members are written in a different way with respect to conventional matrix approaches. With reference to Figure 3, the displacement vector $\{\Delta_i\}$ of a member lists only the displacement of the member end, relative to the member start. Accordingly, only the force vector $\{F_i\}$ at the member end appears in the stiffness equation for the member, because the corresponding vector at the member start can be obtained from rigid-body equilibrium conditions:

$$\{\bar{F}_i\} = [L_i]\{F_i\} \tag{5}$$

By this mean, rigid-body modes are eliminated from the local equations:

$$\{F_i\} = [K_i]\{\Delta_i\} \tag{6}$$

that become directly invertible, and restrained degrees of freedom are automatically eliminated during assembly of the global stiffness matrix.

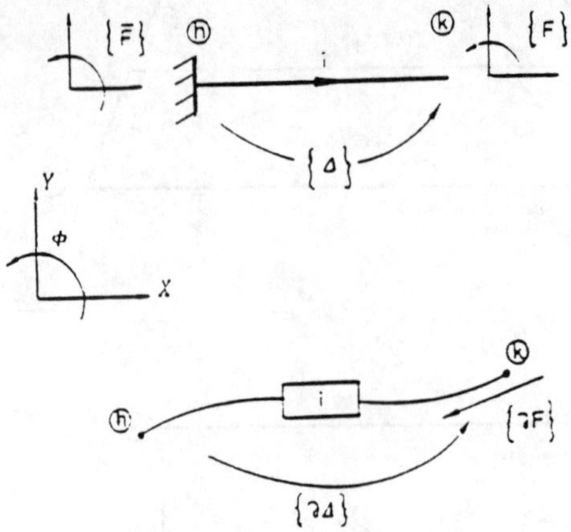

Figure 3. Behavioral model of the single member

It is noted [15] that a topological formulation leads to a perfect procedural duality between the force and the displacement methods.

In order to formulate the global system equations fully consistently with the topological approach, it is intended that every local member relationship, (5), (6) is referred to the global reference system, i.e. the rotation matrix expressing vector transformation from local to global coordinates or viceversa is a property of the member, depending on member orientation with respect to global axes, and is therefore another piece of information associated with the branches of the graph.

In conclusion, the equations of the topological model take the following form:

nodal equilibrium: $[N_g]^t \{F_g\} = \{P\}$ (7)

constitutive law : $\{F_g\} = [K_g] \{\Delta_g\}$ (8)

compatibility : $\{\Delta_g\} = [N_g] \{\delta_g\}$ (9)

where g indicates reference to global axes, $[N_g]$ is the incidence matrix, $[K_g]$ contains all the member stiffness matrices as diagonal submatrices, $\{\delta_g\}$ is the nodal displacement vector, and $\{P\}$ is the vector containing the applied nodal loads.

In order to obtain the corresponding qualitative model, qualitative variables are chosen to represent variations of the behavioral parameters $\{\partial\Delta\}$, $\{\partial\delta\}$ and $\{\partial F\}$ with respect to a reference state. Variations are produced by external disturbances represented by a vector $\{\partial P\}$. Qualitative variables and constants belong to the sign space {+,0,-}. By transformation of equations (5),(7),(8), and (9) into confluences, the following qualitative model is obtained:

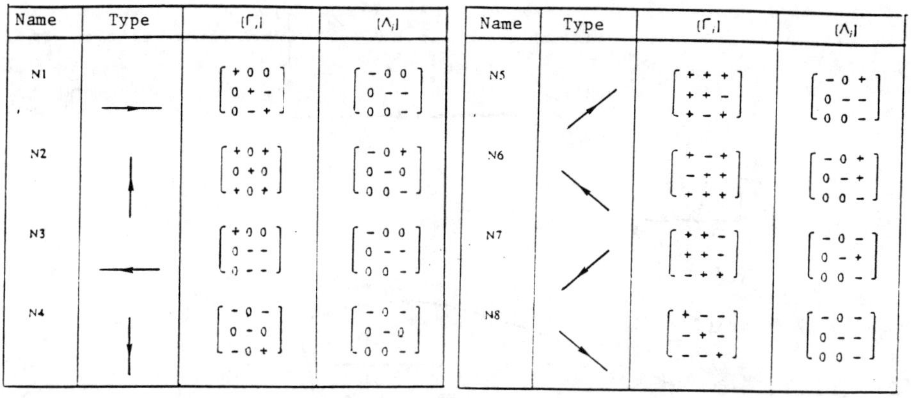

Figure 4. Example of component library for beams

member equilibrium	: $\{\partial\bar{F}\} = [\Lambda]\ \{\partial F\}$	(5′)
nodal equilibrium	: $[A]\ \{\partial F\} = \{\partial P\}$	(7′)
constitutive law	: $\{\partial F\} = [\Gamma]\ \{\partial \Delta\}$	(8′)
compatibility	: $\{\partial \Delta\} = [N]\ \{\partial \delta\}$	(9′)

The qualitative translation transforms all the matrices appearing in the quantitative equations, originally composed by numeric constants, into matrices composed only by signs. Qualitative incidence matrices [A], [N] are directly obtainable from graph topology; qualitative stiffness and local equilibrium matrices [Γ], [Λ] must be predefined by means of a <u>beam element library</u>, where all the situations giving rise to different qualitative matrices are identified. Figure 4 illustrates the component library used in the example for rigidly connected beams. Other libraries have been used for hinged ends.

The qualitative model has been implemented on an EXPLORER lisp-machine, using the KEE language integrated by LISP modules. The application has been divided in two separate sections: the first one interprets the description of the structure and produces the qualitative model, and the second one performs the reasoning scheme, finding all the compatible solutions for a single disturbance applied at one node.

The experiences performed with this application have led to some interesting conclusions.

A first result was that the reasoning scheme usually do not fail in detecting degenerate structural schemes. For example, in the case depicted in Figure 5a the solution process is able to recognize the unstable structural arrangement; in such instances the solution process stops and an empty set of solutions is generated.

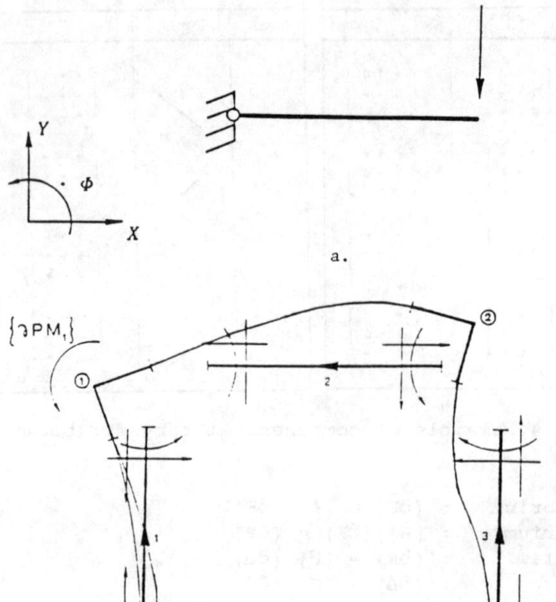

Figure 5 - Sample cases

Figure 5b illustrates the typical results that can be obtained from a complete analysis of a simple case. The figure shows, superimposed, the solutions in terms of signs of member end forces and moments and in terms of deformed shape of a portal frame, subjected to a disturbance composed by a positive moment applied at node 1. It can be seen from the example that the solution is not unique, because both positive and negative moments are possible at the restrained end of the left vertical member (consequently, two different deformed shapes are possible). To proceed further in the analysis of the structure some additional information is needed concerning the relative stiffness of the members; this means that the capability of handling some more quantitative data should be added to the model if unicity of solution for any given disturbance is required.

As a final comment, it can be observed that soundness and completess can be obtained for the model only depending on some factors, as for example redundancy of informations in the topological description, effectiveness of the component heuristics added to resolve ambiguity in the solution process. All these factors generally relate to the solution phase, and substantiate alternative solution approaches as proposed by other researchers (see for example the paper by Dormoy [13]).

EXTENSION TO ELASTIC-PLASTIC BEHAVIOR

As previously noted, linear models only give rise to intrastate analysis. To investigate potential application of the approach to interstate behavior, the qualitative formulation of the displacement method for plane frames has been extended to elastic-plastic analysis of shear-type frames.

Shear-type frames represent a substantial simplification of the general plane frame problem, because they can be described by scalar variables (horizontal displacements at the floors) instead of vectors. Moreover, shear-type frames are of a certain importance in view of the application of qualitative reasoning methods to earthquake engineering, as it will be explained later in the discussion.

In particular, with reference to the single story frame of Figure 6a, the definition of an equivalent graph is straightforward, as shown in Figure 6b. The assumption of infinite rigidity of the horizontal beam, allows indeed the identification of a single degree of freedom system, the governing equation of which is represented by the horizontal equilibrium between the applied external force and the resulting shear force at the top end of the vertical members.

The confluences representing the qualitative model can be directly written from the topology of the graph of Figure 6b, as follows:

equilibrium : $\sum_{i=1}^{e} \partial Fx_i = \partial Px$ (10)

constitutive law : $\partial Fx_i = \partial \Delta x_i \quad , \forall_i$ (11)

compatibility : $\partial \Delta x_i = \partial \delta x \quad , \forall_i$ (12)

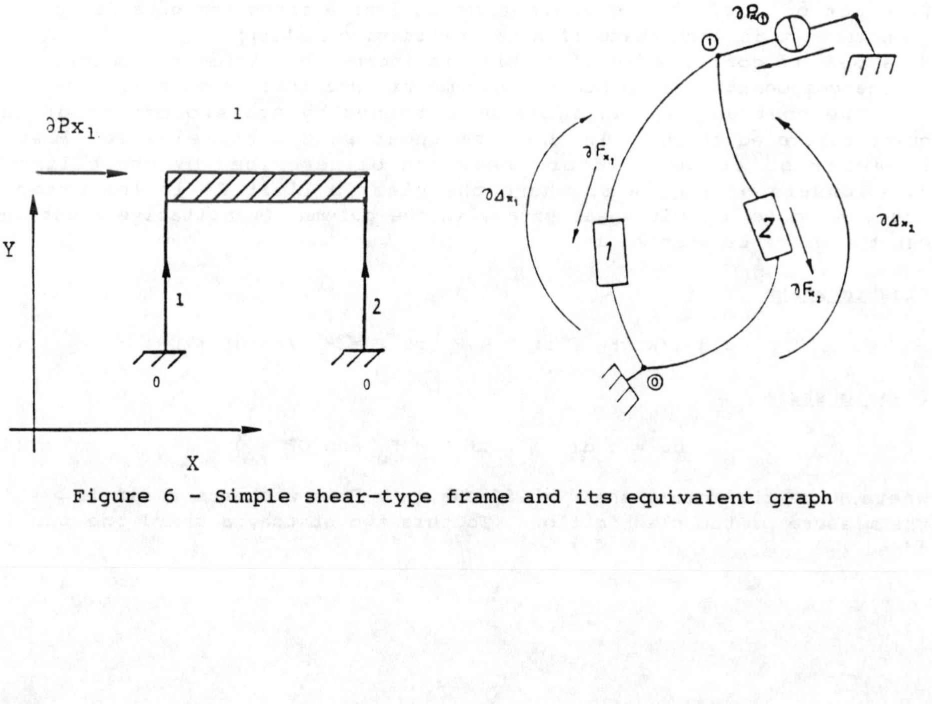

Figure 6 - Simple shear-type frame and its equivalent graph

Due to the inherent simplicity of the model, the development of an INTRASTATE analysis is very simple. Application of a positive disturbance $\partial Px = +$ in equation (10) results in a lock of the solution process if the number of the vertical members is grater than 2. With the aid of some heuristic inference, it can however be found that $\partial\delta x = +$; than, from equation (12) it follows that $\partial\Delta x_i = +$ for any i, and from equation (11) it is concluded that also $\partial Fx_i = +$ for any i. Figure 7 summarizes the results of the intrastate analysis.

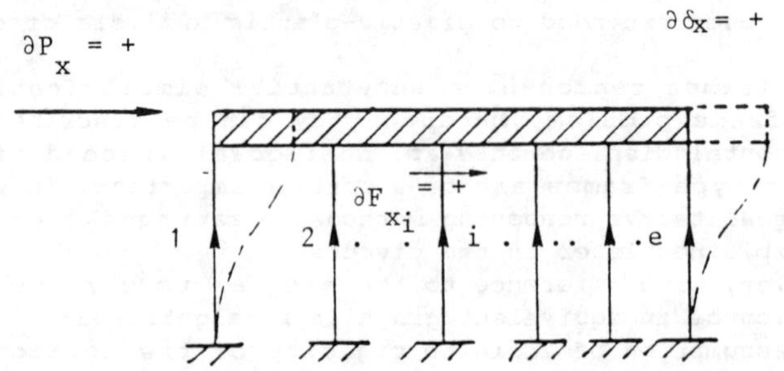

Figure 7. Intrastate analysis of the simple shear-type frame

The same conclusion could obviously have been reached more simply by means of shallow rules; the convenience of a deep, model based reasoning, arises immediatly as soon as attempts are made to extend the validity of the model. A substitution of the appropriate component confluences will indeed be sufficient to model nonlinear behavior of the columns. Again, de Kleer and Brown's theory indicates the procedure to be adopted.

The existence of distinct qualitative states shall be associated to:

1. a set of specifications defining the states in terms of conditions to be verified over the qualitative values of variables (not on their variations);
2. a set of confluences expressing in finite terms the behavior of the component in each state (finite qualitative model);
3. a set of confluences describing in incremental terms the behavior of the component in each state (incremental qualitative model).

The confluences can again be obtained by transformation of the quantitative equations. In the case under study, the elastic-plastic behavior of columns under shear can be described by the bilinear relationship of Figure 8, where the plastic limit F_u is for example associated to a limit shear stress in the column. Quantitative equations can therefore be written as:

<u>Elastic state</u>

$$du = 1/K\ dF \quad , \text{ if } F < F_u \text{ or } F = F_u \text{ and } dF < 0 ; \tag{13}$$

<u>Plastic state</u>

$$du = a\ dl \quad , \text{ if } F = F_u \text{ and } dF = 0 ; \tag{14}$$

where K is the column shear stiffness, a > 0 is a constant and dl > 0 is the measure of the plastic flow. To this two states, a third one can be added :

Failure state

$$F = F_u \text{ and } u = u_u \text{ , (} dF = -\infty \text{ , } du = +\infty \text{).} \quad (15)$$

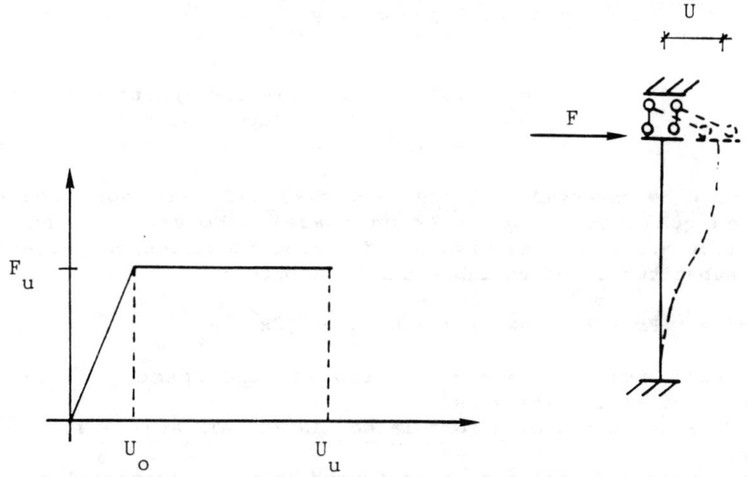

Figure 8. Elastic-plastic behavior of the shear column

In the present case, being the constitutive law anolonomous, the derivative with respect to time is not needed, and to obtain the confluences it is only necessary to substitute qualitative variables to quantitative ones and the corresponding signs to quantitative constants. Finite model confluences can be directly obtained from state specifications. The complete qualitative description of the problem is therefore summarized in Table 4.

Table 4
Qualitative description of an elastic-plastic shear column

State	Specifications	Finite Model	Incremental Model
1) Elastic zero	[Fx] = 0	[Fx] = 0 [Δx] = 0	∂Fx = ∂Δx
2) Elastic	[Fx] = + F_e	[Fx] = + [Δx] = +	∂Fx = ∂Δx
3) Elastic limit	[Fx] = + F_u	[Fx] = + [Δx] = +	∂Fx = ∂Δx ∂Fx = -
4) Plastic	[Fx] = + F_u	[Fx] = + [Δx] = +	∂Δx = + ∂Fx = 0
5) Failure	[Fx] = + F_u	[Fx] = + [Δx] = +	∂Δx = + ∂Fx = -

To obtain the relationships of Table 4, the displacement [Δx] relative to the column base is substituted to the displacement u, and the following relationships for [Fx] are assumed in state specifications:

$$0 < +F_e < +F_u \tag{16}$$

where:

0	=	$[0,0]$	(closed point - interval)
$+F_e$	=	$(0, +F_u)$	(open interval)
$+F_u$	=	$[+F_u, +F_u]$	(closed point - interval)

It should be observed that the values $+F_e$, $+F_u$ are not included in the assumed qualitative space (sign space). However, because state specifications are separately evaluated during the solution process, the following substitution of variable can be adopted:

$$[z] = [Fx - F_e] \Rightarrow [z] = 0 \Leftrightarrow [Fx = F_e] \tag{17}$$

(analogoulsly for $[Fx] = +F_u$), allowing the space {+,0,-} to be suitable for the entire analysis.

Once the model has been formulated, INTERSTATE analysis can proceed as follows:

step 1. an intrastate analysis is performed on the incremental model, for the given disturbance and initial state, in order to determine qualitative values for all the derivatives;

step 2. the derivatives indicate the directions in which system parameters are varying, and are used to determine if some specification is contradicted, inducing a state transition in the corresponding component;

step 3. the possible state transitions for the entire system are obtained as a subset of the cross-product of the possible transitions for its components. If more than one state is possible for the entire system, all the combinations not already analyzed are examined.

As any qualitative analysis, interstate analysis also involves various types of ambiguities, concerning in particular state transitions. Several rules are however available to discern states physically possible in a given situation among states theoretically possible on the basis of parameter combination. Under the fundamental hypothesis that qualitative variables are varying continously, main rules are as follows.

<u>CONTINUITY</u>: variables should vary continuously; i.e. x(t) should be adjacent to x(t'), and ∂x(t) should be adjacent to ∂x(t'), if t is adjacent to t' in the assumed qualitative space.

<u>LIMIT</u>: a state transition can be caused by an increasing (decreasing) variable ∂x = +, if and only if it has an upper (lower) bound x < A.

<u>EQUALITY CHANGE</u>: if ∂x ≠ 0 and [x] = A, where A is a point-interval (k,k), the corresponding state is terminated immediately (momentary state).

<u>CONTRADICTION AVOIDANCE</u> : no transition toward any state is possible, if this state is contradictory with the system. Contradictions are usually revealed by finite models.

EXAMPLE OF NONLINEAR SHEAR-TYPE FRAME

Figure 9 represent the most simple case of a nonlinear shear-type frame. To illustrate the capabilities of interstate qualitative analysis, the case of Figure 9 has been studied by means of the approach presented above.

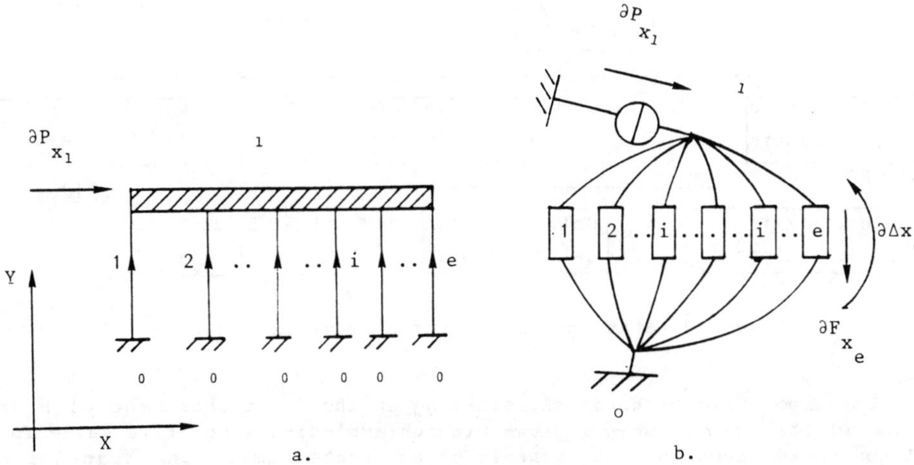

Figure 9. Simple case of shear-type frame

From the graph model, equilibrium and compatibility confluences may be expressed as in Table 5:

TABLE 5
Connection confluences for the simple shear-type frame

INCREMENTAL MODEL	FINITE MODEL
$\partial Fx_1 + \partial Fx_2 = \partial Px_1$	$[Fx_1]+[Fx_2] = [Px_1]$
equilibrium	equilibrium
$\partial \Delta x_1 = \partial \delta x_1$ $\partial \Delta x_2 = \partial \delta x_1$	$[\Delta x_1] = [\delta x_1]$ $[\Delta x_2] = [\delta x_1]$
compatibility	compatibility

The behavior of the components of the system is described by the five states of Table 4. The disturbance is represented by an external load ∂Px_1 applied at the horizontal beam. The analysis is performed by considering successive increments of the applied disturbance (for example all positive), and by developing for each increment the sequence of steps described in the previous paragraph. The analysis will produce the state diagram of Figure 10.

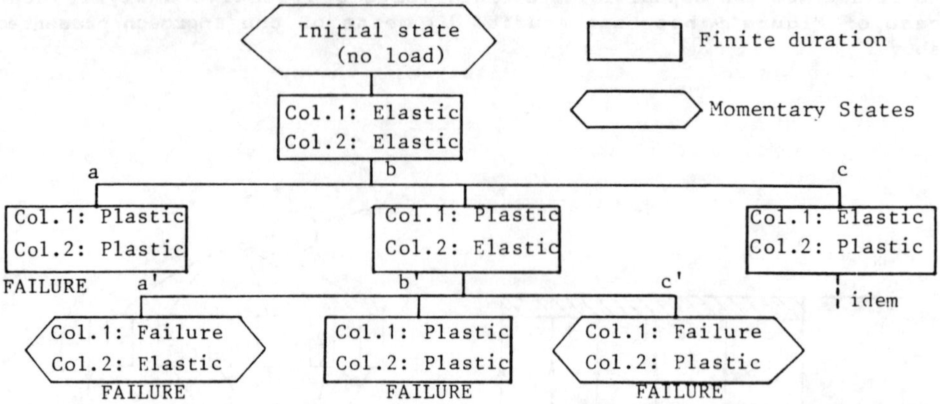

Figure 10. State diagram for the frame of Figure 9

For a positive disturbance, starting at the first increment with an unloaded structure, step 1 gives without ambiguity a positive value for all the derivatives. Step 2, because of the state limits and transition rules, detects (again without ambiguity) that both columns can enter the elastic state. Step 3 recognizes that no contradiction arises with the remaining confluences representing the model.

At the second increment, step 1 gives the same result as the first increment, but step 2 recognizes that both colums can enter the plastic or the elastic limit, or the failure state: for every component, this makes a total of four states mathematically possible (2,3,4,5 in Table 4). Step 3 has to consider a total of 15 new states for the entire system (16 minus the state obtained in the first increment), and reveales that for every column states 3 and 5 are contradictory with the continuity rule. As a consequence of the second increment, three states are possible for the system (a,b,c in Figure 10).

Application of the third increment proceeds with three separate intrastate analyses conducted in parallel, starting from each one of the states determined at the second increment, but no discernment can be made about which one of the three possibilities can actually take place. This ambiguity reflects the characteristic of the model, because no information is given concerning the relative values of the limit points on the constitutive laws for the members. The same reason is leading to the impossibility of determining actual transitions into the failure state.

For the case a), the third increment is immediatly leading to a failure state for the entire system, because of the impossibility of satisfying the equilibrium confluence. Cases b) and c) are substantially identical, interchanging the behaviors of column 1 and 2. Considering case b), the intrastate analysis for the third increment produces three possible states for column 1 (3,4,5), and four for column 2 (2,3,4,5). Successive steps reveal that only three new states are equally possible for the system (a′,b′,c′ in Figure 10), because state 3 in column 1 and states 3 and 5 in column 2 would result in a contradiction. It is observed that all the three states give rise to the failure of the frame.

DISCUSSION

The approaches and the examples presented in the previous paragraphs have shown that qualitative physics formulations can be used to construct procedures capable of simulating expert reasoning on both linear and nonlinear structural systems. These procedures can effectively be implemented into computer applications in order to provide capabilities of forecasting or interpreting the structural response, without the need of numerical simulation programs and, at the limit, even without any quantitative knowledge of the structural system.

The cases examined, very simple in nature, have put into evidence that the effectiveness of the approach should be investigated in more detail. At least two aspects need to be the object of further studies: the first one is concerning the development of real-scale applications, and the second one is concerning the improvement of the precision of the informations that can be used in a qualitative analysis.

The complexity of the analysis seems to increase rapidly as soon as topologically complex structural systems are considered, especially if nonlinear behaviors are to be taken into account. Discussion of the first aspect therefore involves consideration of the interaction between the formulation of qualitative models and computer architectures or programming tools; the problem of ambiguity seems also to have a bigger importance in modelling of real-scale structural systems.

A major problem from the conceptual point of view is however represented by the second aspect, and by the way in which formulations may be adapted to treat more precise informations.

In principle [10], the same qualitative formulation should be able to handle informations of different precision in a sort of continuous transition, in the sense that more information result in a wider class of deductions, not changing details of conclusions previously drawn.

Evolution of the qualitative physics theories described in this paper toward semiquantitative models has been faced in some recent papers [16,17,18]. Modified approaches normally substitute the simple sign space {+,0,-} with interval spaces, rendering the absolute and the relative values of qualitative variables and constants more meaningful and precise. This also reduces ambiguity, although ambiguity can never be eliminated from qualitative analyses.

Introduction of interval spaces (or range variables) introduces into the formulation mathematical aspects of a different nature, rendering the structure of the model and the solution procedure closer to approaches already known in the mathematical literature [19].

The Authors are presently developing an application of a similar type, consisting in the simulation of the nonlinear monotonic behavior of shear type frames, where qualitative variables and parameters are defined into interval spaces. The application is intended for use in seismic vulnerability analysis of masonry buildings (that can be represented by one or more connected shear-type frames), where geometrical and mechanical data are estimated only as ranges of equally possible values.

The first experiences gathered from a general formulation seem to be promising. The computer application is being implemented under the NEXPERT shell on a 486 personal computer.

CONCLUSIONS

The paper has illustrated a number of general approaches to the formulation of qualitative models of physical systems and has shown their applicability to also represent the structural behavior.

Although applications in the domain of structural mechanics can be considered at an embrional stage, the experiences presented and discussed in the paper seem to indicate that the procedure could be effectively implemented in real-scale problems of forecasting and interpreting the structural response both in the linear and in the nonlinear fields, without the need of precise numerical determination of system parameters. Further studies are however needed to validate effectiveness and reliability of qualitative models for the mechanics of structures.

REFERENCES

1. Lancaster, K., The scope of Qualitative Economics. Rev.Econ.Stud., 1962, 29.

2. Lancaster, K., The Theory of Qualitative Linear Systems. Econometrica, 1965, 33.

3. Adorni, G., Del Grosso, A., Burdese, M., Loddo, R. and Zucchini, A., A qualitative approach to Structural Mechanics. In Proc. 2nd Workshop on Qualitative Physics, Paris, 1988.

4. Del Grosso, A., I modelli qualitativi nella Meccanica delle Strutture. In Atti Conv. Naz. sulla Meccanica dei Materiali e delle Strutture, Rome, 1989, pp. 337-346.

5. Adorni, G., Del Grosso, A., Burdese, M., Loddo, R. and Zucchini, A., Un modello qualitativo per l'analisi di travature elastiche. Internal Report n.17, Istituto di Scienza delle Costruzioni, University of Genoa, 1990.

6. Zucchini, A., Modello qualitativo del comportamento elastoplastico del telaio shear-type. Internal Report n.1, Istituto di Scienza delle Costruzioni, University of Genoa, 1991.

7. Hayes, P.J., The Naive Physics Manifesto. In Expert Systems in the Microelectronic Age, ed. D. Michie, Edinburgh University press, 1979.

8. Hayes, P. J., The Second Naive Physics Manifesto. In Formal Theories of Commonsense World, eds. J.R. Hobbs and R.C. Moore, Ablex, 1985.

9. Various Authors, Artificial Intelligence, 1984, 24.

10. Struss, P., Mathematical aspects of Qualitative Reasoning. Artificial Intelligence in Engineering, 1988, 3.

11. Raiman, O., Order of magnitude reasoning. In Proc. 5th Natl. Conf. on Artificial Intelligence, AAAI, 1986.

12. Travé-Massuyès, L. and Piera, N., The orders of magnitude as Qualitative Algebras. In Proc. 11th Int. Joint Conf. on Artificial Intelligence, Detroit, 1989.

13. Dormoy, J.L., Controlling qualitative resolution. In Proc. 2nd Workshop on Qualitative Physics, Paris, 1988.

14. De Kleer, J., An Assumption-based TMS. Artificial Intelligence, 1986, 28, pp.127-162.

15. Spillers, W.R., Automated Structural Analysis: an introduction, Pergamon Press, 1972.

16. Berleant, D., A unification of numerical and qualitative model simulation. In Proc. AAAI Workshop on Model Based Reasoning, Detroit, 1989.

17. Stratton, R.C. and Jarrel, D.B., Developing integrated quantitative and qualitative models for reasoning about physical systems. In Proc. AAAI Workshop on Model Based Reasoning, Detroit, 1989.

18. Tang, D.W., Jolly, H. and Zalondek, K.C., Compiling quantitative model-based diagnostics for embedded applications. In Proc. AAAI Workshop on Model Based Reasoning, Detroit, 1989.

19. Moore, R.E., Interval Analysis, Prentice-Hall, Englewood Cliffs, New York, 1966.

EXPERT SYSTEM MODULES FOR SYSTEM CONTROL

[1] L. Faravelli [2] P. Gherardini

[1] *Professor, Institute of Energetics*
University of Perugia, via F. Cairoli 24, I06100 Perugia, Italy

[2] *Senior Researcher, Institute M. Picone*
Viale del Policlinico 137, I00161 Roma, Italy

ABSTRACT

The paper overviews the application of expert systems prototypes in system control. Some results obtained for off-line control are illustrated. The basic problem, for both on-line and off-line control is identified in the diagnosis of the state of the system. A probabilistic network scheme for a coherent development of this diagnosis is presented.

INTRODUCTION

Process control and system diagnosis, among similar problems in different disciplines, made manifest in the late Seventies [1] the poorness of general reasoning models, working on symbolic data structures but without specific links with the problem under investigation. Artificial Intelligence scientists realized at that time the need of incorporating in an efficient device all the pieces of structured information that are used by a human expert. The same efficient device should also contain the conscious or inconscious reasoning process by which the expert makes use of its knowledge for achieving a solution [2]. Ten years of research produced the capability of reproducing the expert behaviour by a computer code, called 'expert system' [3] [4]. The possibility of accessing great data banks and, mainly, the possibility of enlarging these data banks by a continuous monitoring of both input and output measurements, could cause a learning from service experience which could lead expert systems to be much more confident than the original pool of experts. The weak point in this evolution from 'expert system' to 'super expert system' is the treatment of uncertainty. After a brief review of the potential applications of expert system modules into the control of intelligent systems, the second part of this paper will be completely devoted to the illustration of a probabilistic network scheme and its application in diagnosis problems.

CONTROL MODULES

Figure 1 provides the scheme of a general purpose expert system prototype the authors are building within a plan of preservation, from seismic risk, of Italian monuments. The plan is sponsored by the Ministery of Cultural Goods and the National Group of Earthquake Mitigation (GNDT). The scheme consists essentially of five parts:

1. module 1 collects the data on the building under investigation and the site where it is located;

2. module 2 consults the available data banks; it consists of two submodules: module 2a queries the data bank of the seismic events at the site according to Figure 2 [5]; module 2b consults the data bank of the damages to retrieve the recordeed damages, if any, occurred during the life of the building (see Figure 3);

3. module 3 makes the diagnosis on the 'state of health' of the building on the basis of the available pieces of information. This is presently made by expert system prototypes computing the vulnerability index [6] [7] [8] [9]. Ongoing studies are aiming at introducing in the modules the possibility of performing structural analysis modelling (module 3');

4. module 4 supervises the monitoring activity and the updating, by these measurements, of the data banks exploited in 2.;

5. module 5 builds the appropriate countermeasures everywhen modules 3 and 4 produces an alert. The development of this module is still at an early stage.

All the previous modules were developed within the working environment NEXPERT [10]. This shell makes use of an object oriented view of data with:

- objects and sub-objects;
- classes and subclasses;
- definition of properties at class level;

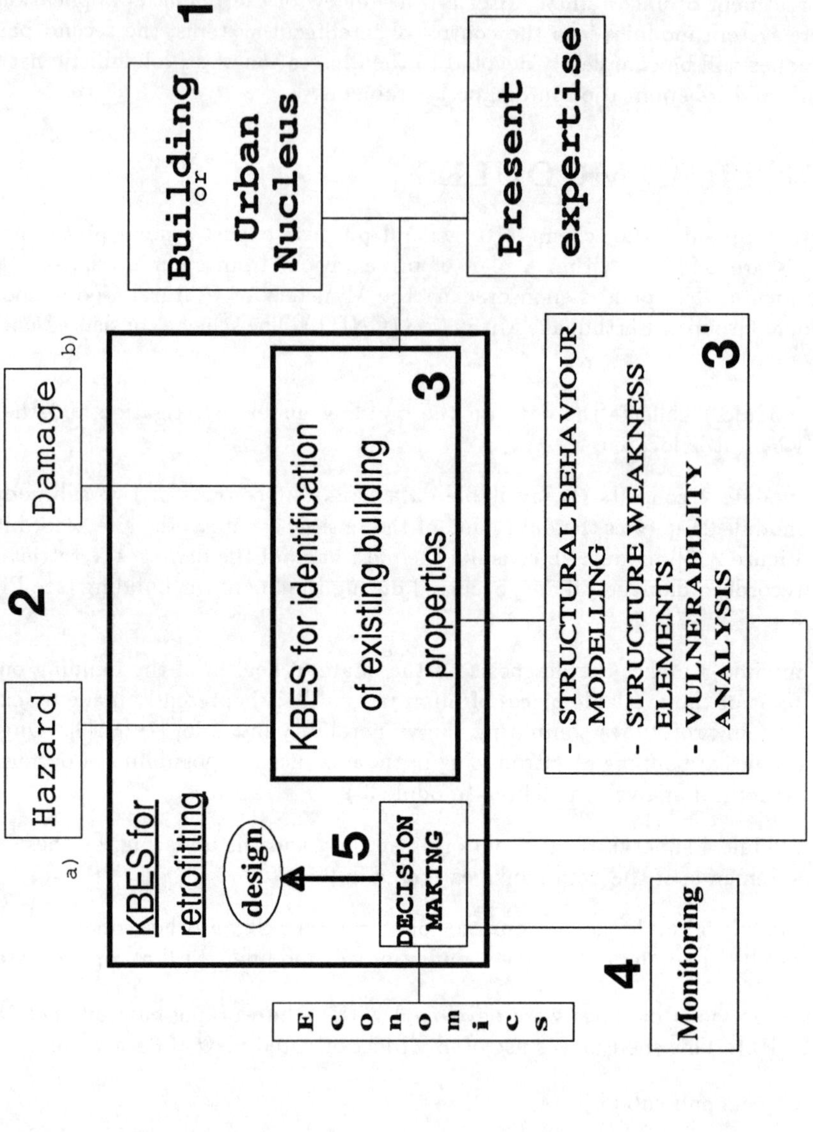

Figure 1: Scheme of a general purpose expert system prototype for seismic risk mitigation.

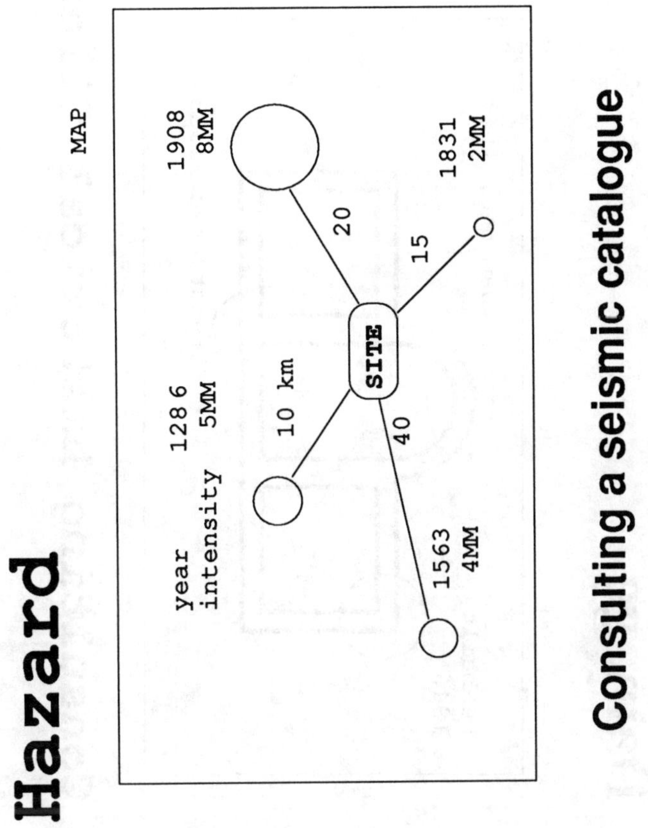

Figure 2: Scheme of an expert system module consulting the data bank of the seismic events.

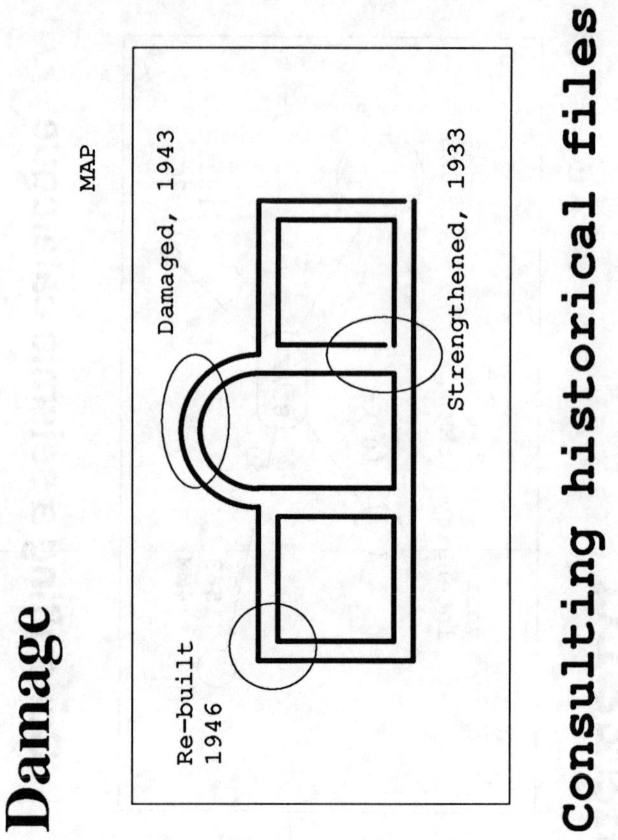

Figure 3: Scheme of an expert system module consulting the data bank of structural damages.

- inheritance of properties and/or values;
- procedures definable at the property level (either as source of values or as 'demons');
- dynamic creation/deletion of objects and links;

The reasoning representation is pursued by rules in the form **'IF premises (LHS) THAN hypothesis and DO action (RHS)'**. All the hypotheses are Boolean variables referrable to in both the LHS and RHS of any rule. Backward and forward reasoning is supported and inference strategies are definible in a dynamic way. In this scheme rules become relevant as a consequence of something happening which is linked to their LHS ('event driven architecture'). Nexpert application programming interface, then, allows both:

- embedding the Nexpert environment into a user application;
- calling user procedures sharing the Nexpert working memory (i.e. the data) from within Nexpert rules.

In this sense one says that Nexpert has an'open architecture'. Moreover, compatibility at the source level across several hardware platforms is ensured.

The set of modules previously discussed were realized for a off-line control scheme. Nevertheless, also on-line control systems require that a decisional analysis is performed. In this case pieces of evidence are provided both by monitoring and by data bank consultation. They are provided to a 'brain' which must take the decision of activating or not activating the actuators. Several sources of uncertainty interact with this decisional process. The pieces of evidence (from monitoring measurements and/or data bank consultation) must be filtered through a propagation network of a probabilistic nature to enter the decision criterion. One assumes that:

- the network is identified;
- the probabilistic structure of the relevant conditional probabilities is known.

The next Section presents a way for propagating pieces of evidence from the input to the filter response.

CAUSAL PROBABILISTIC NETWORKS

A causal network (or Bayesian network, as in [11]) is defined as a directed graph $G = (V, E)$, i.e. a set of nodes V and a set of directed links E. Adopting such graphical display for representing knowledge bases, we assume that the nodes represent conceptual entities coming from a specific knowledge domain, while the links express a causal relationship between these entities: if, say, directed links go from entities a and b to entity c, then we say that c is caused by a and b. In this context the causality relation has to be thought of in a very broad sense: the terminology only refers to the implicit assumption being made that knowledge related to entities a and b (the parents) can lead to modify our belief on the possible state for entity c (the child).

In a probabilistic framework these entities will be in some way associated to random variables which, for the sake of simplicity, we now assume to be binary, i.e. representing events which can or cannot occur with a given probability p ($1 - p$, resp.).

To represent the causality links we state that the probability distribution sitting on each node is in fact a conditional probability distribution specifying, for each possible configuration of states of the parent variables, the probability of the child variable being in each of its two states.

So, a CPN (Causal Probabilistic Network) can be thought of as having two component: the qualitative structure, specifying the causality dependencies, and the quantitative behaviour, i.e. a probabilistic assessment of these dependencies. In this setup, which can be regarded as the basic structure of a probabilistic expert system, the problem arises of defining a procedure, both correct and computationally feasible, for the propagation of uncertainty, as expressed by probabilities. We think of procedures which strictly obey the probability calculus axioms, and so guarantee a correct answer in any circumstance.

But what do we mean by propagation of uncertainty and why is it needed ? A CPN is in a sense a static model for a certain knowledge domain: we want to use the model in situations evolving dynamically, and so to be able to answer questions such as "how would the probability of event x occurring be modified, when it is available new evidence concerning other events linked to x in the graph, the last ones being either close neighbours of x or distant relations ?".

In other words if, at a certain stage, we collect data which completely specify the value of a subset of variables in the network, so that no longer uncertainty exists about them, how this knowledge would modify our conditional probability assessments for the variables which have not been observed ?

Correct computational procedures to give this problems an answer can become very unefficient in large networks, because in that case one would end up with functions of hundreds of variables. A detailed introduction to this problem, and indeed its solution in many general cases, can be found in [12]: very briefly, a correct and efficient probability updating can be performed exploiting local relations between nodes in the graph, so that an a priori cumbersome computation can in fact be reduced to a sequence of very simple steps, involving only small subsets of variables at a time. Further, the updating algorithm can be described in an object oriented framework: each time new information is to be included, which means modifying our knowledge about one or more nodes (objects), updating messages need to be sent to (and only to) their adjacent nodes (see also [13]).

At the end, when no further updating is necessary, a new equilibrium is found that reflects the new state of the world: from this, revised marginal probabilities for each unobserved variable can be easily obtained.

The Nexpert implementation of both the CPN structure and the probability propagation algorithm proved to be very convenient, especially when integrated with 'ad hoc' numerical routines (details can be found in [14], [15] and [16]). Basically it consists of two steps:

- the original network is transformed into a junction tree, i.e. a tree whose nodes are well defined complete subsets C_i, called cliques, of the original node set V; these cliques are ordered in such a way that, if any two of them, say C_i and C_j, have common elements, than these elements can be found in the intersection set of any two adjacent cliques on the path that connects C_i and C_j in the tree. This property garantees that any new information, concerning individual node, entering the network, be propagated in a consistent and non redundant way;

- each clique is assigned a belief table, derived in a proper way from the original conditional distributions, which is in fact (up to a normalizing constant) the joint probability distribution on the nodes belonging to the clique. The propagation scheme is then described in term of two basic operations to be performed locally: collection of evidence, in which the new information entered from different branches is propagated to a single node acting as a root, and distribution of evidence, which means that updating factors, summarising the needed modifications, are propagated back from the root to all other cliques.

At the end of this procedure an equilibrium state is reached, i.e. a state such that marginal distributions on single nodes can be derived from any belief universe containing that node, with the guarantee that the results obtained are the same (adjacent belief universes give consistent results on common nodes).

NUMERICAL EXAMPLE

A control device is simplified schematically by:

1. three input sensor detecting smoke levels higher than a given threshold, wind velocity v higher than a level v_u and ground acceleration a higher than a_u, respectively;
2. one output sensor detecting values of the displacement u higher than the limit value u_u. A correct design of the structure should regard v_u and a_u as the values of the actions wind and earthquake, respectevely, which, acting separately, cause u_u.

Now, the displacement thereshold exceeding may be due to a dangerous state, induced by wind or earthquake, so as to require the actuators to be activated. But it may also be due to fire or accident or none of these four causes or more than one of them. Evidence of smoke, wind and earthquake, should come from the corresponding sensors, but their potential black-out must also be considered and represented as a random quantity. The structure of this knowledge-base is represented by the directed graph in Figure 4, where a logical link (either accident or fire) is also included. In summary the following events are considered:

SMOKE (Σ) = smoke is detected;

WIND (Ω) = significant wind is detected;

EARTHQUAKE (E) = significant ground motion is detected;

$\imath$ (iota) = the instruments are not in black-out;

α = accident occurs;

σ = smoke occurs;

ϕ = fire occurs;

η = either fire or accident occurs;

ω = significant wind occurs;

ϵ = significant earthquake occurs;

δ = dangerous state: the actuators must be activated;

x = significant displacement is detected.

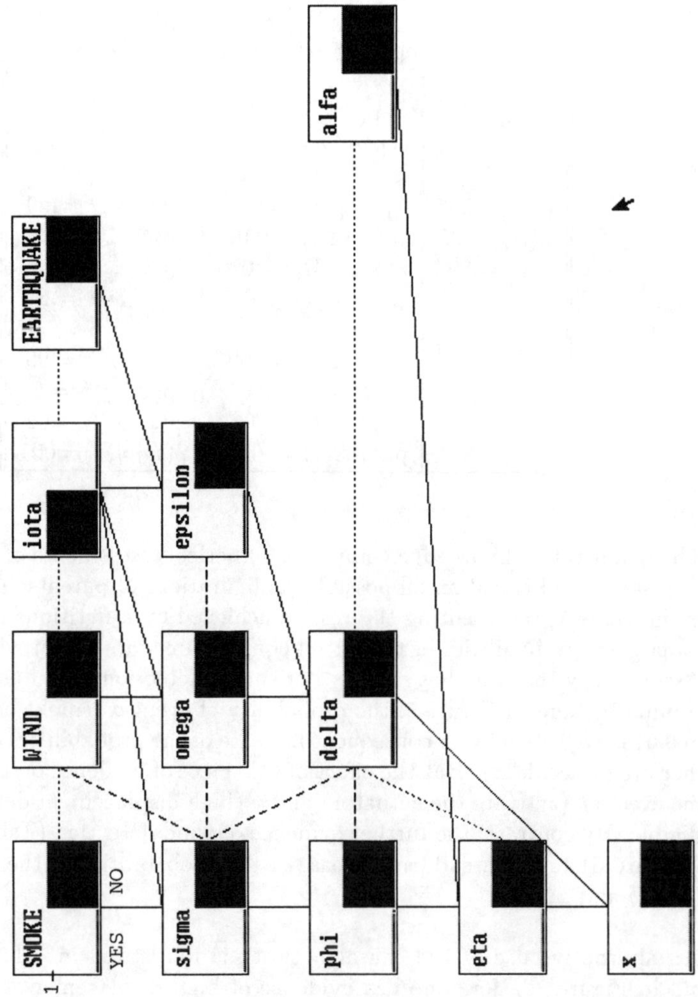

Figure 4: Causal network for the structural control example. Each node has two possible states: yes and no. Direction of causality is from top to bottom. For each node, the bar diagram provides the marginal probability distribution in the initial equilibrium state.

Table 1: Conditional probability table for the numerical example

	$p(\sigma \mid \Sigma, i)=0.9$ $p(\sigma \mid \bar{\Sigma}, i)=0.01$ $p(\sigma \mid \Sigma, \bar{i})=0$ $p(\sigma \mid \bar{\Sigma}, \bar{i})=0.009$	$p(\eta \mid \alpha, \phi)=1$ $p(\eta \mid \bar{\alpha}, \phi)=1$ $p(\eta \mid \alpha, \bar{\phi})=1$ $p(\eta \mid \bar{\alpha}, \bar{\phi})=0$
$p(\Sigma)=.01$ $p(\Omega)=.005$	$p(\phi \mid \sigma)=0.6$ $p(\phi \mid \bar{\sigma})=0.01$	
$p(E)=.005$ $p(\alpha)=.001$ $p(i)=.99$	$p(\omega \mid \Omega, i)=0.9$ $p(\omega \mid \bar{\Omega}, i)=0.01$ $p(\omega \mid \Omega, \bar{i})=0$ $p(\omega \mid \bar{\Omega}, \bar{i})=0.045$	$p(\delta \mid \omega, \epsilon)=.99$ $p(\delta \mid \bar{\omega}, \epsilon)=.9$ $p(\delta \mid \omega, \bar{\epsilon})=.8$ $p(\delta \mid \bar{\omega}, \bar{\epsilon})=.01$
	$p(\epsilon \mid E, i)=0.09$ $p(\epsilon \mid \bar{E}, i)=0.01$ $p(\epsilon \mid E, \bar{i})=0$ $p(\epsilon \mid \bar{E}, \bar{i})=0.0045$	$p(x \mid \eta, \delta)=.99$ $p(x \mid \bar{\eta}, \delta)=.9$ $p(x \mid \eta, \bar{\delta})=.02$ $p(x \mid \bar{\eta}, \bar{\delta})=.001$

The quantitative knowledge comprises numerical assessments of the probability of each state conditional on all possible configurations of parents. Assessments are given in Table 1, representing the results achieved by a fictitious pool of experts. The marginal probabilities in the initial equilibrium state are provided in Table 2.

Assume now that one has evidence of smoke detection. As illustrated in Table 2 (column b)) and in Figure 5, the probability of event σ (smoke occurs) increases from 0.0188 to 0.8910, with consequent increase of the probabilities of ϕ (fire) and η (either fire or accident). But the effect of this piece of evidence on the probabilities of the events δ (activate the actuator) and x (high displacement detection) is quite negligible. By contrast, the further evidence of wind detection (Table 2, column c) and Figure 6) causes a sudden increase of the probabilities of the events ω (wind occurs), δ and x.

An alternative use of the causal network is illustrated in column c) of Table 2 and in Figure 7. Here one has evidence of high displacement detection (event x) with a consequent sharp increase of the probability that the actuators must be activated (event δ).

Table 2: Marginal probabilities for the nodes of the causal network in Figure 4 for the following four cases: a) initial equilibrium state; b) evidence of Σ (smoke detection); c) evidence of Σ and Ω (smoke and wind detection); d) evidence of x (significant displacement).

event	a)	b)	c)	d)
i	0.99	0.99	0.9899	0.9843
Σ	0.01	1.00	1.00	0.0138
Ω	0.005	0.005	1.00	0.1111
E	0.005	0.005	0.005	0.0157
α	0.001	0.001	0.001	0.0017
σ	0.0188	0.8910	0.8903	0.0267
ω	0.0148	0.0147	0.8937	0.3667
ϵ	0.0103	0.0104	0.0109	0.2888
ϕ	0.0211	0.5357	0.5351	0.0361
η	0.0221	0.5361	0.5356	0.0377
δ	0.0308	0.0308	0.7195	0.9527
x	0.0291	0.04	0.6859	1.00

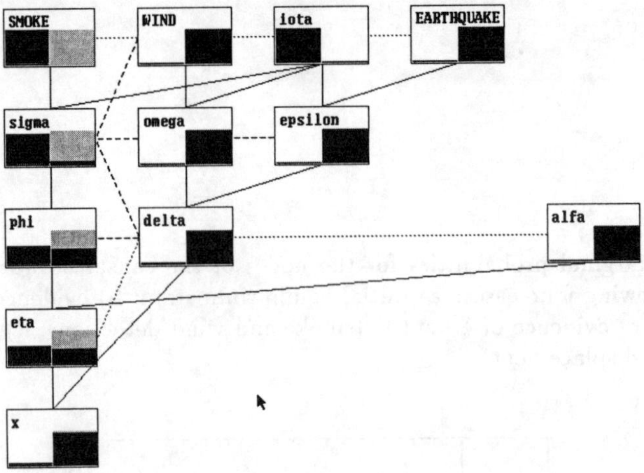

Figure 5: Marginal probabilities in the nodes of the causal network after evidence of smoke detection (Σ). The shaded area denotes the initial condition before the new evidence was acquired.

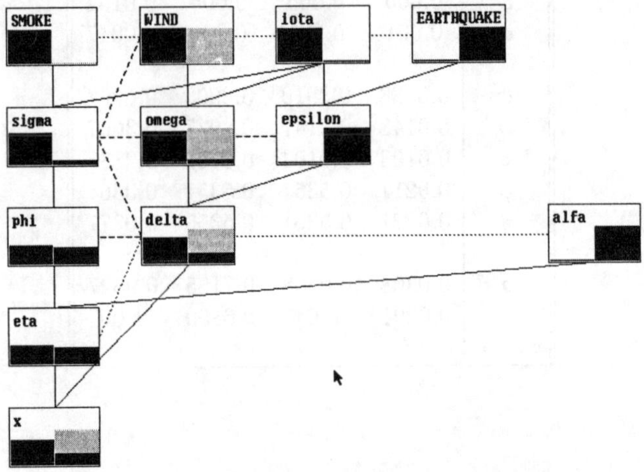

Figure 6: Marginal probabilities in the nodes of the causal network after evidence of smoke and wind detection (Σ and Ω). The shaded area denotes the initial condition before new evidence on Ω was acquired.

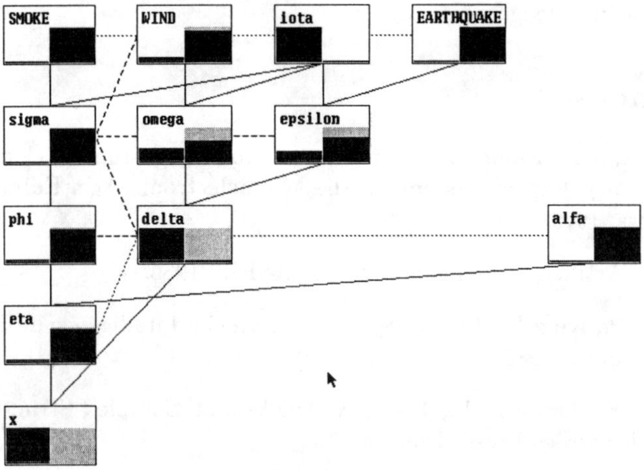

Figure 7: Marginal probabilities in the nodes of the causal network after evidence of high displacement detection (x). The shaded area denotes the initial condition before the new evidence was acquired.

CONCLUSIONS

This paper illustrates a probabilistic treatment of uncertainty which was implemented in the AI working enviroment Nexpert. In synthesis one can achieve a probabilistic assessment of the 'state of health' of the structure on the basis of pieces of evidence collected by the monitoring sensors and by querying historical data bases. In particular, the diagnosis is reached by absorbing and propagating evidence about potential causes or observed effects in a network representing the structural behaviour and the monitoring features. In our first prototype each component of the reasoning network is presented by a Boolean nature. Future developments of this causal network scheme, presently in progress, include:

1. extension to discrete variables for each component;
2. implementation of an automatic sensitivity study for identifyng, among the conditional probabilities which characterize the network, the ones to which the result is more sensitive.

ACKNOWLEDGEMENT

This research has been supported by a grant from the National Research Council (CNR). A grant from the National Group of Earthquake Mitigation (GNDT) is particularly acknowledgement.

References

[1] Young R.M., Production Systems for Modelling Human Cognition, in D. Michie (ed.), Expert Systems in the Microelectronic Age, Edinburgh University Press, 1979

[2] Rich E., Artificial Intelligence, Mc Graw Hill, 1986

[3] Harmon P., King D., Expert Systems: Artificial Intelligence in Business, John Wiley & Sons, 1985

[4] Casciati F., Faravelli L., Fragility Analysis of Complex Structural Systems, Research Studies Press, Taunton, 1991

[5] Faravelli L., Gherardini P., Seismic History at Monumental Sites by AI Working Environment, in Repair and Maintenance of Historical Buildings, Seville, (Spain), 1991, Vol. II, pp. 65-76

[6] Casciati F., Faravelli L., Seismic Vulnerability via Knowledge Based Expert Systems, in Brebbia C.A. (ed.), Structural Repair and Maintenance of Historical Buildings, Computational Mechanics Pubbl., Southampton, 1989, pp. 229-307

[7] Casciati F., Faravelli L., Uncertainty Treatment in a Vulnerability -Assistant Expert System, IABSE Colloquium, Vol 58, 1989, pp. 97-106

[8] Casciati F., Faravelli L., D. Yachaya, An AI Development Enviroment for the Maintenance of Existing Buildings, ASCE Conference '91 , Columbus, Ohio, (USA), 1991, pp. 83-87

[9] Casciati F., Yachaya D., Vulnerability of Masonry Buildings Exploiting Nexpert Utilities, in Repair and Maintenance of Historical Buildings, Seville, (Spain), 1991, Vol. I, pp. 235-246

[10] Nexpert Object, Reference Manual (Rel. 2.0), Neuron Data , Palo Alto, CA (USA)

[11] Pearl J., Probabilistic Reasoning in Intelligent Systems. Morgan and Kaufman, San Mateo, 1988.

[12] Lauritzen S.L., Spiegelhalter D.J., Local Computations with Probabilities on Graphical Structures and Their Application to Expert Systems (with Discussion), J. R. Statist. Soc. B, 50, 1988, pp. 157-224

[13] Jensen F.V., Lauritzen S.L. Olesen K.G., Bayesian Updating in Recursive Graphical Models by Local Computations, Research Report R-89-15, Dept. of Mathematics and Computer Science, Aalborg University, Denmark, 1989

[14] Gherardini P., Uso di tecniche di programmazione ad oggetti in alcune operazioni su grafi (in Italian), Quaderno n. 20, Ist. Applicazioni del Calcolo, CNR, Roma, 1989

[15] Gherardini P., Implementazione object-oriented di un modello di calcolo per reti causali (in Italian), Proceedings of the XXXV Riunione Scientifica della SIS. Cedam, Padova 1990

[16] GherardiniP. and Mayer C, Integrating Numerical Routines into a Nexpert Knowledge Base, Quaderno n. 2, Ist. Applicazioni del Calcolo, CNR, Roma, 1991

TOWARD KBES APPROACH FOR THE MANAGEMENT OF THE PROBLEMS CONCERNING SEISMIC RISK REDUCTION OF BUILT UP AREAS

C. Gavarini
Universita' degli Studi di Roma "La Sapienza", Italy
National Group for the Earthquake Loss Reduction
National Committee for the Seismic Protection of Monumental Buildings and Italian Cultural Heritage

ABSTRACT

The studies aimed to seismic risk reduction deal generally with a great amount of data: it is indeed necessary, in most cases, to consider in a first phase entire populations of buildings, possibly constituting districts, villages or towns; and the data are of any kind: numerical and graphical, related to materials and typologies; historical and artistical; related to human beings and their activities; and so on. One must first gather such data, then load them in suitable Data Bases, then elaborate them in order to define some indexes regarding the measure of seismic risk; after this, i.e. the assessment of risk, there will be the phase of risk reduction, to obtain through several kind of interventions: modifications in occupancy, strengthening of buildings, demolition and reconstruction of buildings, on the same spot or on a new one, application of artificial devices aimed to passive or active control. In order to obtain the best results it is very important to give a great attention to the studies at a global level, before starting with punctual design on the single object, and for such a purpose it is essential to have at disposal appropriate informatic tools as:

- well defined and structured Data Bases;
- friendly softwares for the systematic examination of data;
- suitable tools for the study of possible designs for intervention and the quick verification of their effects in terms of risk reduction;
- methodologies and criteria for decisions.

It seems quite evident that most of the above mentioned tools might be built up as Expert Systems, within the so called KBE (Knowledge Based Engineering). But if the final goal may be conceived, the problem consisting in the definition of the general environment and the implementation of the tools is not an easy one.

Some initiatives have been taken till now in Italy, and the actual presentation will try to give an idea of the general approach we have in mind and present some results obtained.

Four sectors of interest have been considered, with the following characteristics and results:

1) Assistance to non specialized engineers in the emergency condition assessment of buildings damaged by an earthquake: the KBS "AMADEUS" provides a detailed guide to the survey and the usability evaluation of buildings, to be used on the spot on a battery powered Lap Top.

2) Vulnerability assessment of masonry and reinforced concrete buildings using forms or Expert Systems; quick evaluation of the lateral resistance of rc buildings by means of program "PORTAM" and study, with conversational technique on the computer (graphically supported), of the possible retrofitting designs.

3) Preliminary definition of a general system/environment for the survey (architectural and structural), the control and the maintenance of monuments (CA.RI.S.MA); definition of a methodology and a form for the vulnerability survey at level I: in alternative to the form an informatic program (CADISM) enables the direct collection of data on the spot and their loading in the data base, another program (REDASM, written in CLIPPER, like CADISM), enables the friendly consultation of the collected data and finally an Expert System enables the seismic risk evaluation of the single monument on the basis of a simplified typological model.

4) Global organization of the vulnerability data concerning the buildings of an historical center or village, in order to allow the global evaluation of the situation, in terms both of individual risk of the buildings and urban risk, with the possibility to study and to evaluate quickly mitigation provisions.

The studies undertaken are far from being completed, but the approach seems to be promising; and a general principle comes out evident: the informatization of methods and tools in every sector of knowledge cannot be a matter of informatic specialists only, but it does involve the active partecipation of the experts in that very sector.

The KBS AMADEUS has been presented in [01].

The proposed approach for Monuments is the object of a paper under press on the international Review "Microcomputers in Civil Engineering" [02].

The results obtained in sectors 2,4) will be presented in a paper submitted to the 10 WCEE, Madrid 1992 [03].

REFERENCES

[01] T. Pagnoni, C. Gavarini, Z. Tazir: AMADEUS: a knowledge-based system for the assesment of earthquake damaged buildings, IABSE Colloquium: Expert Systems in Civil Engineering, Bergamo, october 1989, pag. 141-150.

[02] C. Gavarini: CA.RI.S.MA: A systematic approach regarding the cataloguing, the survey, the control and the planned maintenance of monumental buildings. General approach and first hypotheses of development, under press on "Microcomputers in Civil Engineering", Elsevier Science Publishing Co., New York.

[03] C. Gavarini: Toward the systematic use of expert systems in seismic risk reduction, submitted to the 10 WCEE, Madrid, July 1992.

Neural Computing for Intelligent Structural Systems

Daniel R. Rehak
Professor, Department of Civil Engineering
Carnegie Mellon University
Pittsburgh, PA, 15213 USA

and

James H. Garrett Jr.
Assistant Professor, Department of Civil Engineering
Carnegie Mellon University
Pittsburgh, PA, 15213 USA

Abstract

Active control of large-scale, flexible, reconfigurable structures is problematic. While a conventional approach might develop a control strategy from the governing equations of the plant, there are problems associated with this method. A neural network based control paradigm, inspired by biological systems, provides an alternative. This paper characterizes the classical neural computational model in terms of the control problem and describes prior work in developing neural applications for control, as well as for related problems of interest within the structures domain. The paper addresses the issues of applicability and use of neural computing for structural control and monitoring, and considers the impediments to building such a system. Conclusions argue that a neural computing approach to structural computations merits further exploration.

A Motivating Vision

Envision a large-scale flexible structure (thousands of components), such as the proposed US or Japanese space stations. Consider the following (futuristic) scenario. A large number of sensors are incorporated into the structure to monitor the entire state of the *plant*, for example, the deflection and acceleration at joints (or at other points), or the strain along members. Technology (e.g., *smart skins*) is available to continuously sample the entire state of the plant (virtually at any point, at nanosecond intervals). In addition, actuators or other active control devices are attached to some members (possibly a significant percentage) of the structure. These actuators and active members are used to control the stiffness and damping characteristics of the structure. A control system is in place to take the sensor readings and manipulate the actuators, in real time, to control the behavior of the structure when subjected to dynamic loads. In addition, the structure's configuration is not static: the entire plant can be reconfigured over time. Given this scenario, what mechanisms can be used to provide the required control?

The vision outlined above is not far from reality. Large structures, such as the space station, have such scale and complex real-time control demands, and are reconfigurable. Similar applications exist in bridges and buildings subjected to wind and earthquake forces, and providing control during the construction process presents the reconfiguration problem. Concepts for active control of such flexible structures have been put forth [Housner 90]. Imbedded sensor technology is under development, as are concepts for active members. What is still lacking are the control mechanisms appropriate for such a plant.

Current control approaches do not appear to be feasible for the active control of large-scale reconfigurable systems. For example, condition assessment for the large number of sensor readings is in itself a complex problem. Computing the behavior of such a large dynamic, non-linear system requires extensive computing, while real-time control demands on such a scale approach the limits of our current computing technology (gigaflop supercomputers). Dedicating such large-scale, general-purpose computing resources to individual structures is not feasible or cost effective.

The following presentation introduces a different approach to control, based on the neural computing paradigm. To date, all of the ideas in this approach are as open as are those of the problem outlined above. However, the concepts are close to realization for smaller, more conventional plants. The presentation begins by outlining the conceptual approaches to control computations, introducing neural computing and contrasting it to other techniques. A more detailed discussion of the mechanisms of neural computing and a summary of prior work in neural-based control follows. The presentation concludes with a discussion of the benefits, problems and future directions of such an approach to structural control.

Alternative Control Mechanisms—Introduction to Neural Computing

Active structural control consists of four major processes: (1) gathering the sensor data, (2) computing the state of the plant based on the sensor readings, (3) determining a set of control forces based on the control regime, and (4) driving the actuators. The computationally difficult problem lies in converting the raw data to the forces (steps (2) and (3)). The following discussion of the mechanisms of the control process concentrates not on the mathematical details of how control is performed, but rather on the computational concepts behind the process.

The conventional approach to control is rooted in the use of the system of fundamental equations which govern the behavior of the system. For anything but the simplest plants, a closed-form solution is infeasible. For structures, a number of behavioral assumptions are made: the result is a variational statement of the energy of the system. Some mathematical techniques (e.g., Ritz, Galerkin) are applied, followed by a discretization. The interim result is a set of non-linear equations describing the dynamic behavior of the plant. Due to the change in representation (continuous to discrete), the complete solution at any arbitrary point cannot be computed (i.e., results are only directly available at the selected discretization points).

Having converted from a variational formulation to an algebraic one still leaves a problem which cannot be solved directly. Numerical techniques, however, can be used to convert the equations into a form which can be solved. The result, though, is computationally intense (iterative solutions of large sets of non-linear equations). The mechanism to solve the equations is to specify an operational procedure (e.g., Newton iteration using Gaussian elimination).

This approach, along with most other computational methods, is based on specifying the "procedure" used to "solve" the problem. Such an approach is thus denoted *procedural.* The procedural paradigm has long been the mainstay of engineering computing. Before discussing alternatives, several points about this approach to control must be restated.

The computational problem is large, and current hardware limits real-time solutions. Alternatively put, when the problem is within the limits of computing technology, the "price" for the solution (e.g., flying a supercomputer on a spacecraft) may be unreasonable. The assumptions limit the accuracy of the solution: real structures, materials, sensors and actuators do not behave as their mathematical models. Also, the assumptions presuppose the locations and essential values needed for control (which accelerations, strains, etc.). Through discretization, only a single structure is modeled. Changes or reconfiguration of the structure require reformulation of the model, a process generally done by hand.

As outlined above, this conventional approach has numerous drawbacks: (1) extensive computational requirements, (2) built-in modeling and control assumptions, (3) limited applicability for reconfigurable systems. These problems are only aggravated as the size of the structure increases. Problems such as dealing with redundant or malfunctioning sensors, and having alternative control actions (i.e., different sets of actuators capable of providing the needed control) only complicate the situation.

An alternative to the mathematical approach is to apply *qualitative* reasoning. For example, Newton's first law states that force is positively related to acceleration. Similarly, an increase in mass will lower frequency. Given such statements, we *could* reason about behavior from first principles.

The approach here is to start with a model, in symbolic form, which describes the plant. The model is augmented with a set of declarative statements about behavior, again in symbolic form, starting from first principles or some other high level of abstraction. A reasoning or inference procedure uses this description to derive information and compute results. The method could be implemented directly in a model-based reasoning system, or the symbolic statements could be restructured for use by a rule-based (i.e., expert) system. The basic approach, coming from the arena of artificial intelligence, is to solve the problem by manipulation of meaningful *symbols*, not by performing detailed computations. Such an approach is classified as a *symbolic* paradigm.

While the idea of computing directly from the basic principles is appealing, there is no way to realize the operations in an effective or efficient manner. Going from simple statements of principles to actual control behavior requires a significant inference process. For circumscribed problems, deriving sufficient relations for a rule-based system is possible, but for general control, the qualitative approach is not feasible—developing the relations is onerous, and using conventional machines requires extensive computations. Until recently, however, the symbolic paradigm was the only alternative to numerical computing. Thus, in terms of problem-solving perspectives, it cannot be ignored.

Given this brief review of existing control approaches, an alterr-ative can be introduced. The approach is inspired by a biological basis (though we will argue that once presented, biological arguments need not apply).

Animal systems are good examples of complex active control plants. Running, jumping, flying, etc., are all complex coordination and control problems. Animal systems sense position and orientation of individual components (e.g., limbs, digits), along with sensing global equilibrium and acceleration. For higher-level vertebrates, the scale of the sensor system is quite large. In our example, the skeletal system corresponds to the structural members, while muscles correspond to the actuators.

These biological systems have significant redundancy, both in sensing and in actuators (e.g., picking up an object can be done by hand-eye coordination only or via touch alone. In either case, the path of motion is not single valued). Under normal conditions, the systems generally operate only within a safe range (e.g., actions which result in breaking a bone are accidental, not planned—the control system will not drive the body out of range).

The brain and central nervous system provide the necessary control mechanisms. These systems can perform complex real-time control activities (e.g., gymnastics), and are adaptable (e.g., walking, walking while carrying loads, walking on crutches).

Control in biological systems is based on neither a procedural paradigm or a symbolic paradigm. Our brains are too slow to solve the control equations. Even if they were fast enough, there is nothing to indicate that the equations of motion or the needed qualitative concepts are inbred, and such behavior is not explicitly taught.

These systems work, however, by electrochemical propagation of signals, in parallel, through a network of highly interconnected simple neurons. Based on primitive inbred capabilities, such as being able to move a muscle, and through exposure to the environment and problem solving (i.e., by trial and error) these systems are able to learn how to control themselves and how to adapt, without external "programming" or "teaching" (e.g., learning to talk is not taught, but is done by trial and error, generating speech, listening, and correcting the way to speak), and these systems perform within the real-time demands of their environment.

Tasks such as walking or running require several footfalls per second. Given the electrochemical make up of neurons, the time required to fire a neuron is on the order of 10 milliseconds (range of 1 to 100). Thus, our brains function at a serial rate of about 100 operations per second. The real-time behavior we exhibit requires massive parallel "computations".

From such a biological inspiration, a system based on a parallel network of computations can be structured. The general form of the paradigm is classified as *neural* computing.

Model of Neural Computing

The neural computing paradigm goes under several names, including *neural networks*, *connectionism*, *massive parallelism*, and *parallel distributed processing*. Inspired by biological models (and often used as a computation model of biological systems), computation is modeled as a very large network of interconnected simple processors (*neurons*). Each processor accepts a set of inputs from other processors and computes an output which is propagated to the connected processors. Programming, per se, does not exist. Networks are trained (*learning*) to recognize input patterns and produce appropriate output responses. Due to the distributed nature of the neural computing architecture, computations are robust and useful results can be obtained (and accuracy maintained) given partial or erroneous input, or when one or more individual neural processors fail. This distributed structure also results in a system where the actual problem-solving representation used is not discernable. Rather, these systems are judged on how well they work, not how they solve the problem.

The first attempts to build neural computing systems came in the 1950's with the introduction of the digital computer, while the biological basis for neural computation dates from the last century. When the fundamental technique for learning in multilevel networks became widely available in the late 1970's, it eliminated a major technical barrier which had limited the methods applicability [Minsky 69]. Since the 1970's the number and range of neural computing applications and investigations has exploded.

A neural system maps an input *vector* (defined in its broadest sense as a collection of elements) from one *space* to another space (e.g., mapping sensed accelerations to control forces). The translation is not specified, but is learned. In learning, the explicit statement of what to learn is also not specified. A network is given a set of patterns (*inputs*) and their associated classifications (*outputs*). As illustrated in Figure 1, the network consists of several *layers* of interconnected neural elements (the network structure is patterned after its physiological equivalent). Just as electrochemical signals propagate through a set of neurons, in the neural model, a set of *activation* values pass through the network, transforming the inputs to the outputs.

Connections are directed paths in the network: the output from one processor passes to all the downstream processors. The layering structure simply reflects the length of a path, counted from either the input or output side. In many neural architectures

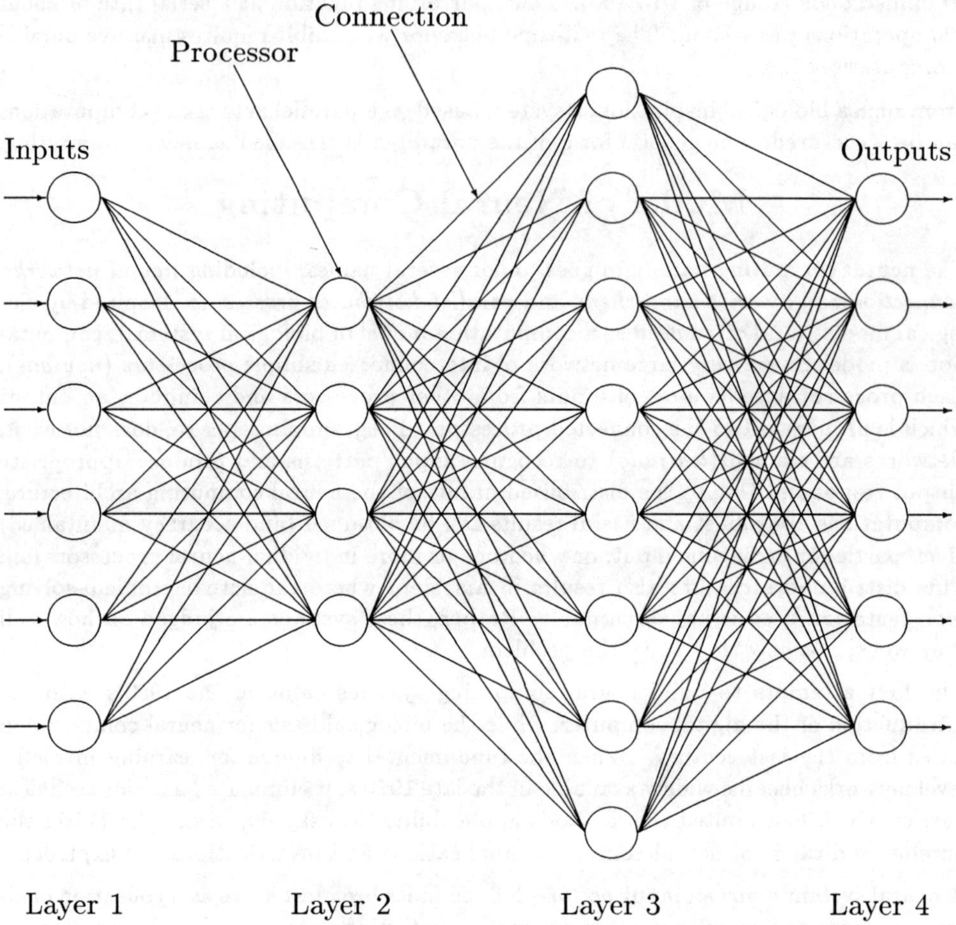

Figure 1: Neural Model

the lengths of all paths through the network are identical, but in general this is not required.

Generally, each processing unit in the network maintains only one piece of dynamic information (its current *activation* level, a scalar), although, a more complex state is permitted. As outlined below, the processor performs simple computations, taking its inputs and its activation value, computing a new activation, and propagating an output.

Neural networks compute by propagating activation through the network. Propagation starts when the network is presented with an input pattern, i.e., a set of processors receive an excitation from an external source. These *input* processors produce outputs which are transmitted to other processors via their connections. Level-by-level from the source, the processors take their input, compute an activation level, and propagate the output. The output of the processor is computed, via some function (e.g., sigmoidal, identity) from its activation level, and is propagated to all *receiving* processors via the connections. Associated with the connection path is a measure of the strength of the connection—a *weight*—which is used to attenuate the signals. A positive strength in the connection from processor A to processor B implies that if A is active, B should be active (e.g., biologically, one neuron firing causes others to fire with increased frequency). A zero (or near zero) strength implies that the activity of A should not influence B, while a negative value implies that A's activity suppresses B's. The strength of the connection is a filter, in the form of a multiplicative constant. The individual processors combine their inputs (generally a simple summation of the weighted values) to compute their activation value, and use this and the predefined activation function to compute a new activation level for output. The "result" of the neural computation is the vector of values produced by the terminal nodes in the network.

Given an input pattern, a *correct* computation will produce the expected output pattern. For a given neural architecture (i.e., number of processors, connection pattern, activation functions), it is the weights of the connections between the processors which determines the output. Thus the *program* is embodied within the connection strengths. Changing the program requires changing the weights, the interconnection pattern or the activation functions.

Thus, in the neural network the knowledge (i.e., what to compute) is maintained in the strengths of the connections, and the process of learning is one of determining these connections. As noted, the problem-solving strategy is not explicitly represented. Simply examining the weights and connectivity will not reveal how the system works. The representation and strategy are *distributed.*

Manual programming (setting) of the weights is usually not performed. Rather the network is trained to exhibit the correct behavior. Training involves presenting the network with many examples, and for each example, the error (defined as the difference between the actual and the expected outputs) is determined. The error is used to control modifications of the connections. This *learning* process is repeated until the behavior of the network is satisfactory (or the architecture is abandoned). After being trained on a large number of cases, a robust network will perform well on many cases which were not in the training set, i.e., it will *generalize.* Also, due to the structure of the computation, modifying individual weights, breaking connections, etc., will not have major impacts on the computed results. Thus a neural network is robust, and will tolerate fuzzy or incorrect inputs, and will produce reasonable results when individual processors fail.

Several methods for modifying the strengths of connections are a variant of *Hebbian Learning* [Hebb 49]. Hebbian learning can be paraphrased as: if unit u_i receives an input from unit u_j and both are active, then the strength of the connection w_{ij} between the two should be strengthened. For simple Hebbian Learning, the learning rule can be stated as:

$$\Delta w_{ij} = \eta o_j o_i \tag{1}$$

where o_i and o_j are the computed outputs of units i and j, respectively, and η is a learning rate constant (between 0 and 1) representing how rapidly the weight changes in proportion to the scale of the outputs.

Another common form of learning is called the *Delta* rule [Widrow 60]:

$$\Delta w_{ij} = \eta(t_j - o_j)o_i \tag{2}$$

where t_j is the expected (teaching) output for unit j. In English, the rule states that the correction to the weight is based on the product of the input and the difference between the expected and actual outputs. The original delta rule was only applicable to two-layer, feed-forward networks (connections are permitted only between one layer and the next, and all connections must be in the same direction). Such two-level networks are limited in what they can compute (e.g., they cannot compute an *exclusive or* function).

Multi-layer, feed-forward networks (such as that shown in Figure 1) can compute any function, but until the mid 1970's, no learning procedure for these networks was known. For multi-layer, feed-forward networks, with a continuous, differentiable activation function, a generalization of the delta rule, called error backpropagation (*backprop*), was developed. The learning procedure propagates errors seen at any level backward to the prior level, essentially using the Delta rule backward through the network.

Training a network by this procedure involves four steps: (1) present each training set to the network; (2) compare the output pattern produced by the network to the expected pattern; (3) backpropagate the error through the network; and (4) modify the weights of the connections according to the learning rule. The process (an *epoch*) is repeated for each training case, possibly many times, until acceptable behavior is obtained. If the system fails to converge, the overall structure of the network (its architecture) must be modified. Learning may encompass hundreds of training sets and thousands of epochs, requiring significant computations.

The development of a backpropagation neural network includes: (1) selecting a neural representation for the inputs and outputs (how many quantities; encoding—binary, −1 to +1, or unbounded values; etc.); (2) selecting the architecture and form of the network (number of layers, number of processing units in each layer, connectivity pattern between these layers, activation function, learning constants, etc.); (3) selecting the training data; (4) training the network (presenting the training epochs and letting the

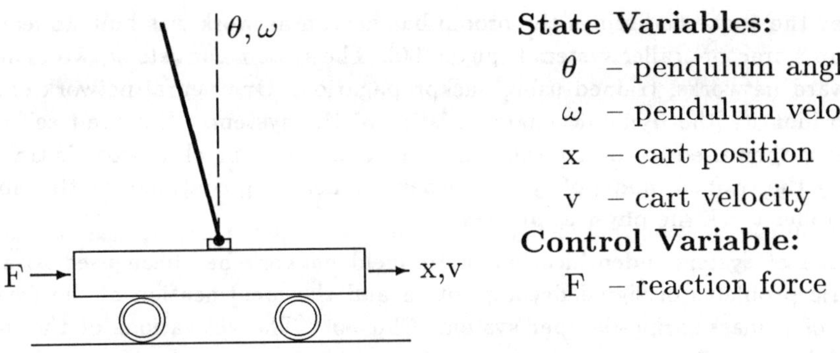

Figure 2: Inverted Pendulum

system learn); and (5) assessing the ability of the network to generalize when presented with patterns on which it was not explicitly trained.

After a network has been developed and trained, it can be used for problem solving. Input data is encoded as required by the representation of the input vector and presented to the network. Via parallel activations and propagation, the network computes a set of outputs which are interpreted as the problem solution.

Control Systems Applications

Control applications of the scale alluded to in the introduction have not been attempted with neural computing, but using neural networks for control applications has received considerable attention. While the vision of a neural controller for structural applications is as much a vision as that of the plant outlined above, fundamental work in applying the concepts has been undertaken, and preliminary work has shown that the ideas do work and can be successfully implemented. Some key contributions are outlined below.

One of the classical problems addressed is control of an inverted pendulum (see Figure 2), known as the "broom balancing" problem [Widrow 63]. As shown in the figure, at any instant, the state of the system is described by four state variables. For *bang-bang* control (the control force is constant, only its direction changes), the required output of the controller is the sign of the force. The collection of state variables forms the input to the system. Given any state, the network must compute a single output. Widrow and others have demonstrated the ability of neural networks to control this and similar dynamic systems (e.g., multi-link pendulums).

More recently, several contributions to control applications have been presented [IEEE 90]. Neural networks are being investigated for control of complex nonlinear systems, with applications: (1) to model the nonlinear behavior of the system, from which a control algorithm can be developed, or (2) through learning, to develop and represent the control algorithm.

Extending the concepts used in the broom balancer, a network was built to learn how to back up a tractor-trailer system [Nguyen 90]. The system consists of two multi-level feed-forward networks, trained using backpropagation. One neural network emulator learns to identify the dynamic characteristics of the system. A second self-trained controller is then used to control the actual system. The control network is trained by simulating the trained model of system behavior. Learning continues as the emulator and controller track the physical process.

In the area of system indentification, a Hopfield network has been used to address parametric problems of signal decomposition and the identification of the frequency response of a mass-spring-damper system [Chu 90]. The activations of the units of the Hopfield network represent the parameters of the system to be identified, and the energy function of the network is defined to be a quadratic of the prediction error of known cases. By using the dynamics of the network (a Hopfield network propagates activitation until it reaches an energy minimum [Hopfield 82]), the parameters are determined so as to minimize the least-mean-square prediction error for the observed system states. The results illustrate system identification in both the time and frequency domains, and show that a neural network can identify the transfer functions of dynamic plants.

In the area of structural control, there have been investigations of system indentification and pulse control [Doo 89]. Several network architectures were shown to perform well in system identification for one-, two- and three-degree-of-freedom linear, time-variant lumped parameter systems. The same models are used to provide pulsed, open-loop adaptive control to supress vibrations in single degree-of-freedom systems subjected to dynamic loadings. Simulations with earthquake excitations showed that a neural network pulsed controller could perform effectively. Assuming control forces can be applied in opposition to large inertial forces developed during seismic excitation, a neural pulsed controller was shown to also be effective for multi-degree-of-freedom systems.

Intelligent Structures

Returning to our initial question—how to control a large, flexible, reconfigurable structure—a scheme utilizing a neural control system can be proposed. Based on the general characteristics of the control problem, the structure and operation of a neural computing system, and the prior work in using neural systems for control, the neural computing paradigm appears to be useful for structural tasks. As noted, the state of neural computing technology is such that the full realization of such a neural controller is not possible today, but the technology is sufficiently well developed to explore this alternative.

Given a set of sensor or data samples for a structure, a neural network can be used to monitor or interpret these values. Monitoring might be used for several purposes. One is to detect sensor failure, an extremely important task for control, independent of the actual control regime and control method. Another application is to perform system identification utilizing the sensed data. An equally important application which has not been discussed is health monitoring (sampling data for signs of structural distress).

In all of these applications, the neural system acts as an input processor for some other control or problem-solving system.

Monitoring, sensor intepretation, system identification, etc., are all problems which are characterized as being a transform function from one state to another. As described, one of the major characteristics of the classical feed-forward, multi-level network is that it is designed to perform such transforms. There have been many successful investigations of neural networks for similar tasks, including vision and monitoring data errors on communications links.

These transfer function problems all share other common characteristics. Explicit methods (suitable for building a procedural solver) do not exist. Thus, applying classical procedural programming techniques to these and similar problems has not been successful. Because a transform is characterized by the relation between the inputs and outputs, the transform can be learned by observing the behavior. The learning methods used to train a neural computing system are analogous to learning by observing the transformation. Thus these problems map well onto a neural computing architecture.

In addition, these structural applications have strong real-time demands, making the parallelism of the neural approach appealing. As the sensing hardware may malfunction (no data from some sensors, bad data from some sensors, response time errors, noisy communcations), a system which can deal with such input is required. Since neural networks are robust for reasonable variations in input (i.e., a small percentage of the input being bad), they are well equipped for effective processing in such domains. Alternatively put, for most of the applications, the "exact" answer is not required, only one which is good enough. Given fuzzy input, the neural system will compute fuzzy output, which is often sufficient for these applications.

An alternative use of neural computing for structures is to provide control. Here the system is given the state of the plant, and it computes the necessary control actions. It is likely that such a neural controller would be coupled directly to sensing elements, but this is not required.

Using a neural system to provide effective control of small systems has been demonstrated [Widrow 63, Nguyen 90]. As with the sensor data interpretation problem, the system is performing a transform from the observed state of the plant to a set of control actions. Implicit in the control actions is the relation between the desired state and the actions. Control is different from interpretation in that there are the additional inputs which take into account the control regime driving the system.

For control, the neural approach is appealing in that it models the actual plant, and does not use built-in modeling or behavioral assumptions. Thus a neural controller would actually control the plant, not the model, and there are no additional requirements of interpreting the real data and actions in terms of the model.

As with interpretation, the neural system can learn how to control the plant by observing the effect of the control actions in a given state. Combining the learning nature of these systems with their transformational characteristics and ability to control the plant directly from observations versus controlling the model makes the idea of a neural system for control appealing. Problems which make control difficult in current approaches, such as time lags, should not be an issue in a neural system—they will be subsumed by the general characteristics of the solver.

As demonstrated [Nguyen 90], a neural system can be self learning. Envision building a neural system to monitor the behavior of a structure. Link that with a control system, and provide feedback between the two. When the system generates a control action, it also generates the expected behavior. Once the action is initiated, the actual behavior can be monitored and compared to what was expected. Differences can be used to correct the controller and the behavior model. Over time, the system will learn to control itself by proposing control actions and observing the behavior. Just as people learn to talk by speaking and listening (the neural model for this task was demonstated with *NETtalk* [Sejnowski 86]), the structure can learn to control itself.

The link between data and expected results could be used to add a "what-if" capability to a system. By inputting fictitious data (i.e., a query), the network would compute the response and actions (i.e., the reply). For example, if a known load were to be applied (e.g., a shuttle docking), this what-if capability could be used to precompute the control actions, which could then be executed while the load was being applied. Thus the controller would only have to deal with the variation in expected versus real loads and actions in real time, rather than responding to a completely unknown situation.

Thus these characteristics of being a transformational problem which requires learning (or lacks an explicit algorithmic solution for real structures), combined with fuzzy data and real-time demands, make neural computing appealing for structural control.

Given these learning characteristics, particularly self-learning, a neural system might be applied to control or monitor a reconfigurable structure. Assuming the change in structure configuration is slow, i.e., slower than learning, the system could adjust its behavior as the structure changes. As neural networks perform reasonably well with network variations, a small configuration change would not result in a serious performance degradation, and after the learning adjustments, performance would improve.

In the extreme of the futuristic vision outlined in the introduction, networks for monitoring and control combined into a self-learning system could be incorporated directly into the structure. The issue of parallel computing will be discussed below, but imagine that the plant itself becomes the computer. Smart skin technologies, in addition to providing an abundance of sensor data, might be extended to incorporate one or more neural processors in each member. The sensor "wiring" used as the communications paths within the structure could also be used to provide the connections between the processors. To utilize the sensing capabilities in a member, wiring interconnections and signal processing are required for each member; the additional cost of incorporating the simple neural processors should not be significant. Thus such a hardware approach might be possible.

Instead of some centralized system which monitors all values and provides control, the plant itself is the computer. The scale of the structure corresponds to the scale of the control system, and there is as much parallelism in the controller as there is redundancy in load path in the structure. Reconfiguring the structure reconfigures the controller, with the self-learning capabilities used to maintain correspondence between the controller and the plant. Failures of either individual members or processors are handled gracefully through the robustness of the systems.

While a biological model provided inspiration for the initial presentation, systems which are ***biologically correct*** or ***biologically feasible*** is not the goal. The concept of a trainable parallel architecture for control is a sufficient basis for exploring neural computing, and a more fundamental biological basis is not required.

There is one fundamental analogy between neural computing and structures. Activation patterns in some network architectures correspond to an energy minimization—the final set of activations is a local minimum of a complex hyperspace. As structural behavior is based on this principle, direct analogies between the energy state of a structure and the corresponding energy state of a neural model of the structure should exist. Such a formal basis has yet to be explored in detail.

The preceeding discussion has painted a bright picture of how neural computing might be applied to structures. There remain, however, several serious (possibly fatal) problems which must be addressed before any major applications of neural computing come to fruition for control, monitoring, system identification, etc.

First, the basis for selection of a neural architecture has not been formalized. No criteria were presented above to suggest what kind of system might be used for these problems—this is an unknown. Given a problem, what is a *good* architecture? While some anecdotal evidence exists, nothing is definitive, and building each new neural application is essentially a trial-and-error process, investigating several architectural alternatives until an acceptable one is found. Clearly, wide-spread applicability of the method requires that more be known about the type and characteristics of a system for a given problem.

Much has been said of the role of parallel computing in the neural model. As structured (multi-level, feed-forward networks), all of the nodes in a layer fire in parallel—the total computation time being of the same order as the maximum path through the network. This is an important biological inspiration for neural networks—given the response time of the electrochemical processes, biological networks are shallow and wide. To obtain the required real-time behavior for complex plants, parallel solutions will be required. Computer hardware, however, remains serial, and, except for some experimental hardware, most neural computing applications simulate parallel computations. Parallel *neural* hardware will be required for the types of problems discussed herein.

Closure

Neural computing has been presented as an alternative approach to providing structural control. The characteristics of the control problem correspond well with the characteristics of the neural computing paradigm, and work to date has shown that neural computing is applicable to simple control and monitoring problems similar to those in the structures domain.

A vision of how neural computing might be developed for structural applications has been presented, along with several major impediments to realizing this vision. While the approach has promise, no formal conclusions can be drawn. At present, the methodology is sufficiently well developed to warrant further investigations for a number of problems, which combined form the basis for large-scale control. This work must move beyond simple experimentation and trial-and-error construction of neural systems to establish a fundamental basis for the use of neural computing (e.g., determination of what fundamental characteristics of the control problem dictate the architectural details of the controller).

As an interim step, and thus independent of this larger vision, neural computing has immediate applicability in the structures domain. These methods are not limited to large plants, and some of the techniques could be applied to many structural systems, especially if the real-time aspect can be addressed. Formal work should go forth to investigate and establish a fundamental basis for applying neural computing applications to structural problems.

References

[Chu 90] Chu, S. R., Shoureshi, R., and Tenorio, M., "Neural Networks for System Identification," *IEEE Control Systems Magazine*, Vol. 10, No. 3, pp. 31–35, April, 1990.

[Doo 89] Doo, L. B., Thewalt, C. R., and Rehak, D. R., *Neural Networks for System Indentification and Pulse Control of Linear Systems*, Technical Report, Department of Civil Engineering, Carnegie Mellon University, Pittsburgh, PA., December, 1989.

[IEEE 90] *IEEE Control Systems Magazine: Special Issue on Neural Networks in Control Systems*, Vol. 10, No. 3, April, 1990.

[Hebb 49] Hebb, D. O., *The Organization of Behavior: A Neuropsychological Theory*, Wiley, New York, 1949.

[Hopfield 82] Hopfield, J. J., "Neural Networks and Physical Systems with Emergent Collective Computational Abilities," in *Proceedings of the National Academy of Sciences*, No. 9, pp. 2554–2558, 1982.

[Minsky 69] Minsky, M., and Papert, S., *Perceptrons,* MIT Press, Cambridge, MA., 1969.

[Nguyen 90] Nguyen, D. and Widrow, B., "Neural Networks for Self-Learning Control Systems," *IEEE Control Systems Magazine*, Vol. 10, No. 3, pp. 62–68, April, 1990.

[Sejnowski 86] Sejnowski, T. J., and Rosenberg, C. R., "NETtalk: a Parallel Network that Learns to Read Aloud," *Technical Report JHU/EECS-86/01,* Department of Electrical Engineering and Computer Science, Johns Hopkins University, Baltimore, MD., 1986.

[Widrow 60] Widrow, B., and Hoff, M. E., "Adaptive Switching Circuits," in *1960 IRE WESCON Convention Record*, pp. 96–104, IRE, New York, 1960.

[Widrow 63] Widrow, B., and Smith, F. W., "Pattern Recognizing Control Systems," in *Computer and Information Sciences (COINS) Symposium Proceedings*, Spartan Books, Washington. DC., 1963.

[Housner 90] Housner, G. W., and Masri, S. F., Eds., *Proceedings of the U.S. National Workshop on Structural Control Research*, University of Southern California, Los Angeles, CA., 1990.

Measurement of System Performance and Adaptive Systems

OPTICAL FIBER SENSORS FOR INTELLIGENT STRUCTURES

RICHARD O. CLAUS
Fiber & Electro-Optics Research Center
Bradley Department of Electrical Engineering
Virginia Polytechnic Institute and State University
Blacksburg, VA 24061 USA

ABSTRACT

Structures of the future may be fabricated from materials into which the functions of environmental sensing, data signal processing and communication, mechanical actuation and control are incorporated. Such "intelligent structures," and "intelligent" or smart materials from which they may be fabricated, have begun to be investigated during the past ten years. For several reasons, optical fiber technology may be a particularly good candidate for many of the environmental sensing and subsequent signal processing and control functions. This U.S.-Italy-Japan Workshop paper reviews work in this area as it applies to intelligent structures of the future, and suggests a wide range of related future research and development concerns.

1. INTRODUCTION

In order to construct, operate and repair large and complicated structures efficiently, sensors are needed to determine the properties of the structure, and the materials from which it is constructed, during its fabrication, operation and damage/repair phases. A recent National Science Foundation Workshop concluded that a wide variety of sensor information is required in such situations [1]. This includes both straightforward mechanical information such as distributed strain, temperature, and vibration, as well as more complicated multi-paramter data such as the degree of corrosion in bridge supports or piping, or the remaining use lifetime for a particular degree of mechanical loading on a damaged composite structure . Clearly, the required numerous sensor measurands and measurement locations will require different types of sensors to be linked together in a global communication and signal processing network. For such applications, optical fiber sensor and communication devices and networks seem particularly well suited.

Optical fiber sensors have been developed during the past fifteen years for the measurement of a wide variety of environmental phenomena. During the past three or four years, such sensors have been demonstrated in large numbers in both aeronautical and biomedical applications, and in a number of specialized industrial applications [2]. The advantages of optical fiber sensors for applications in intelligent structures are several, including their small size and weight, geometrical flexibility, and ability to function under particularly severe environmental conditions. Perhaps more importantly for applications requiring the networking of large arrays of sensors are dielectric fiber's inherent insensitivity to electromagnetic interference and fiber networks' inherent avoidance of instrumentation ground loops, factors which commonly plague similar wire-based sensor systems.

Optical fiber sensors can be designed and configured to measure different physical quantities. As discussed below, fiber sensors can either be implemented to mimic the response of conventional mechanical or electrical transducers, such as strain gauges or accelerometers, or to produce responses to structural properties which cannot conveniently be obtained using conventional sensor elements. Additionally, optical fibers themselves allow data transmission and networking, so fiber sensor networks may lend themselves to multiplexing and multimeasurand configurations, although much work in this area is needed.

Finally, although optical fiber sensors have begun to be developed for material and structural analysis, their application to civil structures has not been emphasized. The following sections discuss how fiber sensor research concepts may be directed toward civil structure needs.

2. FIBER SENSOR ATTACHMENT AND EMBEDDING CONCERNS

In all of the fiber sensor methods to be described below, the fiber sensor consists of a sensor fiber, an appropriate optical source and an optical detector, plus associated instrumentation as required. In order for the fiber sensor to sense the environmental perturbations to be monitored, the fiber needs to be intimately connected to the material, either attached to the surface or embedded inside.

The sensor attachment and embedding problem has been considered for several years with several general conclusions. First, the embedding of fiber sensors, like other sensors, within materials, causes local perturbations of the geometry of the host material and thus creates a potential site for mechanical failure [3]. Although methods have been considered to minimize such deleterious effects, the embedding problem is not solved.

Second, in cases where the sensors are to be attached to the surface of a material, problems which are similar to those associated with the attachement of other types of sensors are encountered. Specifically, adhesives used to form sensor-to-material bonds do not faithfully couple mechanical, electromagnetic or chemical phenomena to the fiber, and their transfer properties vary with time, age and mechanical cycling [4].

Suggested solutions to these problems include 1) the fabrication of fiber waveguides which have the correct combination of coating size and mechanical properties to eliminate stress concentrations at the fiber-to-matrix interface [5,6], and 2) the development of fiber coatings which have specific elastic and inelastic properties, to allow determination of the transfer, aging and creep phenomena in addition to the external phenomenon to be characterized [7].

3. CURE AND FABRICATION MONITORING

Optical fiber sensors have been suggested as elements for monitoring the cure of composite materials and for internal measurements during the fabrication of civil structures. Most attractively, perhaps, would be the use of a single fiber sensor for monitoring the cure of the material, and subsequently for determining other material parameters during the application of the material in a structural system.

The use of optical fibers in the cure monitoring of composites has been documented by several authors. Methods include 1) direct observation of the matrix resin chemistry using infrared transmitting fibers and Fourier transform spectroscopy methods [8], 2) measurement of the local index of refraction of the curing material using a sensor fiber fabricated from the curing resin material itself [9], and 3) local mapping of material modulus using fiber sensor-based acousto-ultrasonic techniques [10]. Much less work has been published concerning the monitoring of the fabrication of other types of materials or structural materials using fiber transducers or transducer arrays.

4. STRUCTURAL LIFETIME MEASUREMENT CONCERNS

The major application of optical fiber sensors in civil structures may be in determining performance during their normal use lifetimes. Two separate but important regimes of sensor operation in this area are discussed in this section, specifically the use of optical fiber sensors as 1) local, short gage length transducers, and 2) global, distributed transducers.

Several fiber sensor methods have been developed for short gage length measurements of strain, temperature and other phenomena. Among these, in-line Fabry-Perot etalon and grating sensors perhaps appear most attractive at this time [11,12]. Intrinsic and extrinsic Fabry-Perot etalons have been fabricated in-line in fiber systems, the intrinsic etalons using internal mirrors [13], and the extrinsic using air gaps between fiber ends in cylindrical hollow core fibers [11]. Both may be operated in either a linear region in which output is proportional to strain in the gage length along the fiber, or a nonlinear region in which counting the number of interference fringe maxima indicates the differential strain which has occurred in the gage length.

Figure 1 shows the design of an extrinsic etalon which has been used to monitor strain in several structures including aircraft wings [14], coal-fired power plant combustor filters [15], and metal cables [16],

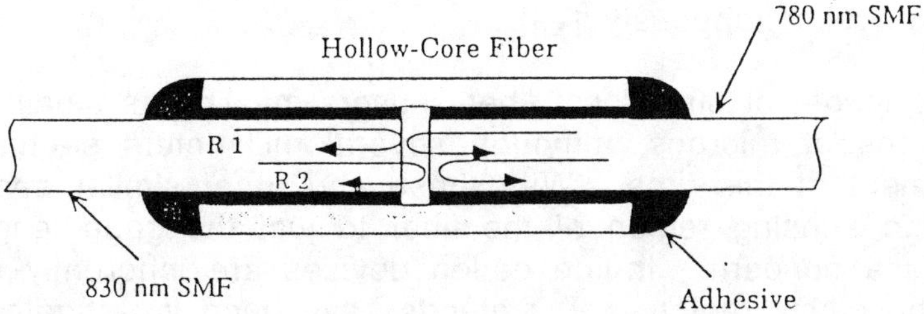

Figure 1. Extrinsic Fabry-Perot etalon sensor geometry. Similar sensors have been used for short gage length sensing applications in civil structures [14, 15, 16].

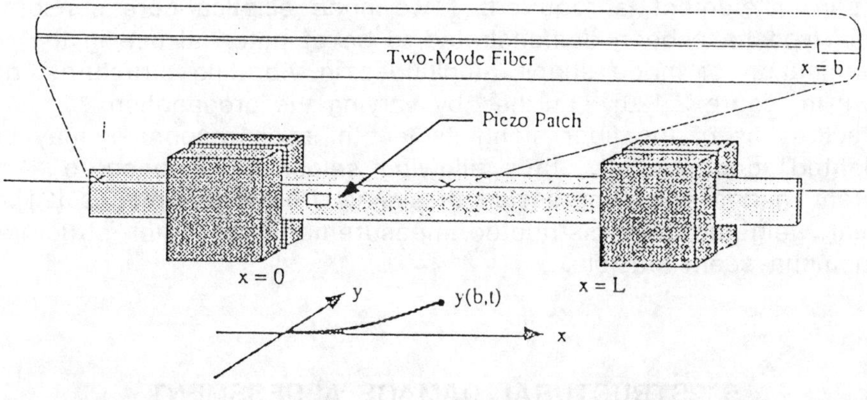

Figure 2. Experimental arrangement of two-mode e-core fiber sensor on clamped-clamped beam for analysis of vibration. Figure adapted from [21.

over a temperature range from liquid nitrogen to approximately 1000°C. By proper design of the sensor element, such sensors may be used for measurements of temperature [17], ultrasonic wave propagation [18], metallic component corrosion and other phenomena. Important here is the possibility of sensor cross-sensitivity to more than one observable, and methods to effectively solve for both phenomena simultaneously [19].

The gage length of such local fiber sensors may be as small as several tens of microns, although several millimeters seems easier to implement at this time. With these and other similar sensors, limiting the sensing region of the fiber to just the gage length section is a concern. In-line etalon devices are inherently sensitive only in the cavity region, and methods have been investigated to limit lead-in and lead-out fiber sensitivity using other measurement schemes [20].

Optical fiber sensors may also be designed to monitor distributed effects over a longer gage length in or on a structural material. In particular, interferometric sensors may be configured to measure the integral of phenomena such as strain and temperature along the gage length of the fiber. Again, control of the gage length by desensitizing the lead-in and lead-out sections of fiber connected to the sensor element is required. Two-mode elliptical-core sensors of this type have been demonstrated to be of potential use in the determination of mode shape amplitudes in vibrating structures, as shown in Figure 2 [20]. Further, by varying the propagation characteristics of the fiber along its length, such responses may be "weighted" geometrically, thus allowing selective response to different modes and suppression of signals from other modes [21]. Similar weighted and distributed measurements of other structural phenomena seem feasible.

5. STRUCTURAL DAMAGE ASSESSMENT

Optical fiber methods may also be used to determine the onset of some types of damage in structural materials; much related work has been published concerning advanced aerospace materials in particular. General approaches here are either to 1) implement a large network of fibers which, when broken, give a qualitative indication of where gross material damage has occurred [22], and 2) utilize an array of sensors placed at strategically important locations to determine damage-related phenomena such as the integrated effects of mapped mechanical loading over a period of time, the occurance of acoustic emission events, especially at critical locations, and the location and characterization of damage-producing impact events [23].

6. INTELLIGENT STRUCTURE FIBER SENSOR SYSTEM CONCERNS

The ability to use optical fibers for both sensing and communication allows an extra dimension in their application to intelligent civil structures. In order to be useful at a system level for the analysis, and perhaps the real time adaptive control, of the structure, sensor information must be combined and processed, and the results of that analysis used to control or repair. Fiber sensors lend themselves to in-line signal processing of sensor data, both through the type of weighted fiber sensing described above and through the incorporation of fixed or adaptive discrete in-line optical signal processing elements [24]. The incorporation of such signal processing capabilities directly into the optical sensor and instrumentation network may significantly reduce the size, power and speed requirements on the associated support computer analysis and control hardware.

Additionally, fiber sensors may be used directly with actuator elements to allow total control system implementation without the need for discrete sensor/signal processing/actuator elements [25]. Such work, albeit preliminary, has been suggested by several research groups.

Finally, the networking of individual fiber sensor elements into local "clusters," with hierarchical communication between clusters via optical fiber local area network methods has been suggested for the global control of large structural systems. Preliminary hardware demonstrations indicated that both communication and sensor data may be transmitted over the same fiber links in such systems, and suggest that local and global control functions may be implemented by the cluster centers and the higher control network node electronics [26].

7. CONCLUSION

Optical fiber sensor and communication technology may be particularly well suited for the measurement of material and structural phenomena, and the implementation of intelligent structure sensing, control and actuation systems. This paper has briefly described current work in several related areas and has suggested ways that this work may be directed toward applications in civil structures in particular.

8. ACKNOWLEDGEMENTS

This discussion has borrowed heavily from the technical papers referenced below. Related research within the Fiber & Electro-Optics Research Center at Virginia Tech has been supported for more than ten years by government organizations and industry. In particular, we acknowledge support from the NASA-Langley Research Center and ONR/DARPA, and cooperation from the Army Harry Diamond Laboratory, for work presented here.

9. REFERENCES

[1] Summary Report, NSF Sensors and Signal Processing for Structural Control Workshop (Washington, DC), February 1991.

[2] **Fiber Optic Sensors**. E. Udd, Wiley (New York, 1991).

[3] "Interferometric measurements of strain concentrations induced by an optical fiber embedded in a fiber-reinforced composite," R. Czarnek, et al., Proc. SPIE O/E Fibers (Boston, MA), September 1988.

[4] "Smart structures for fossil energy applications," K. Murphy, C. Koob, R. Claus and S. Desu, Proc. PETC Technology Workshop (Pittsburgh, PA), September 1990.

[5] "Low profile fibers for smart structures applications," A. M. Vengsarkar, et al, Technomic Press, Proc. Smart Materials and Structures Workshop, (Blackdburg, VA), April 1991.

[6] "Micromechanical strain fields of an optical fiber embedded in laminated composite materials," J. Sirkis, et al, Technomic Press, Proc. Smart Materials and Structures Workshop, (Blackdburg, VA), April 1991.

[7] "Mechanics of optical fibers embedded in fiber composite laminates," A. Dasgupta and J. Sirkis, Proc. SEM Conf. (Milwaukee, WI), June 1991.

[8] "Cure monitoring of high temperature polymeric composites with fiber optics and FTIR spectroscopy," M. Druy, Technomic Press, Proc. Smart Materials and Structures Workshop, (Blackdburg, VA), April 1991.

[9] "Composite cure monitoring using optical fiber sensors," B. Zimmermann, Technomic Press, Proc. Smart Materials and Structures Workshop, (Blackdburg, VA), April 1991.

[10] "Optical fiber-based ultrasonic methods for materials evaluation," W. V. Miller, MS thesis, Virginia Tech, 1991.

[11] "Stabilized extrinsic fiber optic Fabry-Perot sensor for surface acoustic wave detection," T. Tran et al, submitted to Applied Optics, March 1991.

[12] "Weighted optical fiber sensors," A. Vengsarkar et al, Proc. Conf. on Adaptive Materials and Structures (Washington, DC), November 1991.

[13] "Tunable in-line Fabry-Perot interferometer with high-reflectance internal mirror," C. E. Lee and H. F. Taylor, Proc. SPIE O/E Fiber Lase (Boston, MA), September 1991.

[14] "Fabry-Perot fiber optic sensors in full-scale testing on an F-15 aircraft," K. Murphy et al, Technomic Press, Proc. Smart Materials and Structures Workshop, (Blackdburg, VA), April 1991.

[15] "Liquid core optical fiber temperature sensors," M. DeVries et al, Proc. IEEE Region 3 Conf. (Williamsburg, VA), April 1991.

[16] "Embedded fiber optic sensors for cable evaluation," K. Murphy et al, Proc. Review of Progress in Quantitative NDE (Brunswick, ME), August 1991.

[17] "Sapphire fiber interferometer for microdisplacement measurements at high temperature," K. Murphy et al, Proc. Review of Progress in Quantitative NDE (Brunswick, ME), August 1991.

[18] "Cooperative implementation of a high temperature acoustic sensor," E. Baldini et al, Proc. SPIE O/E Fiber Lase (Boston, MA), September 1991.

[19] "Fiber optic dual technique sensor for simultaneous measurement of strain and temperature," A. Vengsarkar et al, submitted to J. Lightwave Technology, January 1991.

[20] "Elliptical-core two-mode optical fiber sensor implementation methods," K. Murphy et al, *J. Lightwave Technology* **8**, no 11, pp. 1688-1696, November 1990.

[21] "Two-mode elliptical-core weighted fiber sensors for vibration analysis," A. Vengsarkar et al, Technomic Press, Proc. Smart Materials and Structures Workshop, (Blackdburg, VA), April 1991.

[22] "Progress towards fiber optic smart structures at UTIAS," R. Measures, Proc. SPIE O/E Fibers (San Jose, CA), September 1990.

[23] "Replaceable and retrofittable fiber optic sensors," R. Claus et al, Proc. SEM Conf. (Milwaukee, WI), June 1991.

[24] "Holographic matched filter for full-field in-line signal processing of optical fiber sensor outputs," G. Indebertouw et al, *J. Lightwave Technology* **8**, no. 7, pp. 1039-1045, July 1990.

[25] "Photofluidics for integrating fiber sensors with high authority mechanical actuators," R. Claus et al, Technomic Press, Proc. Smart Materials and Structures Workshop, (Blackdburg, VA), April 1991.

[26] "Dual MIL-STD 1773 communication and microbend sensor fiber optic lind," J. McHenry, Technomic Press, Proc. Smart Materials and Structures Workshop, (Blackdburg, VA), April 1991.

COMPUTATIONAL ASPECTS OF REAL-TIME DYNAMICS AND CONTROL SIMULATION

K. C. PARK and K. F. ALVIN
Department of Aerospace Engineering Sciences and
Center for Space Structures and Controls
University of Colorado, Campus Box 429
Boulder, Colorado 80309
and
W. K. BELVIN
Spacecraft Dynamics Branch
NASA/Langley Research Center
Hampton, VA 23665

ABSTRACT

The use of adaptive elements for control of space structures presupposes a distributed control, hence requiring real-time computational capabilities involving a large number of controllers and sensors. The paper discusses a computational strategy that consists of: 1) a physically based synthesis of control law that can be interchanged to structural redesign parameters, and vice versa so that the designer can perform trade-off studies between control effort and structural redesign; 2) efficient second-order based control-structure interaction simulation algorithms that can be implemented on a workstation as well as a shared-memory machine; 3) an interface with nonlinear flexible maneuvering and control. Some examples illustrate the viability of the present strategy.

INTRODUCTION

Over the past five years, many investigators have been intensely focusing on the use of adaptive elements for advanced space structural systems [1]. The three major thrusts of research related to adaptive elements for active vibration and shape control have been sensor and actuator developments [2-4], experimental feasibility demonstrations [5-12], and modeling and analytical predictions [6,13-18]. As a result of these studies, several breadboard structures that embody adaptive elements have been designed and tested for their application performance [8,11,19-22]. From the viewpoint of dynamics and control of space structures, adaptive elements are attractive in that they offer non-inertia (or strain) actuation and sensing devices whereas conventional actuators and sensors are mostly inertia-driven devices, viz., proof-mass actuators and accelerometers.

However, in order for adaptive elements to be applicable to wider ranges of structural systems design and to be accepted by the practicing engineers as viable alternatives to conventional structural design with inertia-driven actuators and sensors, several practical aspects of adaptive elements have to be addressed. These include: economical supply of adaptive elements, embeddability into the structures, especially into the layers of composites, sustained performance subject to temperature variations and strain cycles, availability of characteristics for combined structural and adaptive elements, trade-off information between localized and global controller interactions, effects of discrete vs. distributed layout of adaptive elements for stiffness and control, and real-time identification and control. Among these, from the computational viewpoint, the last constitutes a major barrier in the way of practical utilization of adaptive elements for vibration and shape control of structures, especially for space structural applications.

Real-time maneuvering and vibration control of manipulators or crane-like space structures is emerging as a major challenge to the space systems dynamist for the future space mission activities. At present, various real-time control capabilities exist for process control, vibration control of simple structures such as beams. To support such engineering operations, literature dealing with control theory, modeling and control design abounds everywhere. However, as the systems to be controlled become sophisticated from classical rigid models, simple flexible beams, to lattice truss beams under thermal, space disturbances and control actuations, it is becoming increasingly clear that there has to be significantly improved real-time simulation and control capabilities for an adequate operation of space structures. Candidate adaptive space structures that will require real-time control capabilities far beyond the currently available ones include space manipulators made of truss beams, space cranes for future space assembly, and space robots that are capable of performing autonomous maneuvering.

One major computational barrier of existing computational strategies for real-time simulation of *control-structure interaction* (CSI) problems is that the entire interaction systems equations have to be processed as one entity. This usually involves the solution of $(2N \times 2N)$ unsymmetric system even though the governing equations of motion for structure consists of symmetric $(N \times N)$ systems. To this end, the present paper describes a computational procedure that exploits the second-order characteristics in order to expedite the simulation process. Specifically, the procedure employs a partitioned solution of the structural response cast in a second-order form, a stabilized solution of the control force vector in a differential equation form, a second-order based state estimator (observer), and a stabilized estimation error corrector. The resulting computational efficiency and their implementation details will be discussed in detail.

FIRST-ORDER FORM OF CSI MODEL EQUATIONS

The first-order form of the equations of motion for control-structure interaction (CSI) systems with dynamically compensated controllers can be expressed as

$$\begin{Bmatrix} \dot{\mathbf{x}} \\ \dot{\tilde{\mathbf{x}}} \end{Bmatrix} = \begin{bmatrix} \mathbf{A} & -\bar{\mathbf{B}}\mathbf{F} \\ \mathbf{L}\mathbf{H} & \mathbf{A} - \bar{\mathbf{B}}\mathbf{F} - \mathbf{L}\mathbf{H} \end{bmatrix} \begin{Bmatrix} \mathbf{x} \\ \tilde{\mathbf{x}} \end{Bmatrix} + \begin{Bmatrix} \mathbf{E} \\ \mathbf{E} \end{Bmatrix} \mathbf{f} + \begin{Bmatrix} \bar{\mathbf{G}} \\ 0 \end{Bmatrix} \mathbf{w} \tag{1}$$

for which sensor output $\mathbf{z}$ is modeled as

$$\mathbf{z} = \mathbf{H}\mathbf{x} + \mathbf{v} \tag{2}$$

and control force $\mathbf{u}$ is obtained by

$$\mathbf{u} = -\mathbf{F}\tilde{\mathbf{x}} \tag{3}$$

where

$$\mathbf{x} = \begin{Bmatrix} \mathbf{q} \\ \dot{\mathbf{q}} \end{Bmatrix}, \qquad \tilde{\mathbf{x}} = \begin{Bmatrix} \tilde{\mathbf{q}} \\ \tilde{\dot{\mathbf{q}}} \end{Bmatrix}$$

and

$$\mathbf{H} = [\,\mathbf{H}_d \quad \mathbf{H}_v\,], \qquad \mathbf{L} = \begin{bmatrix} \mathbf{L}_1 \\ \mathbf{L}_2 \end{bmatrix}, \qquad \mathbf{E} = \begin{bmatrix} 0 \\ \mathbf{M}^{-1} \end{bmatrix}$$

$$\mathbf{A} = \begin{bmatrix} 0 & \mathbf{I} \\ -\mathbf{M}^{-1}\mathbf{K} & -\mathbf{M}^{-1}\mathbf{D} \end{bmatrix}, \qquad \bar{\mathbf{B}} = \begin{bmatrix} 0 \\ \mathbf{M}^{-1}\mathbf{B} \end{bmatrix}$$

$$\mathbf{F} = [\,\mathbf{F}_1 \quad \mathbf{F}_2\,], \qquad \bar{\mathbf{G}} = \begin{bmatrix} 0 \\ \mathbf{M}^{-1}\mathbf{G} \end{bmatrix}$$

In the preceding equations, $\mathbf{M}(n \times n)$ is the mass matrix, $\mathbf{D}(n \times n)$ is the damping matrix, $\mathbf{K}(n \times n)$ is the stiffness matrix, $\mathbf{f}(t)(n \times 1)$ is the applied force, $\mathbf{B}(n \times m)$ is the actuator location matrix, $\mathbf{G}(n \times r)$ is the disturbance location matrix, $\mathbf{q}(n \times 1)$ is the generalized displacement vector, $\mathbf{w}(n \times 1)$ is a disturbance vector and the superscript dot denotes time differentiation. In (2), $\mathbf{z}$ is the measured sensor output. $\mathbf{H}_d$ is the matrix of displacement sensor locations and $\mathbf{H}_v$ is the matrix of velocity sensor locations. The vector $\mathbf{v}$ is measurement noise. The state estimator in (1) is assumed either to be based on the Kalman filter [23,24] or based on a Luenberger observer [25] if the system is deterministic. The superscript ˜ denotes the estimated states. The actuator output, $\mathbf{u}$, is a function of the state estimator variables, $\tilde{\mathbf{q}}$ and $\tilde{\dot{\mathbf{q}}}$,and $\mathbf{F}_1$ and $\mathbf{F}_2$ are control gains determined for example by pole-zero placement or from the solution of an optimal control problem. The observer is governed by $\mathbf{L}$, the filter gain matrix.

The embedding effects of both the controller and the state observer result in an unsymmetric and nonsparse system matrix of dimension ($4N$ by $4N$), where N is the number of structural degrees of freedom. Solution of (1) would require considerable software modifications of existing structural dynamics analysis programs for large-scale CSI simulation purposes. In addition to losing the computational advantages associated with the finite element based structural dynamic equations, the simultaneous solution approach requires the control law to be embedded into the observer model. If the control law includes actuator, sensor and/or controller dynamics, additional states must be added to the observer. This greatly complicates the observer model and requires significant software development for each class of control law dynamics. The difficulties associated with the *simultaneous solution approach* have prompted development of a *partitioned solution approach* for the CSI equations as described in the next section.

PARTITIONED CONTROL-STRUCTURE INTERACTION EQUATIONS

The equations of motions for control-structure interaction systems stated in the preceding section consist of a $4n$ fully coupled non-symmetric system. Even if one limits the system to be directly fed back, the resulting system would be a non-symmetric $2n$ system. This observation, coupled with the time consuming solution of the associated Riccati equations, indicates that the real-time processing of the closed system equations can become very quickly intractable if the system size increases rapidly. For example, the real-time control of flexible structures whose system size approaching 25 degrees of freedom is presently about a limiting size. For high-performance flexible structural maneuvering in space, it is anticipated that the size of state vectors will most likely far exceed 50 or beyond.

Recently, a novel method for simulating the control-structure interaction systems by a series of divide-and-conquer strategy known as partitioned control-structure interaction (CSI) procedure has been developed [26–30] and applied to a preliminary design of a evolutionary design of an earth observing satellite [31]. A key idea adopted in the partitioned CSI procedure has been to utilize the following component-by-component systems equations:

$$\begin{cases} \text{Structure:} & \mathbf{M}\ddot{\mathbf{q}} + \mathbf{D}\dot{\mathbf{q}} + \mathbf{K}\mathbf{q} = \mathbf{f} + \mathbf{B}\mathbf{u} + \mathbf{G}\mathbf{w} \\ \text{Sensor Output:} & \mathbf{z} = \mathbf{H}\mathbf{x} + \mathbf{v} \quad \text{(no computing is necessary)} \\ \text{Estimator:} & \mathbf{M}\ddot{\tilde{\mathbf{q}}} + \mathbf{D}\dot{\tilde{\mathbf{q}}} + \mathbf{K}\tilde{\mathbf{q}} = \mathbf{f} + \mathbf{B}\mathbf{u} + \mathbf{G}_1\boldsymbol{\gamma} + \mathbf{G}_2\dot{\boldsymbol{\gamma}} \\ \text{Control Force:} & \dot{\mathbf{u}} + \mathbf{F}_2\mathbf{M}^{-1}\mathbf{B}\mathbf{u} = -\mathbf{F}_2\mathbf{M}^{-1}(\dot{\tilde{\mathbf{p}}} + \mathbf{G}_1\boldsymbol{\gamma} + \mathbf{G}_2\dot{\boldsymbol{\gamma}}) - \mathbf{G}_3\dot{\tilde{\mathbf{q}}} \\ \text{Estimation Error:} & \mathbf{Q}_1\dot{\boldsymbol{\gamma}} + \mathbf{Q}_2\boldsymbol{\gamma} = \dot{\mathbf{z}} - \mathbf{H}_2\mathbf{M}^{-1}(\dot{\tilde{\mathbf{p}}} + \mathbf{B}\mathbf{u}) - \mathbf{Q}_3\dot{\tilde{\mathbf{q}}} \end{cases} \tag{4}$$

where

$$\tilde{\mathbf{p}} = \mathbf{f} - \mathbf{K}\tilde{\mathbf{q}}$$

$$\mathbf{G}_1 = \mathbf{M}\mathbf{L}_2 + \mathbf{D}\mathbf{L}_1, \quad \mathbf{G}_1 = \mathbf{M}\mathbf{L}_1, \quad \mathbf{G}_3 = (\mathbf{F}_1 - \mathbf{F}_2\mathbf{M}^{-1}\mathbf{D})$$

$$\mathbf{Q}_1 = (\mathbf{I} + \mathbf{H}_2\mathbf{L}_1), \quad \mathbf{Q}_2 = \mathbf{H}_2(\mathbf{L}_2 + \mathbf{M}^{-1}\mathbf{D}\mathbf{L}_1), \quad \mathbf{Q}_3 = (\mathbf{H}_1 - \mathbf{H}_2\mathbf{M}^{-1}\mathbf{D})$$

Note that the equations for the structure and the state estimator are expressed in a natural second-order form that can be processed for real-time computation purposes by relying on symmetric $n \times n$-matrix. In addition, the control force $\mathbf{u}$ has usually a substantially smaller size than that the structure. Likewise, the estimation error vector, $\boldsymbol{\gamma}$, has the size of the sensors which is also smaller than that of the structure.

TIME DISCRETIZATION OF PARTITIONED CSI EQUATIONS

This section illustrates time discretization of the partitioned CSI equations. In performing the necessary implicit discretization, we have chosen the mid-point version of the trapezoidal rule as it introduces no numerical dissipation and maintains the lowest frequency distortion among the unconditionally stable time integration algorithms [32]. It is noted that, as feedback loops correspond to damping modification and frequency shift, algorithms with no numerical damping and minimal frequency distortion should be preferred for CSI simulation. To this end, we first introduce the following change of variables:

$$\begin{cases} x_1 = x \\ x_2 = M\dot{x}_1 + Dx_1 \end{cases} \tag{5}$$

and the trapezoidal integration formula in the form of

$$\begin{cases} \begin{Bmatrix} x_1 \\ x_2 \end{Bmatrix}^{n+1/2} = \begin{Bmatrix} x_1 \\ x_2 \end{Bmatrix}^{n} + \delta \begin{Bmatrix} \dot{x}_1 \\ \dot{x}_2 \end{Bmatrix}^{n+1/2} \\ \begin{Bmatrix} x_1 \\ x_2 \end{Bmatrix}^{n+1} = 2\begin{Bmatrix} x_1 \\ x_2 \end{Bmatrix}^{n+1/2} - \begin{Bmatrix} x_1 \\ x_2 \end{Bmatrix}^{n} \end{cases} \tag{6}$$

where the superscript n denotes the discrete time interval $t^n = nh$, h is the time increment and $\delta = h/2$.

Upon substituting (5) and (6) to time-discretize (4a), the following discrete equations for the structure are obtained:

$$\begin{cases} (M + \delta D + \delta^2 K)x_1^{n+1/2} = Mx_1^n + \delta x_2^n + \delta^2 Bu^{n+1/2} \\ x_2^{n+1/2} = x_2^n + \delta Bu^{n+1/2} - \delta K x_1^{n+1/2} \end{cases} \tag{7}$$

The state estimator (4c) can be similarly discretized as

$$\begin{aligned} (M + \delta D + \delta^2 K)\hat{x}_1^{n+1/2} &= M\hat{x}_1^n + \delta \hat{x}_2^n + \delta^2 Bu^{n+1/2} \\ &+\delta\{M\bar{L}_1 + \delta \bar{L}_2\}\gamma^{n+1/2} \end{aligned} \tag{8}$$

so that once $\hat{x}_1^{n+1/2}$ is computed, $\hat{x}_2^{n+1/2}$ can be computed by

$$\hat{x}_2^{n+1/2} = \hat{x}_2^n + \delta \hat{g}^n - \delta K \hat{x}_1^{n+1/2} \tag{9}$$

where

$$\hat{g}^n = Bu^{n+1/2} + \bar{L}_2 \gamma^{n+1/2} \tag{10}$$

The time discretization of the stabilized control force and the estimation error can be performed in a coupled manner as they can be solved simultaneously:

$$\begin{bmatrix} (I + \delta \bar{F}_2 B) & \delta \bar{F}_2 \bar{L}_2 \\ \delta \bar{H}_2 B & (I + \delta \bar{H}_2 \bar{L}_2) \end{bmatrix} \begin{Bmatrix} u^{n+1/2} \\ \gamma^{n+1/2} \end{Bmatrix} = \begin{Bmatrix} -\bar{F}_2 \hat{x}_2^n - (\bar{F}_1 - \delta \bar{F}_2 K)\hat{x}_1^{n+1/2} \\ z^{n+1/2} - \bar{H}_2 \hat{x}_2^n - (\bar{H}_1 - \delta \bar{H}_2 K)\hat{x}_1^{n+1/2} \end{Bmatrix} \tag{11}$$

where $\bar{F}_i$ and $\bar{H}_i$ are given by

$$\bar{H}_1 = H_1 - H_2 M^{-1} D, \quad \bar{H}_2 = H_2 M^{-1}$$

$$\bar{F}_1 = F_1 - F_2 M^{-1} D, \quad \bar{F}_2 = F_2 M^{-1}$$

and $\bar{L}_1$ and $\bar{L}_2$ are accordingly defined.

The computational stability of the preceding partitioned solution procedure has been studied and the result shows [33] the procedure maintains its stability provided the stepsize h is restricted to

$$h_{max} \leq 2/|\lambda_{max}| \tag{12}$$

where λ_{max} is the maximum eigenvalue of

$$(\lambda^2 \mathbf{I} + \lambda \bar{\mathbf{F}}_2 \mathbf{B} + \bar{\mathbf{F}}_1 \mathbf{M}^{-1} \mathbf{B})\mathbf{y} = 0$$

It is observed that $|\lambda_{max}|$ is several orders of magnitude smaller than ω_{max} of the structural dynamics eigenvalue problem:

$$\omega^2 \mathbf{M}\mathbf{y} = \mathbf{K}\mathbf{y}$$

In other words, the implicit integration of both the structural and the observer equations removes any stepsize restriction on the magnitude of $\omega_{max} >> |\lambda_{max}|$. On the other hand, the extrapolation of $\gamma^{n+1/2} \approx \gamma^n$ and $\mathbf{u}^{n+1/2} \approx \mathbf{u}^n$ for the solution of the estimation vectors $\hat{x}_1^{n+1/2}$ and $\hat{x}_2^{n+1/2}$ by (8) and (9) imposes the stepsize restriction as dictated by (12).

ILLUSTRATIVE EXAMPLES

The computational efficiency accrued by the partitioned solution procedure as compared with the conventional solution method is sketched out in Fig. 1. Assumed in the construction of the chart are the bandwidth of the stiffness matrix of 0.2, the number of the actuator and that of the sensor to be the same, and the ratio of the structural degrees of freedom and those of the actuator to be 0.1. Note that, given a real-time processing computer that can perform a wallclock rate of 200 samples /second command and control, the conventional method can at most handle the real-time control of a simple beam articulation, whereas the partitioned method can handle the real-time control of complex truss-beam vibrations. For nonlinear problems, the advantages of the partitioned method is more pronounced as can be seen from the chart.

The partitioned CSI simulation procedure as derived in (7) through (11) have been implemented as a stand-alone package. The present software implementation emphasizes the use of the widely available sequential and parallel analysis modules specially developed for the solution of structural dynamics equations. Note that the solution algorithm for both the structural system and the state estimator is the same, hence the software module, provided the right hand terms are treated as apllied forces. Although the stabilized form of the controller $\mathbf{u}$ and the filtered measurements γ are solved in a coupled manner, their size in general is substantially smaller, typically a fraction of the size of the structural system for large-scale problems.

The first example is a truss beam shown in Fig. 2, consisting of 8 bays with nodes 1 and 2 fixed for cantilevered motions. The locations of actuator and sensor applications as well as their directions are given in Table 1. Figures 3, 4 and 5 are the vertical displacement histories at node 9 for open-loop, direct output feedback, and dynamically compensated feedback cases, respectively. The noise distribution matrix **G** is taken to be an identity matrix for the numerical experiments reported herein. Note the effectiveness of the dynamically compensated feedback case by the present second-order discrete Kalman filtering equations as compared with the direct output feedback cases. Figure 5 illustrates a testbed evolutionary model of an Earth-pointing satellite. Eighteen actuators and 18 sensors are applied to the system for vibration control and their locations are provided in Tables 2 and 3. Figures 7, 8, and 9 are a representative of the responses for open-loop, direct output feedback, and dynamically compensated cases, respectively. Note that u_x response by the dynamically compensated case does drift away initially even though the settling time is about the same as that by the direct output feedback case. However, the sensor output are assumed to be noise-free in these two numerical experiments. Although the objective of the present paper is to establish the computational effectiveness of the second-order discrete Kalman filtering equations, we conjecture that for noise-contaminated sensor output for which one would apply dynamic compensated strategies, the relative control performance may turn out to be the opposite. Further simulations with the present procedure should shed light on the performance of dynamically compensated feedback systems for large-scale systems as they are computationally more feasible than heretofore possible.

Table 4 illustrates the computational overhead associated with the direct output feedback vs. the use of a dynamic compensation scheme by the output present Kalman filtering equations. In the numerical experiments reported herein, we have relied on Matlab software package [34] for the synthesis of both the control law gains and the discrete Kalman filter gain matrices. It is seen that the use of the present second-order discrete Kalman filtering equations for constructing dynamically compensated control laws adds computational overhead, only an equivalent of open-loop transient analysis of symmetric sparse systems of order N instead of $2N \times 2N$ dense systems.

DISCUSSIONS

The present paper has addressed the advantageous features of employing the same direct time integration algorithm for solving the structural dynamics equations also to integrate the associated continuous Kalman filtering equations. Therefore, the partitioned procedure offers a substantial computational advantage over the solution methods based on a first-order form when computing with large and sparse observer models.

The formulation, numerical algorithms and software implementation presented herein assumes the actuators and sensors based on lumped parameter models, viz., motors, proof masses, accelerators, strain gauges, etc. For the adaptive elements such as piezoelectric devices and memory-alloy elements which are embedded or distributed over the structural surface, there exists at present no corresponding real-time dynamics and control simulation capability of sufficient generality. The difficulty in modeling the saturation phenomena, localization, filtering characteristics and high-frequency bandwidths makes the realistic real-time control of actively controlled structures with adaptive elements a challenge. Parallel computing [35–38] in conjunction with distributed/hierarchical adaptive control may hold a key for the real-time control of large order structural systems.

ACKNOWLEDGEMENTS

The work reported herein was supported by a grant from NASA/Langley Research Center, NAG1-1021.

REFERENCES

1. Wada, B. K., Fanson, J. L. and Crawley, E. F., Adaptive structures, in: *Adaptive Structures*, ed. B. K. Wada, American Society of Mechanical Engineers, AD-Vol.15, New York, 1989, pp. 1-8.

2. Jaffe, B., Cook, W. and Jaffe, H., *Piezoelectric Ceramics*, Academic Press, London and New York, 1971.

3. Zelenka, J., *Piezoelectric Resonators and Their Applications*, Elsevier Science Publishing Co., Inc., New York, 1986.

4. Cross, L.E., Piezoelectric and electrostrictive sensors and actuators for adaptive structures and smart materials. In *Adoptive Structures*, ed. B.K. Wada, ASME, AD-Vol. 15, New York, 1989, pp. 9-17.

5. Forward, R.L., Electronic damping of vibrations in optical structures. *Journal of Applied Optics*, 1979, **18** (5), 690-697.

6. Crawley, E.F. and de Luis, J., Use of piezoelectric actuators as elements of intelligent structures. *AIAA Journal*, 1987, **25**, (10), 1373-1385.

7. Lee, C.K., Piezoelectric laminates for torsional and bending modal control: theory and experiment. Ph.D. Thesis, Cornell University, Ithaca NY, 1987.

8. de Luis, J. and Crawley, E., Experimental results of active control on a prototype intelligent structure. AIAA Paper No. 90-1163, Proceedings of the 31st AIAA/ASME/ASCE/AHS/ASC Structures, Structural Dynamics and Materials Conference, Long Beach, CA, April 1990, pp. 2340-2350.

9. Baz, A. and Poh, S., Performance of an active control system with piezoelectric actuators. *Journal of Sound and Vibration*, 1988, **126**, (2), 327-343.

10. Anderson, E., Moore, D., Fanson, J. and Ealey, M., Development of an active member using piezoelectric and electrostrictive actuation for control of precision structures. AIAA Paper No. 90-1085, Proceedings of the 31st AIAA/ASME/ASCE /AHS/ASC Structures, Structural Dynamics and Materials Conference, Long Beach, CA, April 1990, pp. 2221-2233.

11. Miller, D., Collins, S. and Peltzman, S., Development of spatially convolving sensors for structural control applications. AIAA Paper No. 90-1127, Proceedings of the 31st AIAA/ASME/ASCE/AHS/ASC Structures, Structural Dynamics and Materials Conference, Long Beach, CA, April 1990, pp. 2283-2297.

12. Bailey, T. and Hubbard, J.E., Distributed piezoelectric polymer active vibration control of a cantilever beam. *Journal of Guidance, Control, and Dynamics*, 1985, **8**, (5).

13. Nailon, M., Coursant, R.H. and Besnier, F., Analysis of piezoelectric structures by a finite element method. *ACTA Electronica*, 1983, **25**, (4), 341-362.

14. Tzou, H.S. and Tseng, C.I., Distributed piezoelectric sensor/actuator design for dynamic measurement/control of distributed parameter systems: a finite element approach. *Journal of Sound and Vibration* 1990, **137**, (1).

15. Burke, S. and Hubbard, J.E., Active vibration control of a simply-supported beam using a spatially distributed actuator. *IEEE Control Systems Magazine*, August 1987, **7**, (6), 25-30.

16. de Luis, J., Crawley, E.F. and Hall, S.R., Design and implementation of optimal controllers for intelligent structures using infinite order structural models. Report No. 3-89, Space Systems Laboratory, M.I.T., Cambridge, MA, 1989.

17. Jia, J. and Rogers, C.A., Formulation of a laminated shell theory incorporating embedded distributed actuators. In *Adaptive Structures*, ed. B.K. Wada, ASME, 1989, pp. 19-24.

18. Hagood, N. and von Flotow, A., Modelling of piezoelectric actuator dynamics for activc structural control. AIAA Paper No. 90-1087, Proceedings of the 31st AIAA/ASME/ASCE/AHS/ASC Structures, Structural Dynamics and Materials Conference, Long Beach, CA, April 1990, pp. 2242-2256.

19. Takahara, K, Kuwao, Shigeshara, M. Katoh, T., Motohashi, S.and Natori, M., Piezo linear actuators for adaptive truss structures. In *Adaptive Structures*, ed. B.K. Wada, ASME, 1989, pp. 83-38.

20. Matsunaga, S. Miura, K. and Natori, M., A construction concept of large space structures using intelligent/adaptive structures. AIAA Paper No. 90-1128, Proceedings of the 31st AIAA/ASME/ASCE/AHS/ASC Structures, Structural Dynamics and Materials Conference, Long Beach, CA, April 1990, pp. 2298-2305.

21. Rogers, C. and Ramaseshan, A., Investigation of embedded actuators using generalized laminated plate theory. AIAA Paper No. 90-1168, Proceedings of the 31st AIAA/ASME/ASCE/AHS/ASC Structures, Structural Dynamics and Materials Conference, Long Beach, CA, April 1990, pp.2234-2241.

22. Tzou, H. and Tseng, C., "Distributed Dynamic Identification and Controls of Flexible Shells: Theory and Finite Element Development," AIAA Paper No. 90-1089, Proceedings of the 31st AIAA/ASME/ASCE/AHS/ASC Structures, Structural Dynamics and Materials Conference, Long Beach, CA, April 1990, pp.2265-2273.

23. Kalman, R. E. (1961), "On the General Theory of of Control Systems," *Proc. 1st International Congress on Automatic Control*, Butterworth, London, **1**, 481-491.

24. Kalman, R. E. and Bucy, R. S. (1961), "New results in linear filtering and prediction theory," *Trans, ASME J. Basic Engineering*, **83**, 95-108.

25. Luenberger, D. G., "Observing the State of a Linear System," *IEEE Transactions on Military Electronics*, 8, 1964, pp. 74-80.

26. Park, K. C. and Belvin, W. K., "A Partitioned Solution Procedure for Control-Structure Interaction Simulations," *J. Guidance, Control and Dynamics*, **14(1)**, Jan.-Feb. 1991, 59-67.

27. Park, K. C., K. Alvin and W. K. Belvin, "Parallel Computations and Control of Adaptive Structures, " in: *Intelligent Structures*, ed. by K. C. Chong, S. C. Liu and J. C. Li, Elsevier Science Publishers, London, 1990, 439-458.

28. Belvin, W. K. and Park, K. C., "On the State Estimation of Structures with Second Order Observers," *Proc. the 30th Structures, Dynamics and Materials Conference*, AIAA Paper No. 89-1241, April 3-5,1989.

29. Park, K. C. and Belvin, W. K., "Discrete Integration of Continuous Kalman Filtering Equations for Time-Invariant Second-Order Structural Systems," *Proc. 1990 Guidance, Control and Dynamics Conference*,22-25 August 1990, Portland, Oregon, AIAA Paper No. AIAA-90-3387.

30. Belvin, W. K. and Park, K. C., "Computer Implementation of Analysis and Optimization Procedures for Control-Structure Interaction Problems," Proc. the 1990 AIAA Dynamics Specialist Conference, Paper No. AIAA-90-1194, Long Beach, Calif., 5-6 April 1990.

31. Belvin, W. K., "Simulation and Interdisciplinary Design Methodology for Control-Structure Interaction Systems," PhD Thesis, Center for Space Structures and Controls, University of Colorado, Report No. CU-CSSC-89-10, July, 1989.

32. Dahlquist, G., "A Special Stability Problem for Linear Multi-step Methods," *BIT*, 3, 1963, pp. 27-43.

33. Park, K. C. and Belvin, W. K. and K. F. Alvin, "Discrete Second-Order Kalman Filtering Equations for Control-Structure Interaction Simulations," *Report No. CU-CSSC-89-08*, Center for Space Structures and Controls, University of Colorado, April 1989 (Submitted for publication in J. Guidance, Control and Dynamics).

34. Wolfram, S., *Mathematica*^{TM}, Addison-Wesley Pub. Co., 1988.

35. Nour-Omid, B. and K. C. Park, "Solving Structural Mechanics Problems on the CalTech Hypercube Machine," *Comp. Meth. Appl. Mech. Eng.*, **61**, (1987) 161-176.

36. Farhat, C., Felippa, C. A. and Park, K. C., "Implementation Aspects of Concurrent Finite Element Computations," in *Parallel Computations and Their Impact on Mechanics*, American Society of Mechanical Engineers, New York, 1987, 301-316.

37. Farhat, C., Sobh, N. and K. C. Park, "Transient Finite Element Computations on 65,536 Processors: The Connection Machine," *International Journal on Numerical Methods in Engineering*, **30(1)**, 27-55 (1990).

38. Alvin, K. F. and Park, K. C., "Implementation of a Partitioned Àlgorithm for Simulation of Large CSI Problems," *Report No. CU-CSSC-91-04*, Center for Space Structures and Controls, University of Colorado, March 1991.

TABLES

TABLE 1a:
Actuator Placement for Truss Example Problem

Actuator	Node	Component
1	2	y
2	18	y
3	9	y
4	9	x

TABLE 1b:
Sensor Placement for Truss Example Problem

Sensor	Type	Node	Component
1	Rate	2	y
2	Rate	18	y
3	Rate	9	y
4	Rate	9	x
5	Position	9	y
6	Position	9	x

TABLE 2:
Actuator Placement for EPS Example Problem

Actuator	Node	Component
1	97	x
2	97	z
3	96	x
4	96	z
5	65	y
6	68	y
7	59	y
8	62	y
9	45	y
10	45	z
11	70	y
12	70	z
13	95	x
14	95	y
15	95	z
16	95	ϕ_x
17	95	ϕ_y
18	95	ϕ_z

TABLE 3:
Sensor Placement for EPS Example Problem

Sensor	Type	Node	Component
1	Rate	97	x
2	Rate	97	z
3	Rate	96	x
4	Rate	96	z
5	Rate	65	y
6	Rate	68	y
7	Rate	59	y
8	Rate	62	y
9	Rate	45	y
10	Rate	45	z
11	Rate	70	y
12	Rate	70	z
13	Position	95	x
14	Position	95	y
15	Position	95	z
16	Position	95	ϕ_x
17	Position	95	ϕ_y
18	Position	95	ϕ_z

TABLE 4:
CPU Results for ACSIS Sequential and Parallel Versions

Model	Problem Type	Sequential	Parallel
54 DOF	Transient	4.5	5.6
Truss	FSFB	9.4	10.2
	K. Filter	13.0	10.7
582 DOF	Transient	98.6	100.3
EPS7	FSFB	190.2	294.5
	K. Filter	284.2	321.5

FIGURES

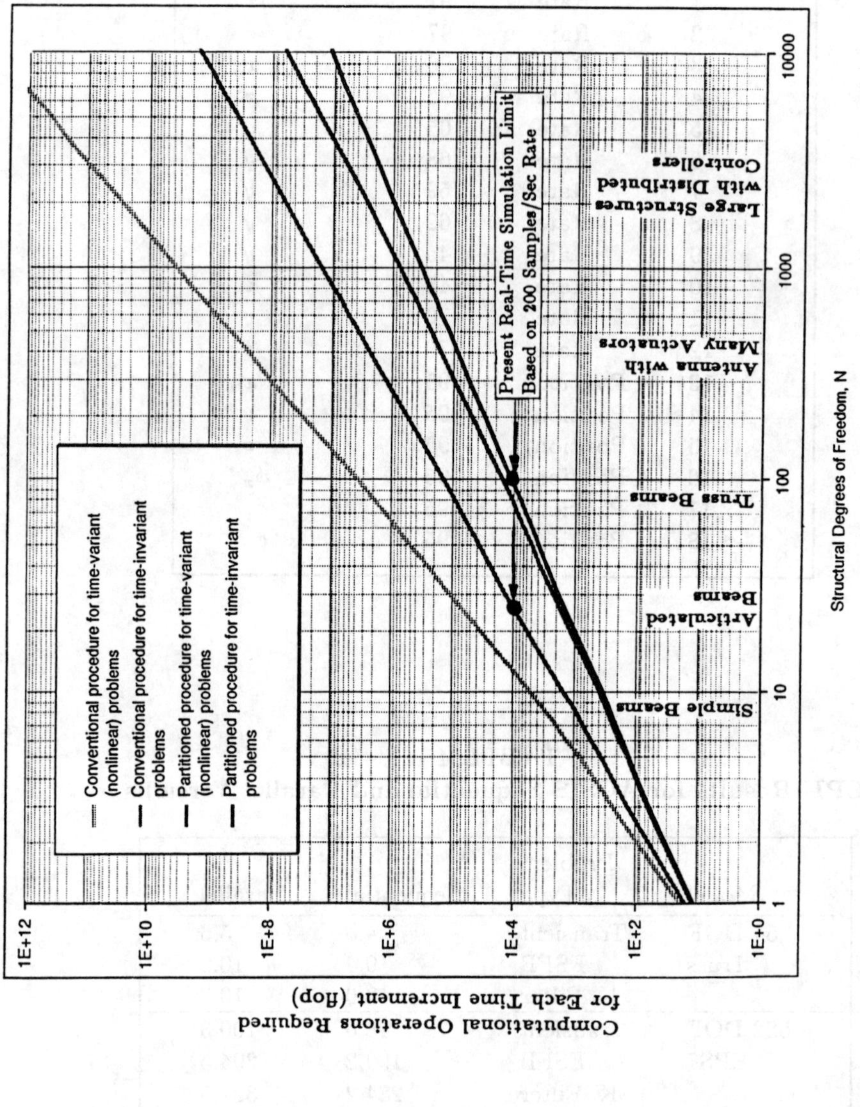

Figure 1. Improvement of computational needs by partitioned procedure.

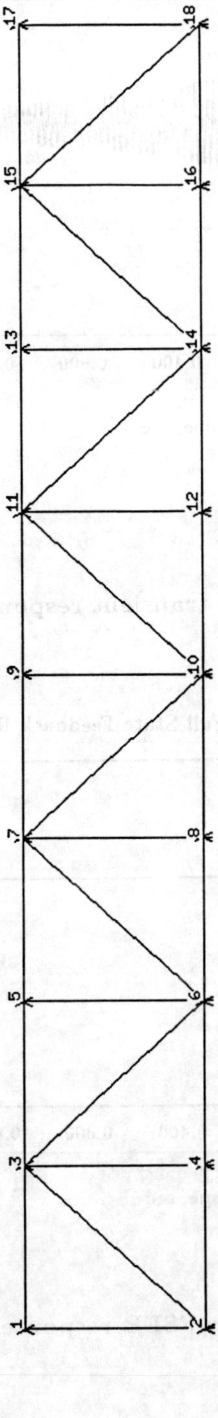

Figure 2. Truss beam problem.

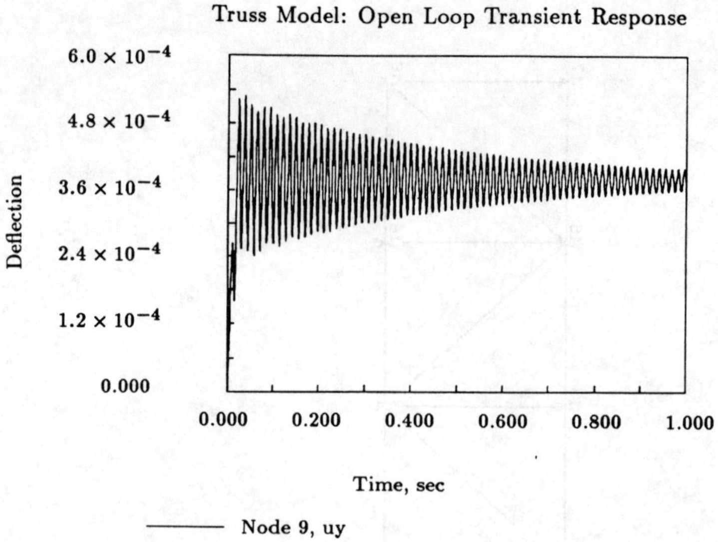

Figure 3. Truss transient response.

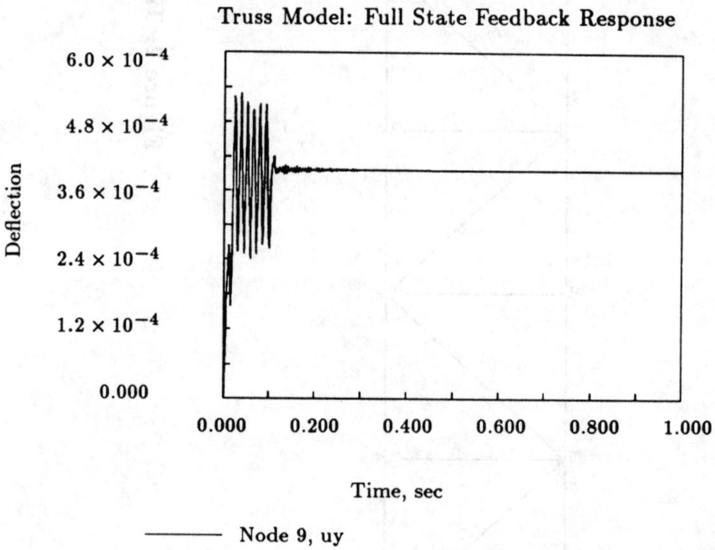

Figure 4. Truss FSFB response.

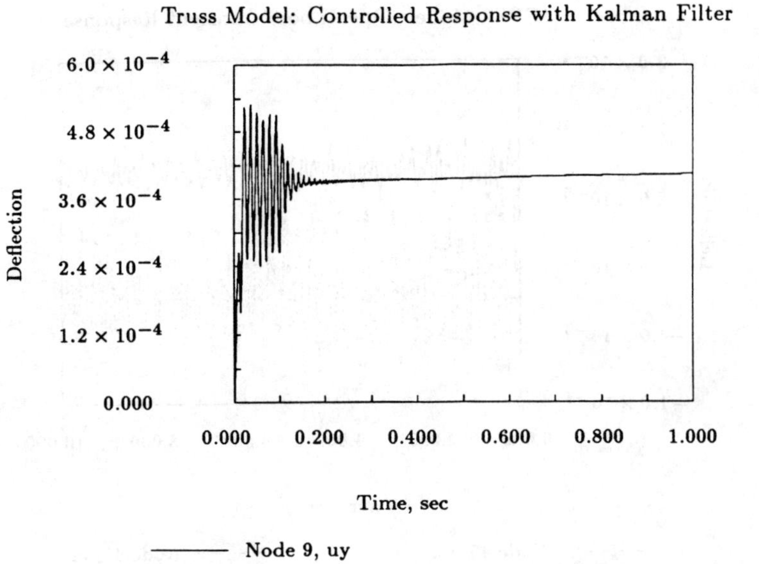

Figure 5. Truss response with filter.

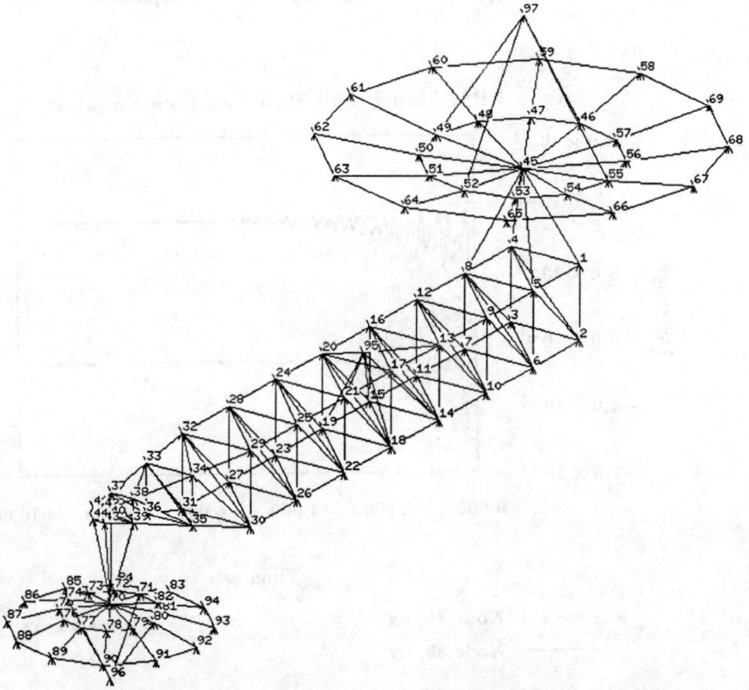

Figure 6. Evolutionary Earth-pointing satellite.

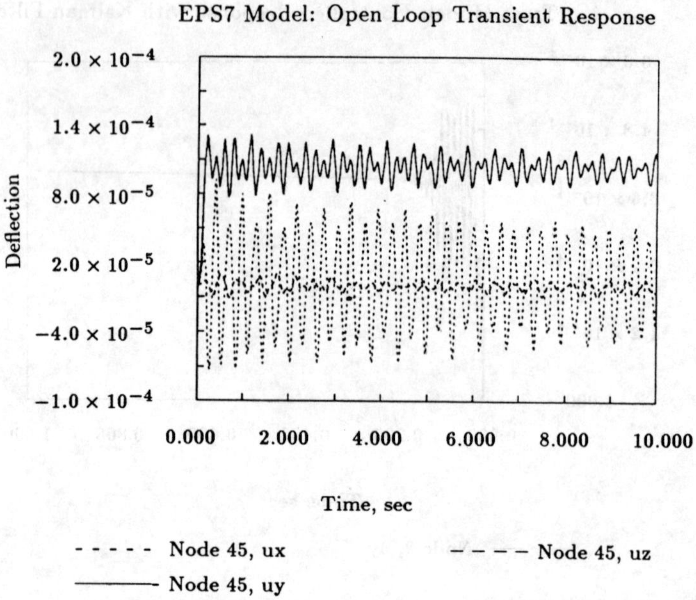

Figure 7. EPS transient response.

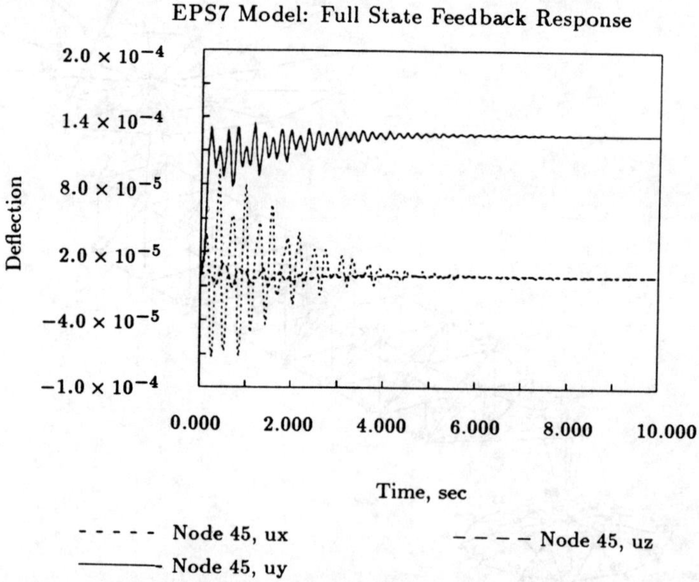

Figure 8. EPS FSFB response.

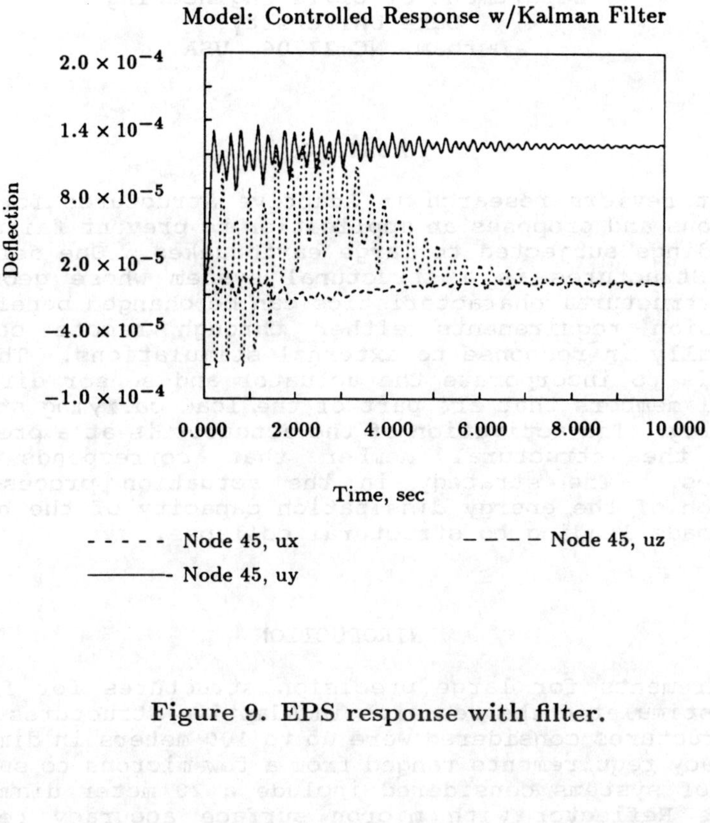

Figure 9. EPS response with filter.

APPLICATION OF ADAPTIVE STRUCTURES CONCEPTS TO CIVIL STRUCTURES

BEN K WADA
Jet Propulsion Laboratory
California Institute of Technology
4800 Oak Grove Dr
Pasadena, CA 91109, USA

SHIVA DAS
Department of Civil Engineering
Duke University
Durham, NC 27706, USA

ABSTRACT

The report reviews research in Adaptive Structures for Aerospace applications and proposes an application to prevent failure of high rise buildings subjected to large earthquakes. One definition of Adaptive Structures is a structural system whose geometric and inherent structural characteristics can be changed beneficially to meet mission requirements either through remote commands or automatically in response to external stimulations. The proposed approach is to incorporate the actuator and sensor directly into structural members that are part of the load carrying structure of the building. The activation of the actuator is at a pre-specified load in the structural member that corresponds to large earthquakes. The strategy in the actuation process is full utilization of the energy dissipation capacity of the building to prevent loads leading to structural collapse.

INTRODUCTION

The requirements for large precision structures for future NASA missions stimulated the research in Adaptive Structures. The size of the structures considered were up to 100 meters in dimension and the accuracy requirements ranged from a few microns to sub-microns. Examples of systems considered include a 20 meter diameter Large Deployable Reflector with micron surface accuracy requirements (Figure 1) to 20-30 meter long Interferometers with sub-micron path length accuracy (Figure 2) requirements. Many structures are truss type structures that are compatible with assembly or deployment in-space. The limited size of future launch vehicle shrouds eliminates the option of fully pre-assembled structures. Using current engineering approaches, the performance requirements can not be reliably achieved and the structure can not be validated through ground tests.

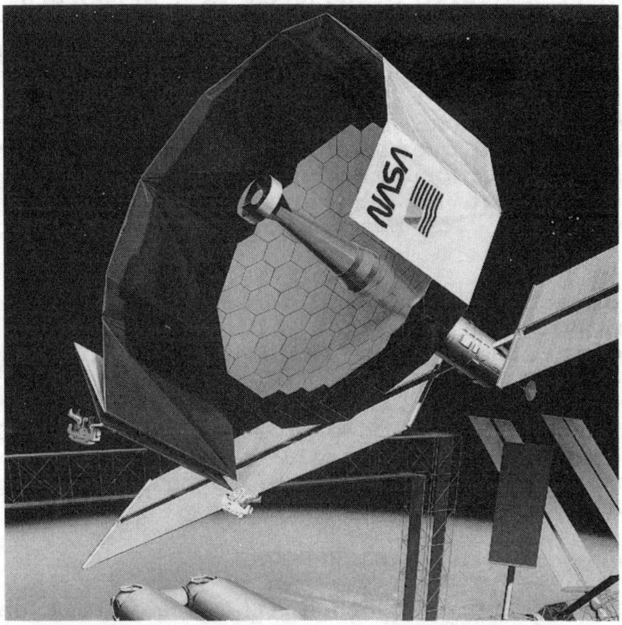

Fig. I Large Deployable Reflector

Fig. II Space Optical Interferometer

One definition of Adaptive Structures is a structural system whose geometric and inherent structural characteristics can be changed beneficially to meet mission requirements either through remote commands or automatically in response to external stimulations. Wada, et al. (1,2) have reviewed the recent developments in Adaptive Structures. Wada, et al. (3) discusses the utilization of Adaptive Structures to overcome the inadequacy of the ground test to validate the structural performance of large precision structures. Garba, et al. (4) summarizes the performance of Adaptive Structures to meet the requirements of precision structures.

Although the objectives for space structures (precision structures with low stresses) differ from civil structures subjected to failure inducing loads (large motion with high stresses), the authors believe that the ideas of Adaptive Structures can be beneficially used for civil structures. Although Adaptive Structures ideas can be used for moderately tall buildings subjected to high winds and small earthquakes, this paper will concentrate on approaches applicable to tall buildings subject to large earthquakes.

BACKGROUND

Even with dramatic improvements in the analytical and test tools over the past 30 years and the goal to design linear structures, occasionally the predictive capabilities are inadequate. Previous test results show that in about 5-10% of the measured modes, the unpredicted (by analysis) characteristics exhibits itself as non-linearities of parameters as a function of amplitude. Even with a good set of test modes, often the comparison of the pre-test analytical model to the test data results in significant discrepancies. Efforts to update the mathematical model to correlate with the test data have resulted in models that have been sufficiently accurate to resolve most engineering requirements to date. Recently many flight mishaps are attributable to the difficulties associated with establishing an adequate ground test program and inadequacies of current analytical predictive technologies.

The data in Figure III shows the accuracy necessary to meet the anticipated requirements. The figure presents the accuracy or changes in the one flexible member of the 20 meter beam, one meter high, which leads to a displacement of one micron at the beam tip. Due to geometric considerations, a 0.1 micron change in the length of the flexible member results in the one micron tip displacement. Included are the estimate of the stress level, temperature change, joint "slop" that are significant to the structure. The requirements are beyond the current fabrication and environmental control capabilities.

Even if the structure can be properly fabricated, the technology to validate the performance of the structure through ground validation tests is not available. Figure IV presents data on the influence

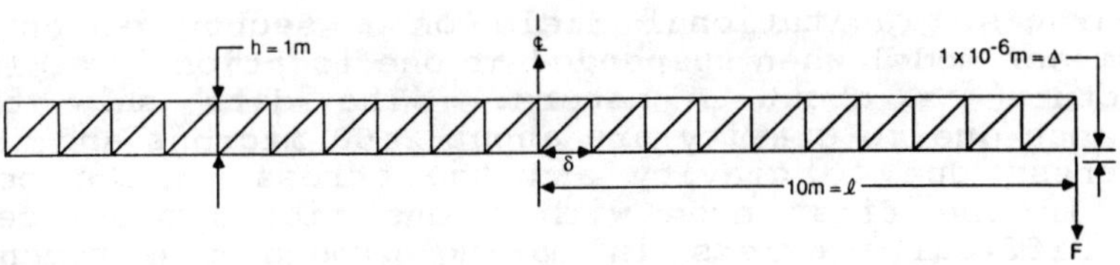

$$\delta = \frac{\Delta = h}{\ell} = \frac{1 \times 10^{-6} m \times 1m}{10m} = 10^{-7} m = 0.1 \text{ micron}$$

SOURCES:	ERROR/CHANGE
• MANUFACTURING ERROR	.000004"
• TEMPERATURE CHANGE	0.1° F
• STRESS	3 PSI
• EXTERNAL FORCE (F)	0.17#

Fig. III Accuracy Requirements

ONE POINT SUSPENSION

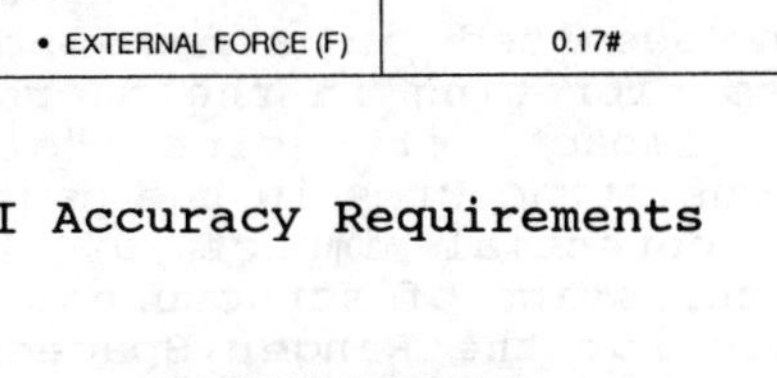

LOAD	MAX DISPL (MICRON)	MAX STRESS (PSI)	RATIO DISPL	RATIO STRESS
-1G Z-DIREC.	2400	5680		
MODE 1	1.0	.25	2400	22,720
MODE 2	1.0	.31	2400	18,323
MODE 3	1.0	.45	2400	12,622
MODE 4	1.0	.52	2400	10,923
MODE 5	1.0	1.09	2400	5,211
MODE 6	1.0	.41	2400	13,854

Fig. IV Optical Interferometer in Earth's Gravity, One Point Suspension of the earth's gravitational field on a sector of an optical interferometer model when suspended at one location. The length of the structure is about 7 meters. The data show that the displacements due to gravity are about 2400 microns and the ratio of the stress due to gravity and the stress in the beam when vibrating in the first mode with a one micron displacement is 22,720. Difficulty exists in making ground measurements with sufficient accuracy, even if the environment can be controlled. The loads in the structure due to gravity will pre-load all the joints that masks any appreciable joint "slop" that may exhibit itself in space. As discussed by Wada et al. (5) current and anticipated technology developments in ground tests are inadequate to validate the structure. Experimentation has revealed that the model accuracy requirements for non-collocated control of lightly damped systems exceed the current predictive capabilities.

A literature search resulted in almost no test data of non-monolithic structures vibrating in the micron displacement range and the potential impact of joint "slop". Experimental observations of various structures in the past 30 years lead to the speculation of the potential impact of joint "slop" on the micro-dynamic characteristics of structures. In the early 1960's, the joints of a truss for the Ranger Spacecraft were loose. An effort to tighten the joint resulted in a more predictable structure with a lower structural modal damping resulting in higher launch loads. The joints were not tightened. Loose trunnion joints during a low level modal test of the Centaur Launch Vehicle resulted in transfer functions that could not be de-composed into normal modes; only data from high levels of excitation represented normal modes. Fortunately, the need was to establish modal characteristics at high levels corresponding to excitations anticipated at launch. Tests of truss structures with loose joints resulted in data that could not be properly de-composed into normal modes. Moon et al. (5) concluded through experimentation that the response of truss structures with "loose" joints is chaotic. Namely, the structure responds randomly when subjected to deterministic inputs. Figure V based upon work by Kuo et al. (6) shows that as the joint is loosen, the micron level response is more "white" in characteristic. Observation of past data suggests that loose joints increases damping and the responses are random in nature; namely the distribution of energy is throughout the frequency range and not concentrated at discrete frequencies. For precision structures, joint looseness must be eliminated.

The current research activities do not alleviate the current deficiencies; the future ground validation test and analytical predictive requirements are significantly more demanding. One approach proposed by Wada et al. (7) is to introduce Adaptive Structures into the spacecraft design. The adjustment or adaptation of the structure to the required shape or characteristics in its operating space environment is through the actuators and sensors integrated into the structure. Adjustment is possible if the structure is linear. Since elimination of joint "slop" cannot be validated by ground tests, Das et al. (8) proposes an approach to eliminate joint "slop" in space. The approach is to induce loads into members of an indeterminate structure.

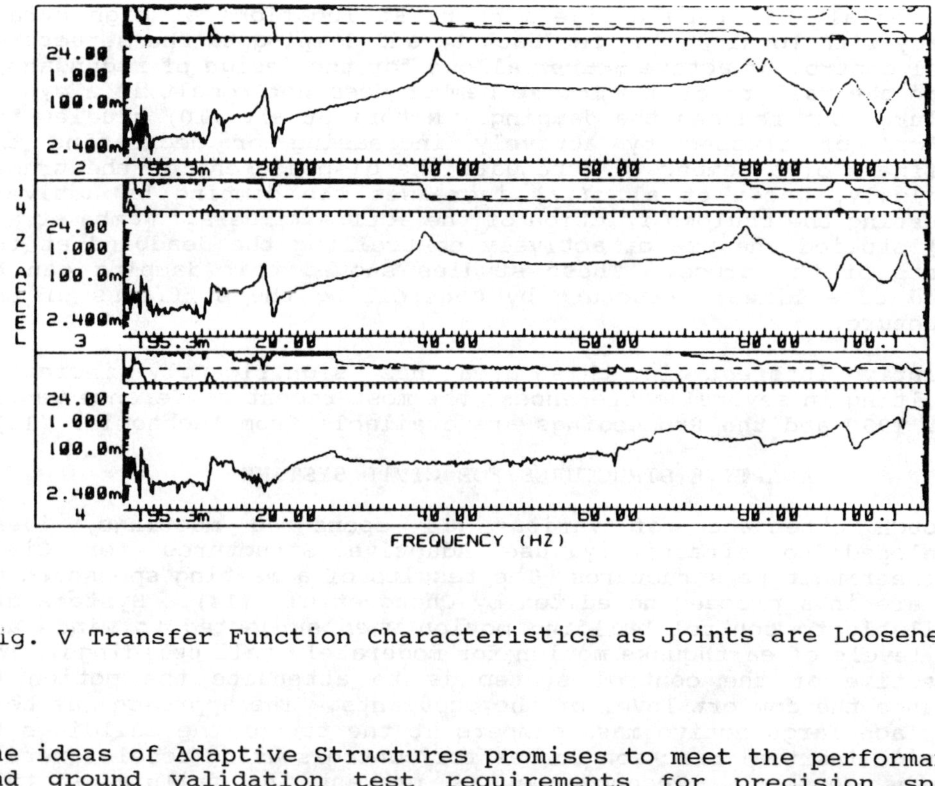

Fig. V Transfer Function Characteristics as Joints are Loosened

The ideas of Adaptive Structures promises to meet the performance and ground validation test requirements for precision space structure. In summary, the approach has been to integrate the sensor/actuator directly into the structure itself and to use local controllers wherever possible to meet the performance and ground validation requirements by the design of a proper redundant system. The classical approach has been to place an inertial/external force actuator at locations of maximum displacement. The reduction in motion is through proper phasing of the forces through an accurate knowledge of the important structural motions. Adaptive Structures uses the structure itself as both the actuator and the sensor and places these active members at locations of high stresses or strain energy in the modes of interest. Considerations to establish the sensor locations include the location of the most sensitive measurements, location in the structure designed for high load capacity, locations where the applied control forces do not closely couple with the controlled modes, locations where the structure itself is most effective to dissipate energy, and locations where redundancies can easily be included. Adaptive Structures seems more effective because of the small additional resources required to control the system, the control hardware is compatible with the structure, the sensitivity of the hardware for control is compatible with the structure, and local controllers can be used to control global motion.

Chen et al. (9) used active members at locations of high strain energy with local analog feedback to add damping to the structure. Local control of active member allows for the design of redundances since the failure of one active member does not result in a system failure, but reduces the damping. Natori et al. (10) studies the control of trusses by actively increasing or decreasing the stiffness of a member to attenuate the displacement of the truss. Similarly, Onoda et al. (11) increases the damping by actively adjusting the coulomb friction of the active members. Namba et al. (12) studied the use of actively controlling the deadband at the joints of the truss. These studies showed that damping can be added to a linear structure by controlling the stiffness of the structure.

Recently interest in this area has significantly increased resulting in several conferences; the most recent conference was in Nov. 1990 and the Proceedings are available from Technomics (13).

ADAPTIVE STRUCTURES FOR CIVIL SYSTEMS

Through attendance of various NSF sponsored meetings, ideas developed to effectively use Adaptive Structures for Civil Engineering type structures. The results of a meeting sponsored by NSF are in a proceeding edited by Chong et al. (14). Systems are available to control building motion when subjected to winds and low levels of earthquake motion for moderately tall buildings. The objective of the control system is to attenuate the motion to enhance the comfort level of the occupants. The approach has been to place large active mass dampers at the top of the buildings to provide inertial reaction forces for controls using accelerometers as the sensors. Technologies to prevent the failure of tall buildings when subjected to large earthquakes is limited.

The extension of these ideas for tall building subjected to large earthquakes where safety is the primary concern must be developed. Although a building seems easy to model accurately, the accuracy of the dynamics of the system is very difficult to predict in the presence of large motion earthquakes. The uncertainty of the dynamics of the foundation/soil that may change with time and the inappropriateness of testing the buildings at high levels of excitation to obtain experimental data leads to the uncertainties. The buildings are often non-linear with the amplitude of motion. Experimentation has shown the performance of control systems using non-collocated sensors and actuators is often dependent on the accuracy of the structural model, especially for lightly damped systems. The current approach requires the placement of large masses at the most undesirable location, the top of the structure. Sensors and control systems are available to control buildings; the technology needs are the requirement for cost effective and reliable actuators with the force and stroke capabilities required for buildings.

Several investigators are using integral structural elements as actuators with sensors to control the motion of the structure. Kobori (15) is using ideas of active bracing systems to alter the dynamic characteristics of the structure to steer the structural frequencies from the dominant (and damaging) frequencies associated with the seismic power spectra at each instant in time. Soong (16) adjusts the tension in the diagonals of the building of structures to provide the actuation required for its controls. Thorp et al. at Cornell University has proposed a cable system distributed throughout a truss type structure that provides a pre-established moment distribution as an actuator; the actuation or moment distribution is through adjusting the cable tension. Nemir et al. (17) has not proposed an actuator, the idea is to alter effectively the stiffness of the structure to transfer energies from lower modes to higher modes.

PROPOSED APPROACH

The proposed approach combines information obtained through work on Adaptive Structures for space applications and information obtained through attending NSF sponsored meetings regarding the technology needs for buildings when subjected to large earthquakes. The main thrust of this paper is to propose an approach to control the motion and thus loads of tall buildings subjected to large earthquakes. The objective is to save lives by preventing the collapse of the structure and allow cosmetic cracks and failures that can be repaired.

The objective is to use the total capacity of the structure itself to absorb energy and to use the structural members itself as an integral part of the actuator and sensor. In most buildings, the largest loads leading to structural collapse results from the excitation of the primary modes of vibration in one or more of the six degrees of freedom. Structural failure will generally occur at the building/soil interface because of the concentration of the high loads or strain energy in the first bending mode. Since damping is often directly related to relative motion of structures, the damping contribution is high from the portion of the structure with the highest concentration of strain energy. For the lower modes, the concentration is at the base of the building. The utilization of the full capacity of the structure to dissipate energy by distributing the energy of the lower modes to higher modes before the loads in the structure exceeds its load bearing capacity is the goal. Local non-structural failures contribute to the energy dissipation mechanism. The approach may be similar to that proposed by Nemir et al. (17). In contrast to the small motion requirements for space structures, large motions are important for tall building subjected to large earthquakes. Thus the introduction of loose joints is beneficial since it adds damping to the structure. Experimental data show the distribution of the energy is spread over a wider frequency range with the introduction of loose joints; this implies a more effective utilization of the structure. The actuators are the structural members with the sensors (strain gages) and release/attach mechanism (simple mechanism that may be activated based on the sensor reading) using local analog controllers. As an example, one simple actuator is a mechanism initiated at a high pre-specified level of member loads (may be approaching its load carrying capacity) to allow the relative motion of both ends of the member. This actuator introduces "slop" in the structures and changes the structural stiffness to attenuate the loads. The actuators will be at the locations of maximum strain energy or loads (base of the building) anticipated during the control process. Member load is the most direct sensor since the objective is to control the member loads. Accelerometers provide only an indirect measure of the significant parameter of interest and the steps required to estimate loads depends upon parameters that are often difficult to estimate accurately. Also the loads in the structure are a direct measure of the modal response of the lower modes with the knowledge of the mode shapes. The loads at the base of the building are a measure of the effective mass of the lower vibration modes.

After a large earthquake, the building can be restored by using local jacks to reposition the members and to reset the pins that prevents relative motion. Then the local cracks can be repaired to restore the building to its original condition.

SIMULATION STUDIES

Incorporating an idea of Adaptive Structures into a simple mathematical model of a building with a numerical simulation established the feasibility of the idea. The simplest actuator is the member itself and simplest sensor is strain gages mounted on the member itself to measure loads. The actuation can be passively initiated by designing mechanical fuses (eg. bolts that fail in shear at prescribed loads) that fail at a pre-specified load or systems which retracts a pin holding the two parts of the member. Upon initiation of the actuator, the change in stiffness of the system to another pre-specified stiffness level, attenuates the loads. A slack member that engages at a prescribed relative displacemnt to change the structural stiffness is feasible. The objective is to reduce the structural loads to prevent structural collapse. The strategies are to attenuate the loads by (1) detuning the structural frequency from the excitation force, (2) redistribute the strain energy more uniformly throughout the structure to absorb energy by shifting the energy from the lower modes to the higher modes, (3) establishing a structure representing a "loose jointed" structure that responses randomly when subjected to a deterministic input, and (4) increasing the modal damping by introducing loose joints.

Consider a spring-mass system as shown in Fig. VI; this system can be used to model the characteristics of a building. Base excitation is equivalent to earthquake excitation.

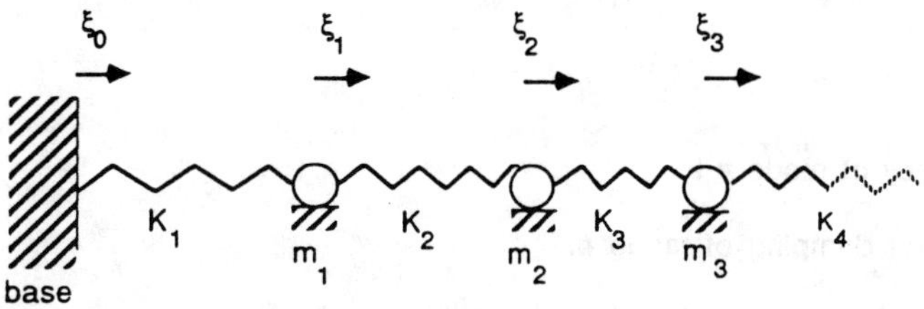

Fig. VI General Spring-Mass System.

In a typical earthquake, a series of "waves" hit the base of the structure, in the form of base excitation. These low frequency waves are composed of higher frequency excitation; thus, a periodic excitation will grow and die in a cyclic fashion. In order to model such an effect we considered the base excitation as being the product of two sine functions, i.e.,

$$\xi_0 = P \sin(\alpha t) \sin(\beta t) \quad (1)$$

where $\alpha >> \beta$

The equations of motion of the spring mass system in Fig. VI are

$$[\mathrm{diag}(m_i)]\, \ddot{\xi} + \mathbf{C}\dot{\xi} + \mathbf{K}\,\xi = \mathbf{p}(t) \quad (2)$$

where

ξ = list of displacements, i.e., $[\xi_1, \xi_2, ...]^T$

C = natural structural damping matrix

K = tridiagonal stiffness matrix (the only non-zero entries in, say, row #i are $k_{i,i-1} = -K_i$; $k_{i,i} = K_i + k_{i+1}$; $k_{i,i+1} = -K_{i+1}$)

p(t)=equivalent forcing function (the only non-zero element of **p**(t) is $p_1(t) = K_1\, \xi_0(t)$)

Assuming that **X** is the matrix of flexible modes of vibration of the model in Fig. VI (so, for example column #i coresponds to mode #i), we can express ξ as

$$\xi = \mathbf{X}\,\eta \quad (3)$$

where η is the list of modal participation factors.

We can now diagonalize eqn. (2) as follows:

$$\ddot{\eta}_i + 2\,\zeta_i\,\omega_i\dot{\eta}_i + \omega_i^2\,\eta_i = P\,x_{1i}\sin(\alpha t)\,\sin(\beta t) \tag{4a}$$

where

ω_i = natural frequency of mode # i

ζ_i = natural structural damping of mode #i

x_{1i} = entry of matrix **X** corresponding to column #i, row #1

Given that the system starts from rest, we have initial conditions

$\eta_i(0) = 0$

$d\eta_i/dt|_{t=0} = 0$

The corresponding solution to the system (4a), (4b) is

$$\eta_i = \frac{P x_{1i}}{2[(\omega_i^2-\gamma^2)^2+4\zeta_i^2\omega_i^2\gamma^2]}\left[(\omega_i^2-\gamma^2)\cos(\gamma t)+2\zeta_i\,\omega_i\,\gamma\sin(\gamma t)\right]$$

$$-\frac{P x_{1i}}{2[(\omega_i^2-\delta^2)^2+4\zeta_i^2\omega_i^2\delta^2]}\left[(\omega_i^2-\delta^2)\cos(\delta t)+2\zeta_i\,\omega_i\,\delta\sin(\delta t)\right]$$

$$+e^{-\zeta_i\omega_i t}\left[\frac{P x_{1i}}{2}\left\{\frac{(\omega_i^2-\delta^2)}{[(\omega_i^2-\delta^2)^2+4\zeta_i^2\omega_i^2\delta^2]}-\frac{(\omega_i^2-\gamma^2)}{[(\omega_i^2-\gamma^2)^2+4\zeta_i^2\omega_i^2\gamma^2]}\right\}\cos\omega_{di}t\right]$$

$$+e^{-\zeta_i\omega_i t}\left[\frac{P\,x_{1i}\,\zeta_i\,\omega_i}{2\,\omega_{di}}\left\{\frac{(\omega_i^2+\delta^2)}{[(\omega_i^2-\delta^2)^2+4\zeta_i^2\omega_i^2\delta^2]}-\frac{(\omega_i^2+\gamma^2)}{[(\omega_i^2-\gamma^2)^2+4\zeta_i^2\omega_i^2\gamma^2]}\right\}\sin\omega_{di}t\right] \tag{5a}$$

where

$\gamma = (\alpha - \beta)$

$\delta = (\alpha + \beta)$

$\omega_{di} = \omega_i (1- \zeta_i^2)^{1/2}$ (5b)

On observing eqn. (5), we can see that if either γ is close to ω_i or δ is close to ω_i, mode #i is going to suffer large amplitude motion. (Usually the forcing frequencies γ and δ will be closest to mode #1. Structural damping of the modes is usually of the order of 2% or 3%.)

If we are to imagine, then, the scenario where γ and δ (see eqn (5a)) are close to, say, ω_i the probability of structural failure is high. In order to avert this possibility, we suggest a procedure to change the stiffness of the building system in a passive sense. To illustrate this procedure, consider the 2 degree of freedom spring-mass system of Fig. VIIa (with springs K_1 and K_2) and Fig. VIIb (with springs K_1+K_{11}, K_2). Spring K_{11} is really a system of springs with the following characteristics: spring K_{11} is engaged at the beginning, but as soon as $|\xi_1-\xi_0|>d_1$ it snaps (only spring K_1 is effective now); if $|\xi_1-\xi_0|$ continues to grow, another spring of stiffness K_{11} will engage at $|\xi_1-\xi_0| =d_{22}$ and will snap when $|\xi_1-\xi_0| =d_2$ ($d_1 < d_{22} < d_2$.....).

[In a practical application to buildings see Fig. VIII. The symmetric system of cables shown (of equal stiffness) are provided with varying slacks (to engage at different diagonal elongations) and they are also provided with connections that snap at the desired stretch.]

In our 2-degree of freedom problem the location of spring K_1 was chosen to add spring K_{11}, since it can be shown that a fixed change in the stiffness at this location has the most change in the 1st natural vibration frequency. Let us say that the first natural vibration frequency of the system in Fig. VIIa is ω_1 and the first natural frequency of the system in Fig. VIIb with spring K_{11} engaged all the time (i.e., $d_1=\infty$) is ω_1^*. By allowing K_{11} to snap and engage, as explained earlier, we are essentially "confusing" the natural frequencies of the spring mass system. Thus if, say, δ is close to ω_1, the snapping and engaging will cause the first natural frequency to alternate between ω_1 and ω_1^*, thus attenuating the peak encountered at a forcing of ω_1. If, on the other hand, δ is close to ω_1^*, the snapping and engaging of K_{11} will cause the natural frequency to alternate between ω_1^* and ω_1; the consequence would be that we create

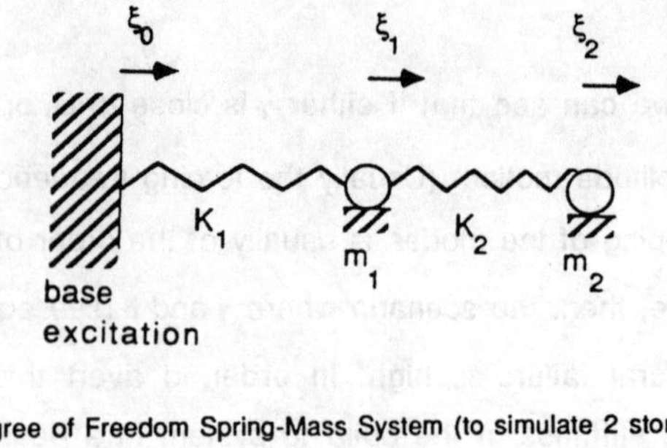

Fig. VIIa 2 Degree of Freedom Spring-Mass System (to simulate 2 storey building).

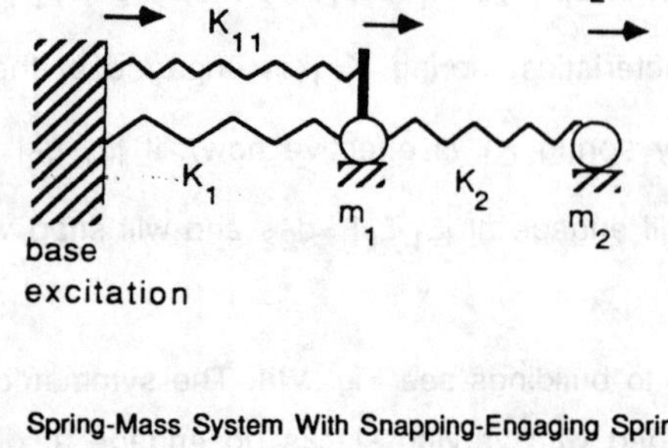

Fig. VIIb Spring-Mass System With Snapping-Engaging Spring K_{11}.

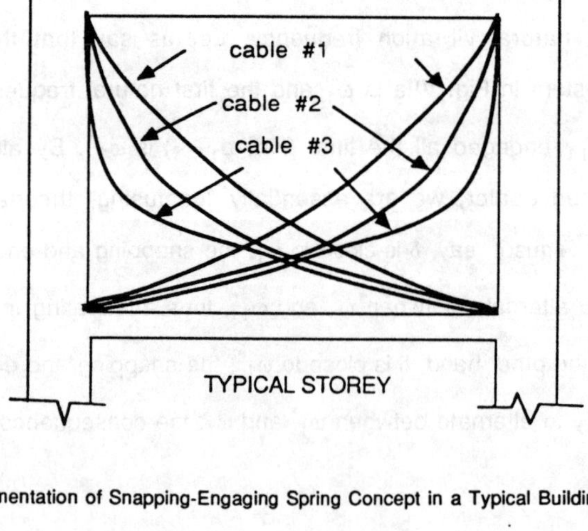

Fig. VIII Implementation of Snapping-Engaging Spring Concept in a Typical Building Storey.

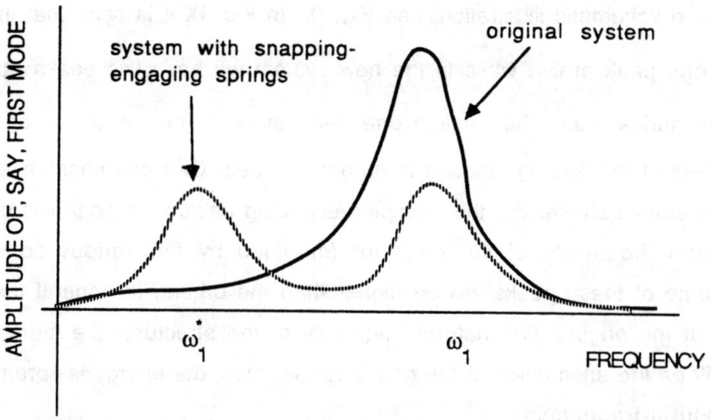

Fig. IX Mode 1 Response of Original Spring-Mass System (Fig VIIa), and System With Snapping-Engaging Spring (Fig. VIIb).

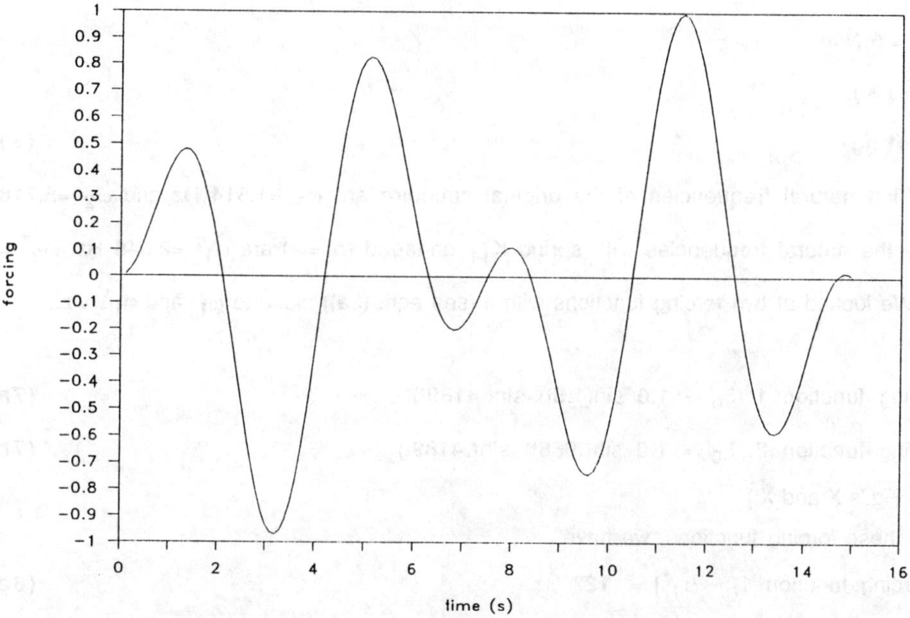

Fig. X Variation of Base Excitation 1 (Forcing Function 1).

a peak at ω_1^*. For a schematic illustration, see Fig. IX. In Fig. IX it is seen that in the original system we have one peak at ω_1, while in the new system we have two peaks at ω_1 and ω_1^* (having lower amplitudes than the original one peak at ω_1). If ω_1 and ω_1^* are sufficiently separated, the effect of the forcing function is further reduced. One can imagine a tall building with ten candidate stories chosen for the snapping-engaging cables - a peak will exist at every possible first natural frequency of the structure (obtained by the various combinations of springs). While some of these peaks will be higher than the others, in general the effect is to reduce the peak at the original first natural frequency of the structure; the more peaks there are the greater will be the attenuation of the original peak. Also, the energy is spread among the different first-natural-frequencies.

For the sake of simulation, in our 2-degree of freedom structure, the following values were chosen

K_1 = 5 N/m

K_2 = 15 N/m

K_{11} = 5 N/m

m_1 = 1 kg

m_2 = 1 Kg (6)

The first natural frequencies of the original structure are ω_1 =1.514 Hz and ω_2 =5.718 Hz, while the natural frequencies with spring K_{11} engaged ($d_1=\infty$) are ω_1^* =2.046 Hz, ω_2^* =5.98 Hz. We looked at two forcing functions with δ (see eqn. (5a)) close to ω_1 and ω_1^*, viz.,

Forcing function 1: ξ_0 = 1.0 sin(1.5t) sin(.4189t) (7a)

Forcing function 2: ξ_0 = 1.0 sin(.968t) sin(.4189t) (7b)

(See Fig.'s X and XI)

With these forcing functions, we have

$|\delta$(forcing function 1) - $\omega_1^*|$ = .127 Hz (8a)

$|\delta$(forcing function 1) - $\omega_1|$ = .127 Hz (8b)

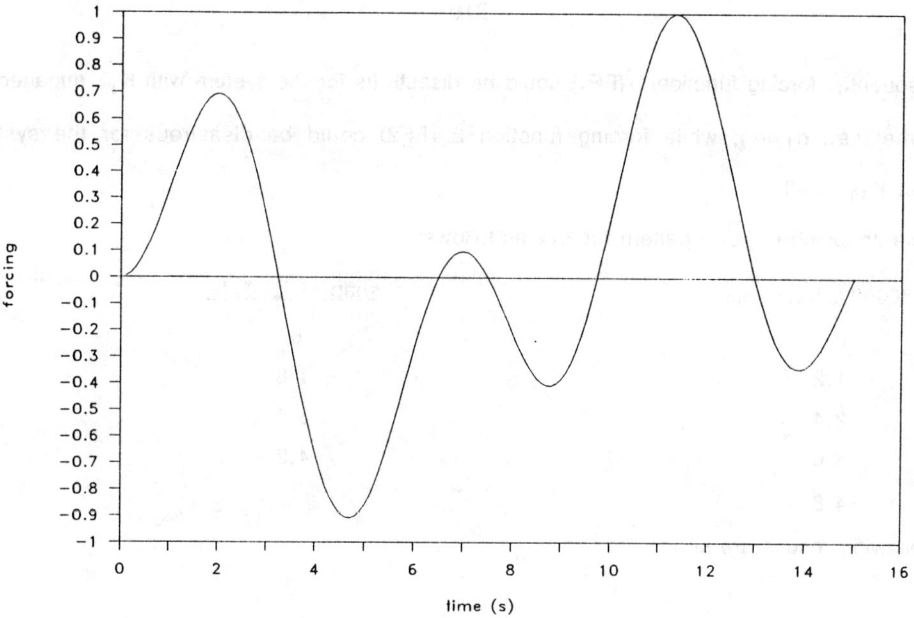

Fig. X1 Variation of Base Excitation 2 (Forcing Function 2).

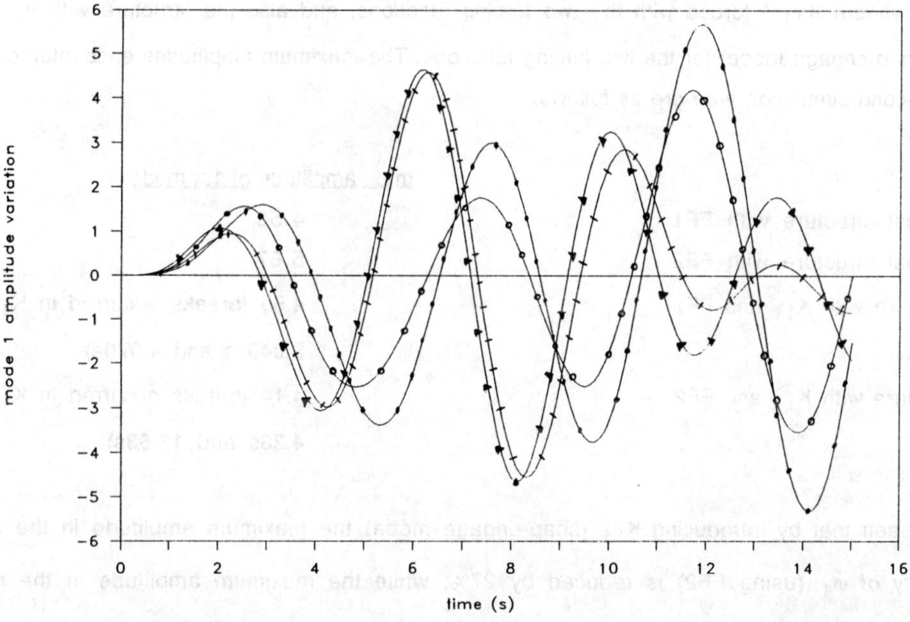

Fig. XII Amplitude Variation of Mode 1 for: —|—|— = Forcing Function 1 with original system; —●—●— = Forcing Function 2 with original system; —▲—▲— = Forcing Function 1 with snapping-engaging spring; —0—0— = Forcing Function 2 with snapping-engaging spring.

Consequently, forcing function 1 (FF1) could be disastrous for the system with K_{11} engaged all the time (i.e., $d_1=\infty$), while forcing function 2 (FF2) could be disastrous for the system without K_{11} at all.

Using a snapping-engaing pattern for K_{11} as follows:

engage at $\lvert\xi_1-\xi_0\rvert=$	snap at $\lvert\xi_1-\xi_0\rvert=$
0	.6
1.2	1.8
2.4	3.0
3.6	4.2
4.8	∞

the following results are seen.

Fig. XII:

This plot shows the amplitude variation of the first mode for the original structure (i.e., without K_{11}) forced with the two forcing functions, and also the structure with K_{11} (in the snap-engage mode) for the two forcing functions. The maximum amplitudes encountered in a 15 second simulation time are as follows:

	max. amplitude of 1st mode
original structure with FF1	4.58
original structure with FF2	5.67
structure with K_{11} and FF1	4.65 (breaks occurred in K_{11} at 2.045 s and 4.070s)
structure with K_{11} and FF2	4.14 (breaks occurred in K_{11} at 4.36s and 13.63s)

It is seen that by introducing K_{11} (snap-engage mode) the maximum amplitude in the near vicinity of ω_1 (using FF2) is reduced by 27%, while the maximum amplitude in the near

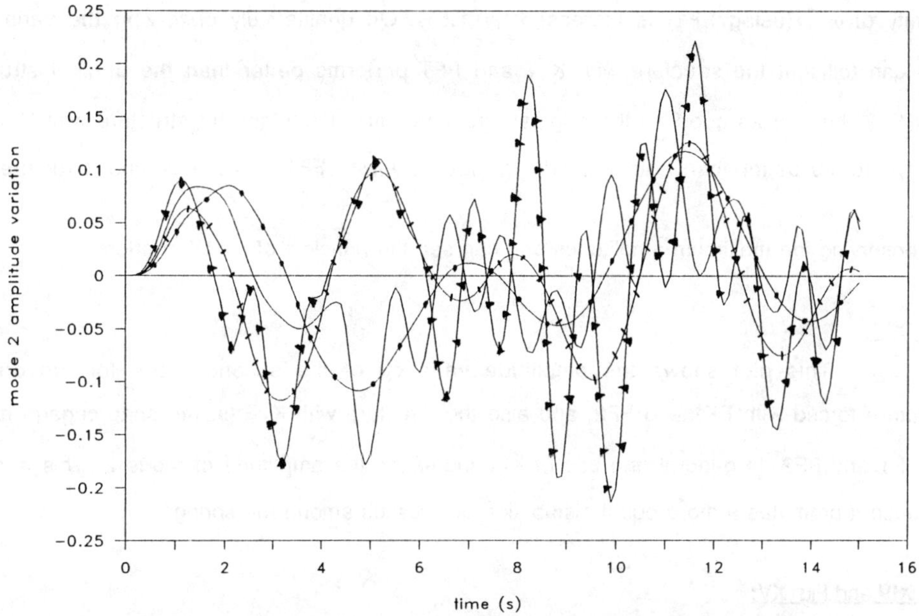

Fig. XIII Amplitude Variation of Mode 2 for: —|—|— = Forcing Function 1 with original system; —●—●— = Forcing Function 2 with original system; —▲—▲— = Forcing Function 1 with snapping-engaging spring; ______ = Forcing Function 2 with snapping-engaging spring.

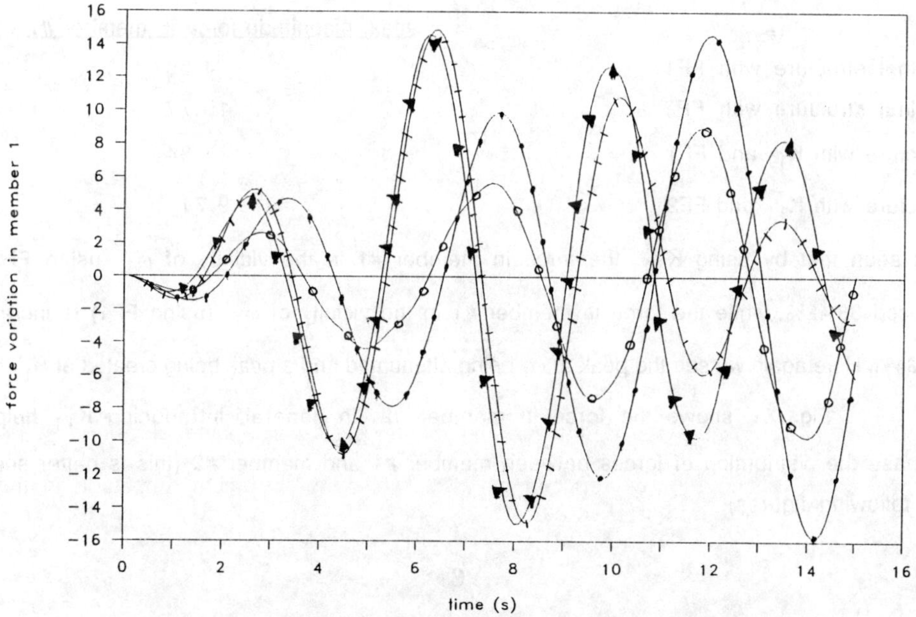

Fig. XIV Force Variation In Member 1 for: —|—|— = Forcing Function 1 with original system; —●—●— = Forcing Function 2 with original system; —▲—▲— = Forcing Function 1 with snapping-engaging spring; —0—0— = Forcing Function 2 with snapping-engaging spring.

vicinity of ω_1^* (using FF1) is increased by 1.5%. On qualitatively observing the variations, one can tell that the structure with K_{11} and FF1 performs better than the original structure with FF2, but not as good as the original structure with FF1. Also, the structure with K_{11} and FF2 performs better than both the original structure with FF1 and the original structure with FF2.

On observing the maximum amplitudes one can see the principle of Fig.IX in effect.

Fig. XIII:

This plot shows the amplitude variation of the second mode for the original structure forced with FF1 and FF2, and also the structure with K_{11} (in the snap-engage mode) for FF1 and FF2. In general, the use of K_{11} increases the amplitude of mode 2. This is good, because it promotes a more equal distribution of stresses among the springs.

Fig. XIV and Fig. XV:

Fig. XIV compares the forces generated in member # 1 for the original system with FF1, original system with FF2, spring K_{11} with FF1, spring K_{11} with FF2. The magnitude of the maximum forces in member #1, encountered in a 15 second simulation period are:

	max. magnitude force in member #1
original structure with FF1	14.89
original structure with FF2	15.77
structure with K_{11} and FF1	14.94
structure with K_{11} and FF2	9.71

It is seen that by using K_{11}, the force in member #1 in the vicinity of ω_1 (using FF2) is reduced 38.42%, while the force in member #1 in the vicinity of ω_1^* (using FF1) is increased by .34%. Here again we see the peak at ω_1 being attenuated and a peak being created at ω_1^*.

Fig. XV shows the force in member #2. In general, introducing K_{11} helps to increase the distribution of forces between member #1 and member #2 (this is better seen in the following figures).

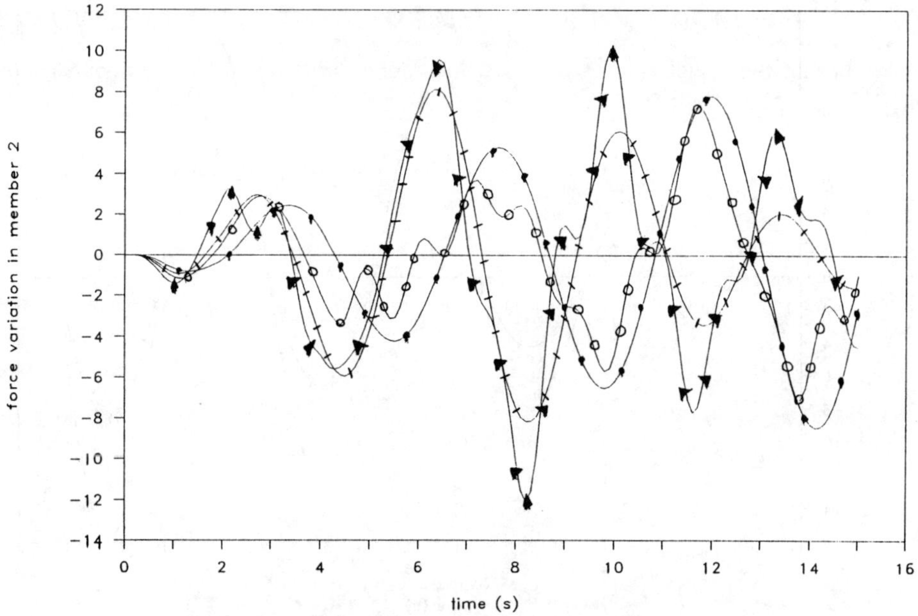

Fig. XV Force Variation in Member 2 for: —|—|— = Forcing Function 1 with original system; —●—●— = Forcing Function 2 with original system; —▲—▲— = Forcing Function 1 with snapping-engaging spring; —0—0— = Forcing Function 2 with snapping-engaging spring.

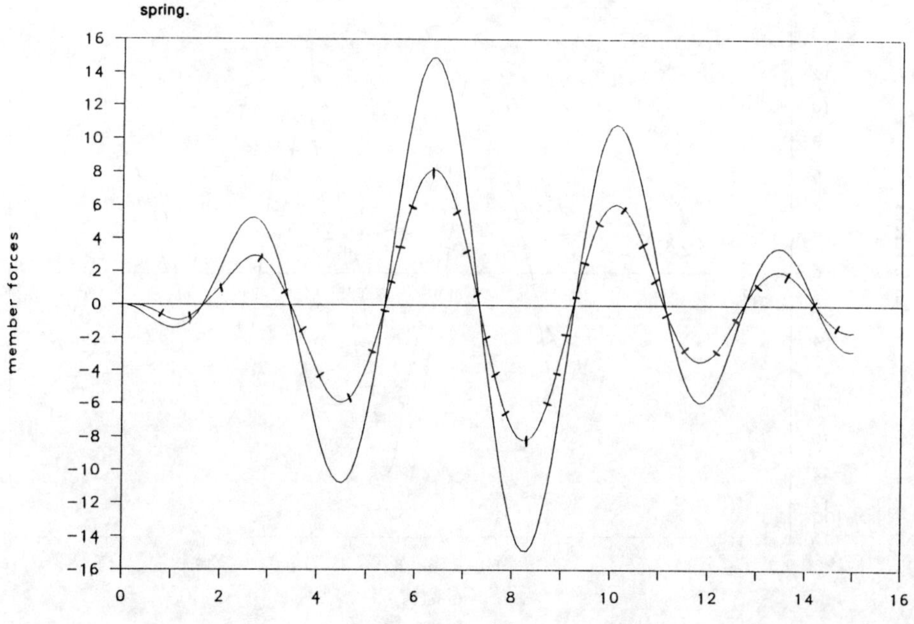

Fig. XVI Force Variation in Members 1 and 2 for the case of Forcing Function 1 with original system. ——— = Member 1, —|—|— = Member 2.

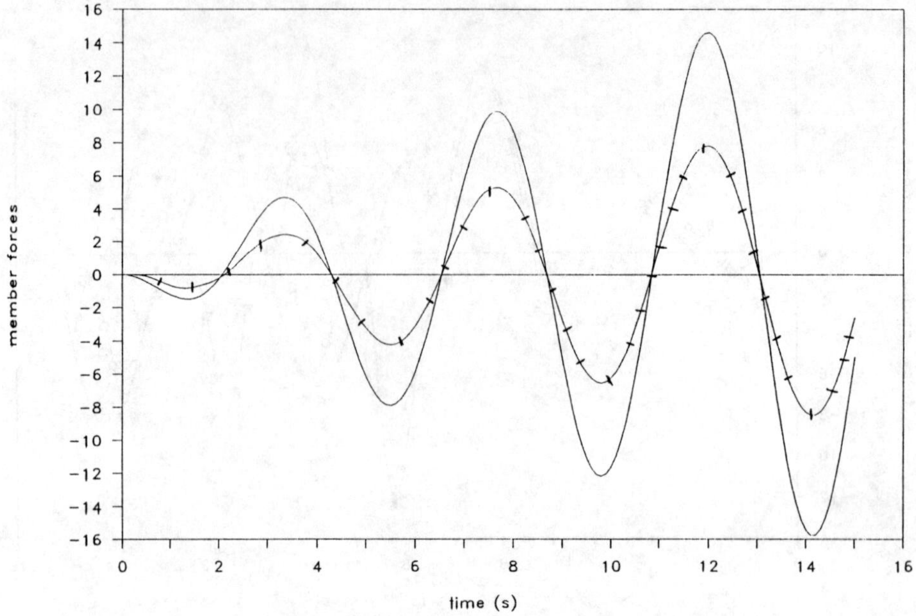

Fig. XVII Force Variation in Members 1 and 2 for the case of Forcing Function 2 with original system; ______ = Member 1, __|__|__ = Member 2.

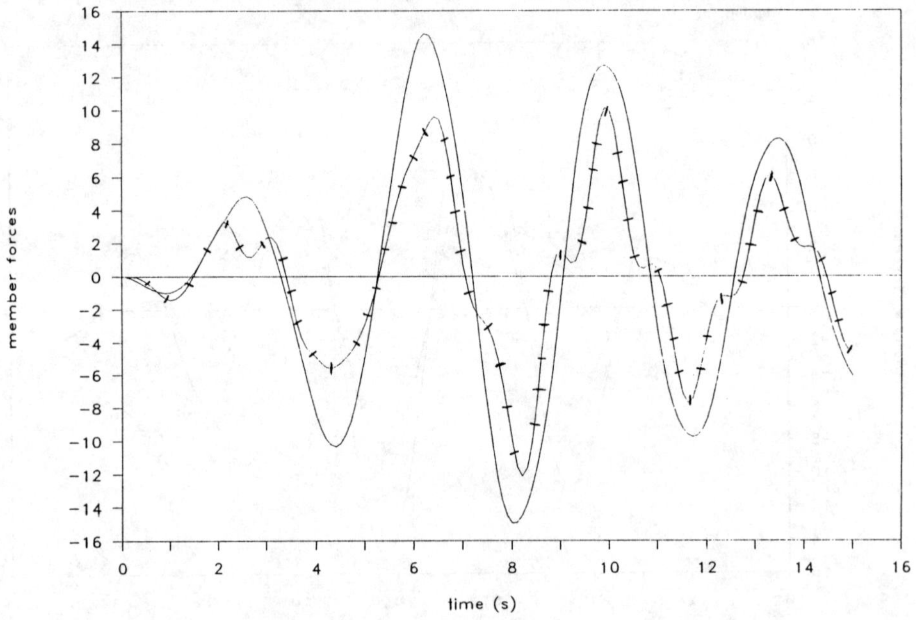

Fig. XVIII Force Variation in Members 1 and 2 for the case of Forcing Function 1 with snapping-engaging spring; ______ = Member 1, __|__|__ = Member 2.

Figures XVI - XIX:

These figures show the variation of member forces for the four cases, i.e., original structure with FF1, original structure with FF2, structure with K_{11} and FF1, structure with K_{11} and FF2. It is seen that using K_{11} (i.e., snap-engage) results in member #1 and member #2 distributing the internal forces more equally than before.

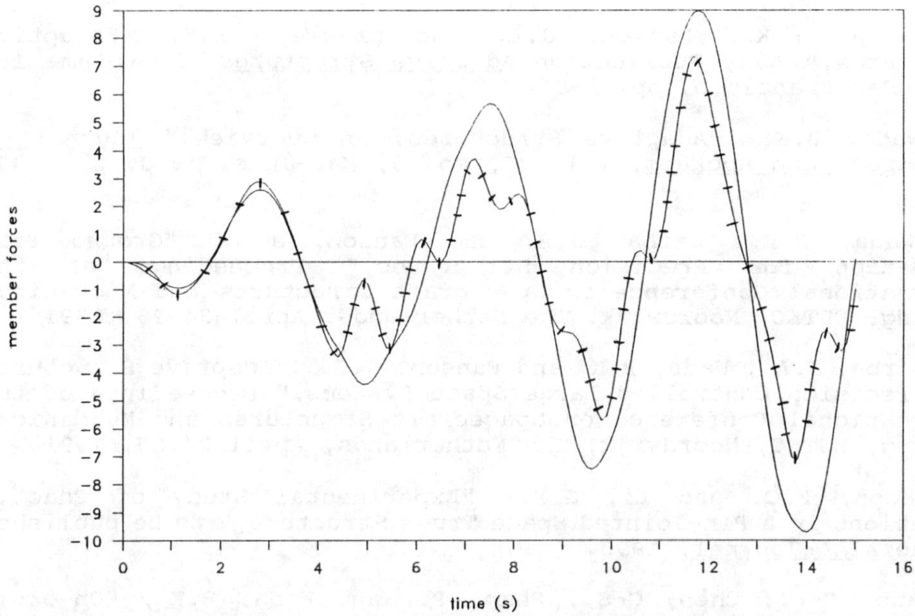

Fig. XIX Force Variation in Members 1 and 2 for the case of Forcing Function 2 with snapping-engaging springs. ——— = Member 1, —┼—┼— = Member 2.

SUMMARY

The extension of Adaptive Structures ideas and the use of the structural members as actuators to introduce stiffness changes in the structure appears to provide an attractive approach to designing tall structures when subjected to large earthquakes. The objective is to prevent structural collapse and to provide a system that can be restored to its original condition. The idea is consistent with reliable low cost actuators that is feasible for civil structures application. The approach can be designed to be redundant; namely when properly designed failure of few active members is tolerable. The approach does not depend on an accurate knowledge of the dynamics of the structure. Preliminary simulations on a simple systems indicates the feasibility of using the ideas of Adaptive Structures for civil structures. In addition to extending the size and scope of the numerical simulations to evaluate other strategies to attenuate the loads, laboratory tests are recommended.

REFERENCES

1.0 Wada, B.K., Fanson, J.L. and Crawley, E.F., "Adaptive Structures," ASME Publication Adaptive Structures, AD-Volume 15, 1989, San Francisco, pp. 1-8.

2.0 Wada, B.K., "Adaptive Structures; An Overview," Journal of Spacecraft and Rockets, Vol. 27, No. 3, May-June, 1990, pp. 330-337.

3.0 Wada, B.K., Garba, J.A. and Fanson, J.L., "Ground Test Validation for Precision Structures," Proceedings of the International Conference on Spacecraft Structures and Mechanical Testing, ESTEC, Noordwijk, The Netherlands, April 24-26, 1991.

4.0 Garba, J.A., Wada, B.K. and Fanson, J.L., "Adaptive Structures for Precision Controlled Large Space Systems," Proceedings of the International Conference on Spacecraft Structures and Mehcanical Testing, ESTEC, Noordwijk, The Netherlands, April 24-26, 1991.

5.0 Moon, F.C. and Li, G.X., "Experimental Study of Chaotic Vibrations in a Pin-Jointed Space Truss Structure," to be published in the AIAA Journal, 1990.

6.0 Kuo, C-P., Chen, G-S., Pham, P., and Wada, B.K., "On-orbit System Identification Using Active Members," Proceedings of the 31st SDM Conference, Long Beach, CA, April 1990.

7.0 Wada, B.K., Fanson, J.L. Chen, G-S, and Kuo, C.P. "Using Adaptive Structures to Enable Future Missions by Relaxing Ground Test Requirements," to be published in the AIAA Journal of Spacecraft and Rockets.

8.0 Das, S.K., Utku, S., Chen, G-S., and Wada, B.K., "A Mathematical Basis for the Design and Design Optimization of

Adaptive Trusses in Precision Control," AIAA Paper No. 91-1160, presented at the 32nd SDM Conference, Baltimore, MD, April 1991.

9.0 Chen, G-S., Lurie, B.J. and Wada, B.K., "Active Member Control for Vibration Suppression in Truss Structures," Intelligent Structures, Ed. Chong, Liu and Li, Elsevier Science Publishing, 1990.

10.0 Natori, M., Murohashi, S., Takahara, K., and Kuwao, F., "Control of Truss Structures Using Member Actuators with Latch Mechanism," ASME AD-Vol. 15, Adaptive Structures, Ed. Wada, New York, Dec. 1989.

11.0 Onoda, J., Endo, T., Tamaoki, H., and Watanabe, "Vibration Suppression by Variable Stiffness Members," AIAA Paper No. 90-1165, presented at the 31st SDM Conference, Long Beach, CA, April 1990.

12.0 Namba, H., Aoki, S. and Natori, M., "Passive and Active Systems of Adaptive Joints for Truss Structures in Space," Proceedings of the 32nd SDM Conference, Baltimore, MD, April 1991.

13.0 Wada, B.K., Fanson, J.L. and Miura, K., Eds, First Joint US/Japan Conference on Adaptive Structures, Technomics Publishing Co., Lancaster, PA, May 1991.

14.0 Chong, K.P., Liu, S.C. and Li, J.C., Eds., Intelligent Structures, Elsevier Science Publishers, Essex, England, July 1990.

15.0 Kobori, T., "Technology Development and Forecast of Dynamic Intelligent Building," Intelligent Structures, Elsevier Science Publishers, Essex, England, July 1990.

16.0 Soong, T.T., Active Structural Control; Theory and Practice, Longman Scientific and Technical, Essex, England, 1990.

17.0 Nemir, D.C., Osegueda, R.A. and Lin, Y.J., "Motion Control of Flexible Structures Using Modal Energy Transference," submitted to the Journal of Engineering Mechanics.

DEVELOPMENT OF TAKENAKA MOBILE LABORATORY
- On-site Vibration Measurement and Analysis System -

TOMOYUKI INUKAI
Technical Research Laboratory,
Takenaka Corporation
5-14, 2-chome, Minamisuna, Koto-ku, Tokyo 136, Japan

ABSTRACT

The "Takenaka Mobile Laboratory" is a large vehicle loaded with state-of-the-art measurement and analysis systems. It plays an extensive role in ground surveys at the planning and design stages for building construction, disaster damage prediction, post-construction vibration characteristics tests and seismic evaluation for existing buildings. In particular, this system allows the expandable capsule to be removed on site so that it can double the area and make the comfortable working space. Measurement is performed on-site and processed in real time for rapid feedback to design. This paper reports on the outline of the development of our system.

INTRODUCTION

The structure design of architectures is done mainly in viewpoint of the safety of the structure against external forces such as earthquakes and winds. A number of vibration surveys and tests are required at all stages of building construction, from the planning and designing stages through construction and long-term maintenance. In the planing and designing stages of super high-rise buildings, mega structures, nuclear power plants, the newest base isolation and structural control systems and so on, all kinds of vibration measurements and analyses are indispensable. Tests are also needed during and after construction to confirm the vibration characteristics of buildings, while seismic evaluation and vibration hazard surveys are often required on buildings that are already existence.

For the rapid feedback of the test results to the structure design, vibration measurement, real-time data processing and analysis are required to carry out on-site. Modern trends in vibrational measurement and analysis are as follows;

1) Enlargement of the measurement scale
2) Diversification of the test methods, measurement of the non-linear dynamic behavior of structures for instance
3) Requirement of the various types of analysis on-site such as spectral analysis and modal analysis
4) Visualization of the measurement and analysis results etc.

To meet these needs and trends, the author et al. began the development of the new vibration measurement and analysis system in 1985. An outline of our development objective is as follows;

1) Taking the mobility of new system seriously, adopt a large remodeled vehicle loaded with state-of-the-art measurement and analysis system
2) Installing some devices to prevent the vibration hazard, in moving, to the measurement instruments
3) Easy operations of system check such as data acquisition system and wiring

In 1987 we has developed "Takenaka Mobile Laboratory" satisfied above mentioned objectives. We started the vibration measurement and analysis works used the newest system "Takenaka Mobile Laboratory" in 1988. Figure 1 shows the state of the measurement on-site, The Yokohama Arena, in 1988.

This report describes about the development of Takenaka Mobile Laboratory and shows the description and features of our system.

Figure 1. State of measurement on-site (The Yokohama Arena in 1988)

<u>Large Remodeled Vehicle</u>

Taking the mobility of the vibration measurement and analysis system seriously, a large remodeled vehicle was adopted. This vehicle is 8.25m in length, 2.45m in width and 3.57m in height during the moving so that this is not enough to install many measurement and analysis instruments. However, our system allows the expandable capsule to be removed by hydraulic and electric deployment on-site, including our parking lots, so that it can double the area, a $24m^2$ mini-laboratory, and make the comfortable working space, as shown in Figure 2. It is even flexible enough to cope with various tests, analyses and presentations.

Outside view

Inside view

Figure 2. Outline of remodeled vehicle

Vibration Isolation System

In the vibration isolation system for the measurement and analysis instruments loaded on vehicle, there are many important problems. One is the protection of instruments against the impulse load caused by sudden stop and start, and driving rough road. Another is the elimination of high cycle vibration generated by engine etc. Due to settle these problems, the vibration isolation system is required the non-linearity vibration characteristics such as soft and hard stiffness against high cycle vibration and impulse load, respectively.

In the development our system, the rubber vibration isolator with a quadratics con shape (by Tohsho Eng. Co., Ltd.) is adopted as the vibration isolation system. Figure 3 shows the outline of the rubber vibration isolator. Table 1 shows the efficiency of the vibration isolation system when driving speed is 35-45km/hr. It is confirmed that the vibration isolation system is efficient, especially against vertical vibration. In the case of high driving speed, 90-100km/hr, similar efficiency is confirmed.

TABLE 1
Efficiency of vibration isolation system

	Moving Direction	Right Angle with Moving	Vertical Direction
On Floor of Vehicle	97dB/0.75m/sec²	96dB/0.66m/sec²	118dB/8.65m/sec²
On Isolation Bed	90dB/0.35m/sec²	93dB/0.46m/sec²	103dB/1.54m/sec²
Damping	7dB	3dB	15dB
Ratio of Acc.	0.48	0.7	0.18

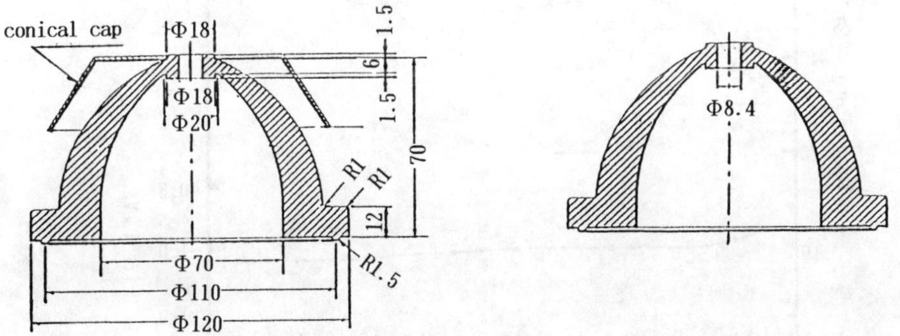

Figure 3. Outline of rubber vibration isolator with quadratics con shape

Programmable Signal Selector

In usual vibration measuring, using a matrix pin board, the input signals from all kinds of sensor-amp systems are connected to the data acquisition system such as A/D converter system. Therefore, in proportion as the total number of measuring channels increases, the work for connecting and checking of wiring becomes rather complex and take time. In our system, due to settle these problems, the programmable signal selector (by Tohsyo Eng. Co., Ltd) was adopted.

By using analog multiplexor, electric switching, this selector can choice the arbitrary 48 channels among 144 channels of input signals, and output them to A/D converter system and analog data recording system. The selection of signals is programmable and its connection pattern is able to load from and save to floppy disk arbitrarily. And this programmable signal selector has automatically self-checking function. As shown in Figures 4 and 5, this selector has good frequency characteristics such as wide flat range and the drift of output voltage as heat and time is very small.

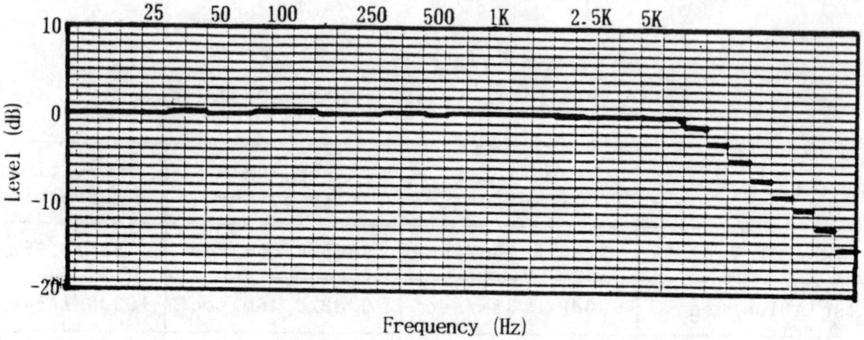

Figure 4. Frequency characteristics of programmable signal selector

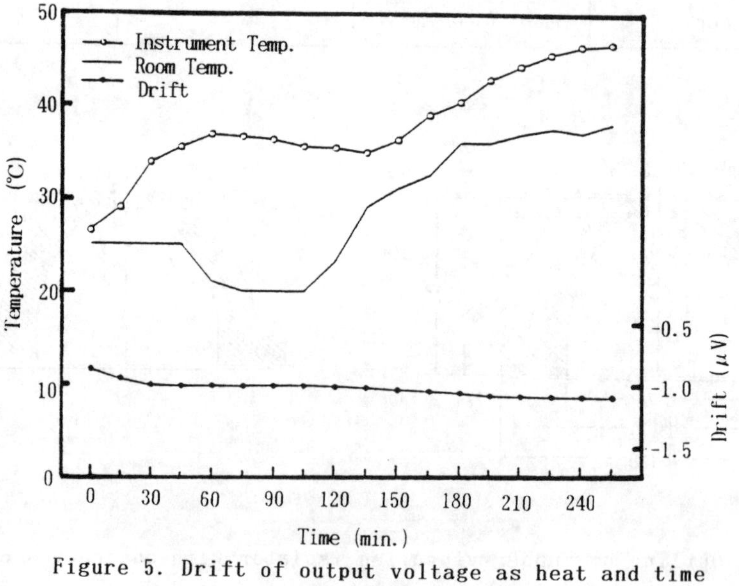

Figure 5. Drift of output voltage as heat and time

Outline of Hardware System

Due to deal with the diversification of the test methods, dynamic strain amplifiers and BNC input terminals for optional signal were installed in addition to standard displacement/velocity amplifiers. In selecting the kind of measuring instruments, it was taken note of that they could allow programmable control by data processing system.

While in selecting the kind of data processing and analysis system, computer system, we took note of real-time data acquisition, high speed and accuracy sampling, high speed processing for computationally intensive scientific and engineering applications, high performance graphic function, large capacity mass storages etc. And we introduced MC5600 system that was upgraded to MC6600 system in 1990 (by Concurrent Nippon Corp.).

As shown the layout of the measurement and analysis system in Figure 6, the expandable capsule of the remodeled vehicle can make the wide and comfortable working space. Figure 7 shows the block diagram of hardware system. The features of this hardware system are as follows;

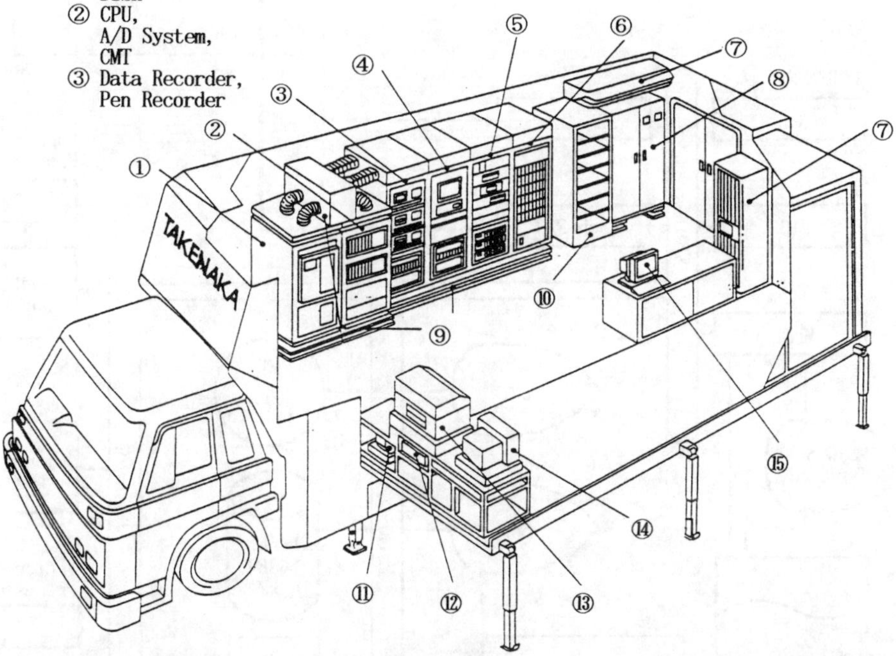

① Open MT, Disk
② CPU, A/D System, CMT
③ Data Recorder, Pen Recorder
④ Signal Selector, Monitor Board, Pen Recorder
⑤ Frequency Counter, Founction Generator, Dynamic Strain Amp.
⑥ PCM Telemeter Unit
⑦ Air Conditioner
⑧ Power Supply Stabilizer
⑨ Vibration Isoration System
⑩ Utility Rack
⑪ Video Monitor, Video Recorder
⑫ Video Making System (NTSC)
⑬ Printer/Ploter
⑭ Color Graphic Display
⑮ Character Display

Figure 6. Layout of measuremant and analysis instruments

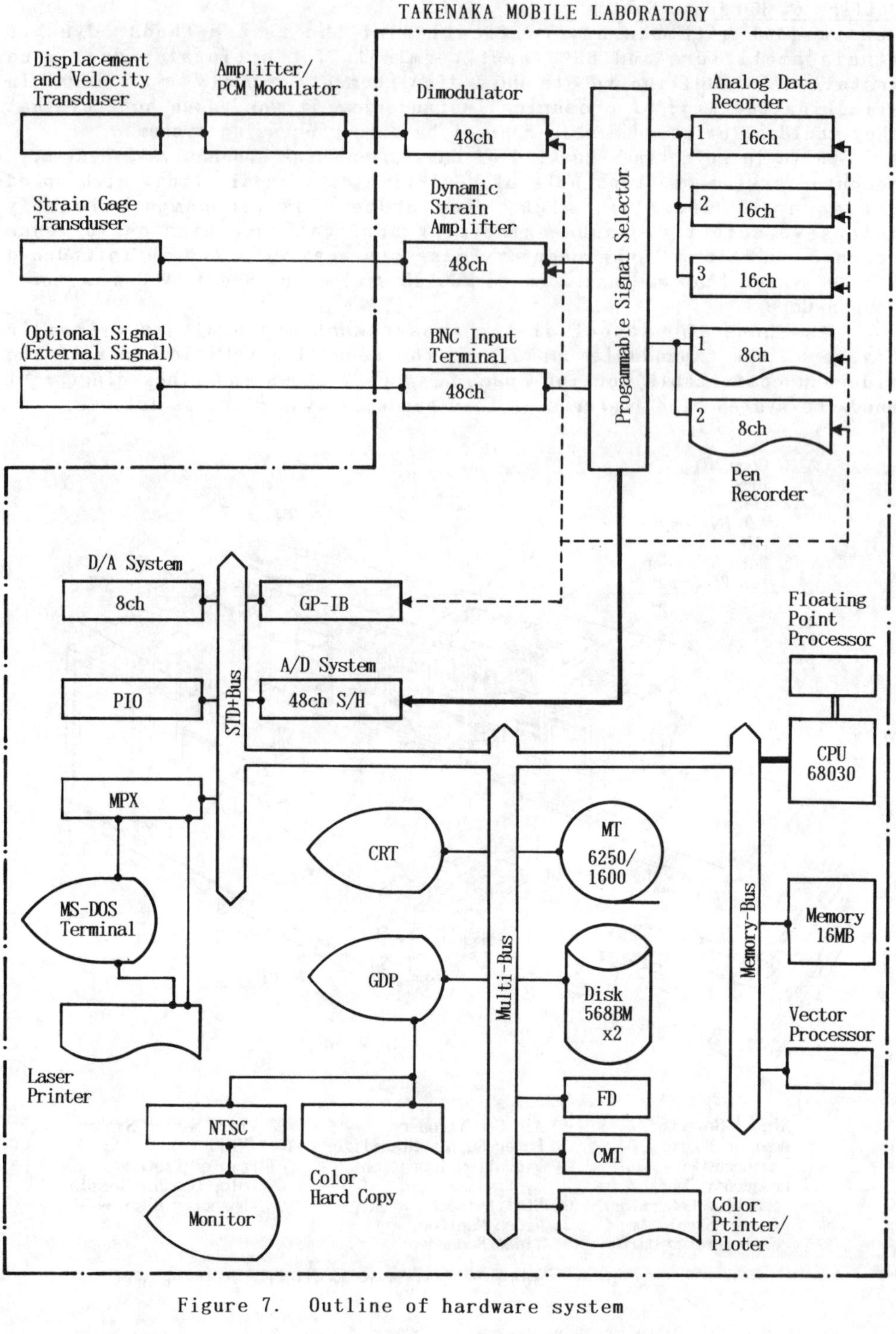

Figure 7. Outline of hardware system

1) Computer
   ·32-bit virtual CPU (MC68030, 25MHz) with 16MB physical memory
   ·High speed data processing utilizing high speed floating point co-processor and built-in vector accelerator
2) Software
   ·Operating system; RTU, real-time UNIX compatible with AT&T system V and Berkeley 4.2BSD
   ·Compilers; FORTRAN 77, C etc.
3) Mass storage
   ·Disks; 8"winchester 568MBx2
   ·Tapes; 150MB cartridge MT, 6250/1600bpi open MT
4) Graphic function
   ·High speed/resolution color graphics utilizing independent graphic processor
   ·Presentation of graphic results by video making system (NTSC)
5) Data acquisition
   ·A/D convertor; 48 channels sample and hold, 12bit
   ·High data transfer rates of up to 2MB per second
   ·Programmable signal selector capable of measuring a maximum of 144 channels, in steady state
6) Communication with vehicle exterior
   ·Radio telephone (equipped with FAX)
   ·Communication net working (preparing)

Outline of Software System

The measurement objectives of this system are steady state forced vibration test, random vibration test such as shaking table test, impulsive excitation test, seismic prospection by explosion, microtremore etc. In developing the software system, the points to which attention should be paid are as follows;

1) Cording the software by using FORTRAN language mainly to upgrade easily in the future
2) Introducing interactive operation
3) Automatic control the measuring instruments due to be able to carry out self- and programmed-measuring
4) Easy handling of data acquisition system check, A/D converter, connection of channels etc.
5) Visualization of the measurement and analysis results such as mode animation and real-time monitoring
6) Substantial application softwares

Figure 8 shows the outline of the software system.

The measuring programs (by XYZYX Co. Ltd.) can be classified roughly into three types such as steady state vibration measurement, random time domain measurement and vibration monitoring. Each program has original data file and measuring condition file, and is able to restart as the need arises. Finally, all measured and corrected data are managed by using the Takenaka standard data format so that the installation of application programs, used on the another system at present, is rather easy.

The steady state test program has a role which is the measurement of amplitude and phase-lag from excitation force. The data sampling rate is controlled corresponded to the excitation frequency. Utilizing

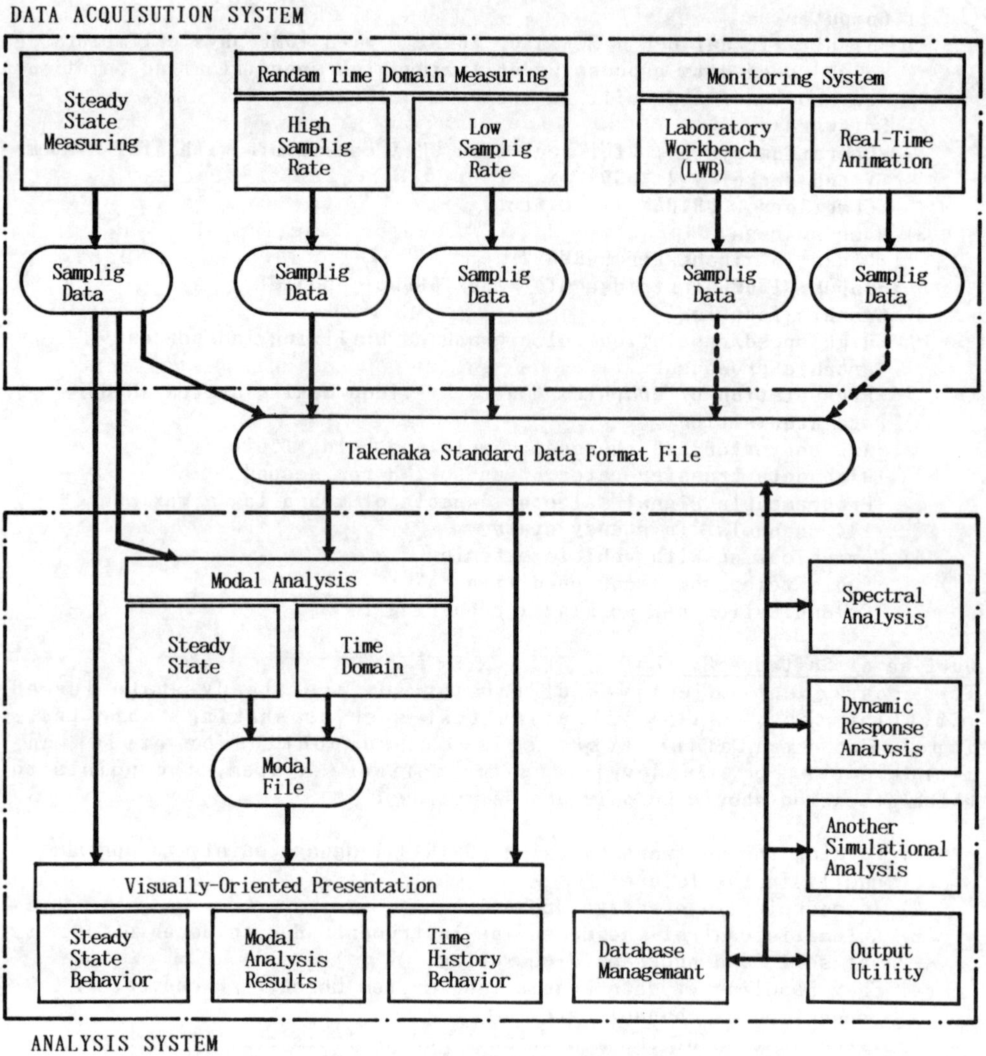

Figure 8. Outline of software system

the programmable signal selector, this is capable of measuring a maximum of 144 channels. By presenting the steady state vibration animation at the need arises, the vibration behavior of measurement object is able to confirm.

The programs for random time domain test can be classified into two types by data sampling rate such as high and low sampling program which maximum of sampling rate is 1000Hz and 200Hz respectively. The low sampling program can do real-time monitoring of wave form, orbit such as x-y plot, running spectrum, transfer function etc., and still self- and programmable measurement. Each program is capable of measuring a maximum of 48 channels.

Laboratory Workbench (LWB, by Concurrent Computer Corp.) is visually-oriented tools for configuring data acquisition and signal processing systems. Utilizing LWB and the programmable signal selector, the check of measuring system can be done easily and well reliability. Real-time animation program is very effective for checking the measuring condition before measurement and confirming the dynamic behavior of object. But this can't present high speed behavior with the present capability of our system.

In our software system, the modal analysis, identification of modal parameters, and the animation of the analysis results can be done corresponded to each measuring program.

DISCUSSION AND CONCLUSIONS

Since we started the vibration measurement and analysis works used the newest system, Takenaka Mobile Laboratory, in 1988, the good results by introducing this system are as follows;

1) The preparation and measuring works of the vibration tests were helping to reduce man power and spending time so that the efficiency of our works were improved fairly well.
2) The dealing with the enlargement of the test scale and the diversification of the test method had improved considerably.
3) All kind of data for vibration phenomena are accumulated so that the rapid feedback of the measurement and analysis results to the structual design can be done.

ACKNOWLEDGMENT

The author et al. gratefully acknowledge the great work of the members of Mitsui Engineering & Ship Building Co. Ltd., XYZYX Co. Ltd., Tohsyo Engineering Co. Ltd, Concurrent Nippon Corp. and Tsuboi Tokusyu Syatai Co. Ltd. for the development of the Takenaka Mobile Laboratory.

Structural Control

U.S. RESEARCH IN STRUCTURAL CONTROL: PROGRAM EVALUATION AND INITIATION

S.C. LIU, H.J. LAGORIO, K.P. CHONG, and W.A. ANDERSON
Engineering Directorate
National Science Foundation
Washington, D.C. 20550

ABSTRACT

This paper assesses the current status of control technology and its potential application to structural safety against seismic, wind and other dynamics loads. An intensive effort underway to develop a national research program is described. Technology assessment and research recommendations made in three major planning and program development workshops are reported. Research needs and important problems remaining from both the technological and practical standpoints for structural control are suggested.

INTRODUCTION

Bridges, buildings and other structures have traditionally relied on their strength and ability to dissipate energy through engineering design to resist dynamic forces such as strong seismic events, extreme winds, ocean waves, and heavy traffic or machine loads. In recent years, however, the use of passive and/or active systems to control and reduce structural and architectural damage caused by earthquakes, winds, and other environmental loads has become an area of considerable research interest.

Historically there have been a number of successful applications of control concepts wherein the objectives, limits and criteria for the controller are essentially established by the specific design requirements. However, the pace of practical application in a global sense and technological advancements in the field of intelligent materials and control systems for civil structural design and construction, have been falling short of expectations. Control technologies required for such civil structural applications clearly need further development and enhancement through use of advanced sensors, computers, and actuators to: (a) detect changes in the system's internal state and environment, (b) monitor quantitative models of system performance by means of multiple sensor inputs, and (c) generate real time actions to modify or alter system behavior, performance, and operation due to earthquake, extreme winds, and other natural or man-made hazard conditions.

In connection with passive systems, much research has been accomplished in specific areas related to base isolation techniques and a variety of mechanical energy dissipators such as structural bracing systems, friction dampers, viscoelastic dampers, and other mechanical devices. In the area of active systems, active mass dampers, variable stiffness devices and active bracing or tendon systems represent some of the devices being developed, tested, or installed both in the laboratory and, in certain cases, in prototype full-scale structures. Recently, the combined use of active and passive systems, often referred to as hybrid systems, have been suggested for some specific structural applications.

DYNAMIC CONTROL OF STRUCTURES

Logically, what is attempted to address here specifically is the application of intelligent control system technology for the safety protection of land-based structures, such as buildings, bridges, roadways, and tunnels. An active or hybrid controlled structure can be viewed under the general context of "intelligent structures" or "intelligent materials". The intelligence can be programmed or built into the structure by material composition and processing, real-time measurement and monitoring of dynamic response, detection of microfracture, yielding, or other failure, or conditioning through computerized servo system to adapt in a controlled manner to external forces or other stimulus. In a general sense, a "controlled" structure usually contains, through ingenious engineering design, its own sensors, data acquisition and reduction, actuators, computational/control, and feedback capabilities.

It appears that such technologies, when sufficiently developed and broadly implemented, possess an excellent potential to impact the safety design and construction of new structures in the future. Furthermore, the dynamic control and intelligence technology also offers a viable, alternative solution to the nation's enormous problems in the rehabilitation of existing structures which are deficient in their seismic, wind, or other resistant capacities.

It is generally agreed that dynamic control and intelligence technology holds considerable promise in the design applications of civil engineering structures, new and existing. By its interdisciplinary nature, such technology also has the potential to impact the future development of diverse fields of scientific or technical subjects, such as material science, vibration, biotechnology, neural network, and new devices.

To forge ahead in the development and practical implementation of structural control technology, a coordinated research program covering these diverse but interconnected subjects is needed and should be developed. In the following sections the development process of such a program by the National Science Foundation (NSF) is described and the results of the current technology assessment and future research needs in this field of emerging importance as identified through a series of special workshops is presented.

RESEARCH PROGRAM DEVELOPMENT

Over the last few years, intensive development efforts have been made for the purpose of charting a long-term national research agenda on structural control. An important capstone of such an agenda is to foster, on a coordinated basis, multidisciplinary research and development of passive, active and hybrid control technology for application to structures and nonstructural building systems under dynamic loads. Collaborative efforts among researchers in interdisciplinary areas involving academics, industrial experts and professional practitioners are considered extremely essential.

The development effort, which is expected to lead to an official announcement of research initiative by NSF, can be described by the following series of technical events that have taken place to date:

(1) In 1989, a U.S. Panel on Structural Control Research was established under the auspices of NSF. The Panel has the responsibility of:

(a) Facilitating the transmission of information concerning state-of-the-art developments in the field.

(b) Identifying and prioritizing needed research and development.

(c) Developing preliminary plans for analytical and experimental advancement in the field.

(d) Developing plans for the performance of full-scale testing and demonstration.

The Panel's organization consists of an Executive Committee with Professor George W. Housner as its chairman, and seven (7) Working Groups in: (1) Analytical Methods, (2) Experimental Methods, (3) Building Applications, (4) Non-Building Applications, (5) Interdisciplinary Approaches, (6) International Coordination, and (7) Information Dissemination. In the meantime, a Japanese counterpart panel on Seismic Control Research was also established by the Science Council of Japan and Dr. T. Kobori is the chairman of the Japanese Panel. Both the U.S. and Japan panels have been working closely in the development of a joint US-Japan research agenda in the areas of active and hybrid control.

(2) On July 23 to 26, 1990, an International Workshop on Intelligent Structures was held in Taipei, Taiwan. The emphasis of the workshop was on the development, application, and modification of active/hybrid control theories, and through innovative materials processing and system concepts, to achieve designed intelligence of civil engineering structures. Three major theme areas for the technical program of the workshop were established: (a) Sensing and monitoring techniques, (b) Structural control, and (c) Intelligent systems [1].

(3) On October 15 and 16, 1990, a workshop was held at the Electrical Power Research Institute (EPRI) under the joint auspices of NSF and EPRI. The objective of this workshop is to evaluate the status of and research needs in the area of Intelligent Control Systems [4]. The technical program was structured on four technical areas: (a) General theoretical foundations assessment, (b) Knowledge-based systems, (c) Application related research, and (d) Software/hardware requirements.

(4) On October 25 and 26, 1990, the U.S. Panel on Structural Control Research organized and held the "U.S. National Workshop on Structural Control Research" at the University of Southern California (USC), Los Angeles. The workshop identified four specific areas related to earthquake safety for technical discussions. They are: (a) Analytical research, (b) Experimental research, (c) Building applications, and (d) Non-building applications. Proceedings of the Workshop is now available from USC [2].

(5) On February 25 and 26, 1991, NSF held another Workshop on "Sensors and Signal Processing for Structural Control" in Washington, D.C. This Workshop focused on assessing the current state of knowledge and technology concerning sensors and measurement techniques, signal processing, actuator mechanics as they applied to structural control [3].

The essential conclusions and recommendations of the Workshops as summarized below provide a comprehensive technical assessment of the state-of-the-art technology in intelligent control systems and structural control research.

U.S. NATIONAL WORKSHOP ON STRUCTURAL CONTROL RESEARCH [2]

The discussions focused on four (4) primary areas - Analytical Methods, Experimental Methods, Building Applications, and Non-building Applications. The major findings are as follows:

(1) <u>Analytical Methods</u>
Special attention should be paid to nonlinear control theory. Interactive effects between "actuator dynamics" and "structural dynamics" warrants particular study. In connection with

uncertainties in structural modeling and loading, the robustness of control schemes is important and probabilistic models may be required. Other important issues include: redundancy in sensors, development of "smart structures" such as embedding active elements in structural members, time delay effects caused by actuator dynamics or by computer processing, and the potential of adaptive control algorithms. Finally the importance of integrating design of structures and control systems into a seamless process is emphasized.

(2) Experimental Approaches
Relatively little experimental research has been done in active, passive, or hybrid control systems, and an intensive future effort is needed which must occur simultaneously with correlative analytical studies. The following major issues are deliberated:

(a) Component Tests - Standardized tests for each component of the control system need to be developed. These include the performance of different actuators, pumps and servo-hydraulic systems, sensors, computer control systems, etc. The time-delay effects of these elements, their reliability and interactive effects should be evaluated.

(b) Component Development - Sensors need to be tested and developed, response quantities to be measured need to be identified, and to what degree of accuracy measurements must be made should be investigated. More compact and efficient actuators, development of hardware in conjunction with software algorithms, and the use of single component, serial components and parallel systems, should be developed.

(c) Model Tests - Laboratory tests of small-scale and medium-scale control systems and controlled structural models should be conducted utilizing existing experimental facilities.

(d) Full-Scale Tests - Control systems need to be tested on full-scale structures before practical implementation is possible. Such tests are particularly suitable for U.S.-Japan cooperative efforts because of test facilities available in Japan.

In addition to above, the workshop recommends the development of active structural elements such as active braces that induce changes in structural stiffness and passive devices which can function to distribute model kinetic energy or dissipate input energy. Development and deployment of adaptive systems and hybrid systems are encouraged for systems which undergo nonlinear response and to help reduce the external power supply requirement.

INTELLIGENT CONTROL SYSTEMS

An active or hybrid controlled structure might also be looked at from a broad perspective as an "intelligent" controlled system. Structures which undergo dynamic motions due to winds or ground shaking maybe considered as real-time systems whose complexity in control increases with (1) the number of functions to be controlled, (2) the rate at which functions to be controlled, and (3) the amount of information to be considered for a control decision [4]. From the basic research and development viewpoint, it is of increased interest to engineers in intelligent systems which possess improved flexibility in terms of application, operability (in terms of reliability and performance), and availability (in terms of serviceability and maintenance).

At an NSF/EPRI Workshop [4], the following research commendations were made to advance the development of complex dynamical control systems and to introduce new technologies for improvement of quality, safety, and flexibility, operability, and availability of such systems:

- Establishment of theoretical foundation.
- Techniques for technology transfer and implementation.
- Advanced hardware (e.g., sensors and actuators) and software (e.g., information processing and knowledge implementation).
- Baseline technology for system design and implementation.
- Test beds for system verification and validation.

SENSORS, ACTUATORS, AND SIGNAL PROCESSING

Sensor and actuator technologies and signal processing techniques as applied to structural control were discussed at a special workshop at NSF in February, 1991 [3]. The major considerations behind this Workshop were that while these topics have been advanced quite dramatically in their respective fields in recent years, these developments--often of advanced nature--are not familiar to structural engineers. Structural engineers working the control areas need to know the baseline technologies in these areas, products and technique available, and their utilities and limitations. On the otherhand, specialists in sensors, devices, control systems, and information theory need to know what structural/architectural measurements are needed and what kinds of information processing are to be dealt with.

The Workshop was organized into three interest groups: sensors, actuators, and signal processing. Findings in these group meetings are as follows:

(a) Sensors - The discussion ranges from an examination of previous research programs in this topic area to sensor needs by structural engineers, to required measurement parameters of civil structures, to new sensor types.

Specific recommendations were generated in terms of the need to meet research goals as well as NSF program goals:

- Explore issues of integration and installation of sensors in/on intelligent civil structures.
- Pursue advanced techniques for in-service monitoring and control.
- Address the need for application-specific damage assessment and Non Destructive Evaluation for critical structures.
- Investigate frontier technologies which will enable new types of measurements to be made.
- Advance basic research to the proof-of-concept level, integrating developing technologies into quantitative demonstrations of capability.
- Encourage and/or require interdisciplinary cooperation, a path to information dissemination, and a means for technology transfer.
- Establish Centers of Excellence to address the increasingly complex needs for sensing and control in intelligent civil structures.

(b) Actuators - Central to all actuator question was the need to transfer energy from one form to another. As a result, most discussion was directed at questions related to practical constraints, cost/effectiveness, control philosophy and system design. The research needs identified were:

- Develop standards for structural control systems. Specifically, test procedures and performance criteria, including control parameter levels.
- Undertake theoretical studies to contrast the uses of distributed versus discrete actuators.
- Research programs should also focus on actuators and systems for existing structures.
- Research should be undertaken to study the use of actuators as structural elements.
- Research programs should be cross-disciplinary and have multiple principal investigators.

(c) Signal Processing - In signal processing in structural control, the interface with sensors and actuators has to be addressed first so that the input and output of a signal processing system can be defined. This interface relates to the sensor type, number of sensors required and the manner they are deployed. From these questions the type and volume of signals to be processed, dynamic range of input signals, and the accuracy and speed requirements for processing can be determined. Similar analysis for actuators will provide guidelines for output signal processing.

The recommendation of the signal processing groups are:

- Encourage interdisciplinary research team activities with emphasis on close working relationships among team members.
- Identify existing signal processing techniques suitable for structural control in order to avoid duplication of work.
- Formulate signal processing problems that cannot be solved using existing techniques and present the challenge to the signal processing community.

PROBLEMS AND RESEARCH NEEDS

From the intensive investigations conducted under the special meetings and workshops as described above, a clear picture of the current state of structural control technology is now emerging.

While considerable progress is being made in research and development of passive and active structural control systems both in the U.S. and abroad, a number of important technological as well as practical issues of cross-disciplinary nature need to be addressed in order to forge close ties between research and practice and to accelerate the process of implementation. Such issues by their disciplinary nature include:

- Determination of quantitative parameters of the input forcing functions as well as structural response parameters that could be measured and incorporated in the control algorithms;
- Development of high-tech sensors as well as effective use of conventional sensing devices currently available for sensing and measuring the desired forcing and response parameters which

would minimize the time-delay and spill-over problems;

- Investigation of actuator mechanics which allow the desired real-time actions to be generated reliably within practical power supply limitations for safety or for comfort control according to the established performance specifications through system design and integration;

- Use of, or use through modifications of, existing control concepts, theories, and systems developed or applied in other fields such as aerospace, navigation, mechanical or electric systems to civil engineering structures;

- Development of new materials and sensors which possess their own intelligence;

- Development of effective signal processing techniques and algorithms for instantaneous, feedback/feedforward, adaptive, or optimum on-line control operation;

- Development of advanced computer technology and systems incorporating artificial intelligence, fuzzy logic, and/or neural networks for specific structural control application;

- Reduction and control of chaotic vibrations in mechanical and transportation systems;

- Development of innovative structural, architectural and nonstructural systems which are amenable to the application of structural control;

- Modelling and robustness of structural control under uncertainties; and

- Reconciliation of the conventional concepts of structural safety and controlled safety.

It is with such a background that a coordinated national research initiative in this emerging area of significant application potential is being formulated. This initiative is designed to facilitate and accelerate research and development in structural control. Through this effort a solid technological base is expected to be established which will make possible for broad, practical implementation of structural control systems to safeguard structural and nonstructural systems against unpredictable natural hazard events.

The fundamental objectives of the program initiative under development are identified as:

- encourage innovations in passive, active, and hybrid control systems.
- advance sensor and actuator technology and signal processing techniques and their applications.
- develop intelligent robots, other devices and computer systems for structural safety.
- develop innovative systems for energy absorption, added damping, and variable stiffness.
- study the robustness and system reliability of control systems.
- develop guidelines, standards and practical approaches and techniques for design, fabrication and field installation.
- investigate emerging technology subjects in innovative structural systems, and smart materials, sensors, and devices for detection, sensing, monitoring, and diagnosis.

In line with the above objectives, the following key research problems are identified:

- New materials.
- Performance evaluation, analytical and experimental verification.

- Standardized test procedures and performance criteria.
- Integrated structural and architectural systems design techniques and standards.
- Control algorithm development, robustness, and reliability, adaptivity, and optimization.
- Sensors, actuators and monitoring system development and optimal deployment.
- Advanced computer technology incorporating artificial intelligence, neural network, fuzzy logic, etc. for on-line targeting, detection, monitoring, diagnosis, and control.
- Intelligent robots and devices.
- Modelling, signal processing, and computer simulations.
- Practical methods and techniques for design, manufacturing, fabrication, and implementation.
- Laboratory and field experiments of systems and devices under actual or simulated dynamic excitations.
- Design and demonstration projects of control systems as applied to new or existing structures.
- Innovations in base isolation, energy dissipation devices, and other passive techniques.

In some of the research areas identified above, collaboration among researchers, practicing engineers, and manufacturers is necessary. Results obtained from such a coordinated research effort are expected to generate new knowledge, techniques, and a better understanding of the working principles as well as practical limitations inherent in these control systems so that conditions of applicability and design limits of each class of systems can be established and evaluated. It should be emphasized again that the ultimate goal of this multi-year research effort under planning is to bring about technological advances so that control technology can be realistically implemented in a timely fashion to provide safety resistance to protect structures such as buildings, bridges, tunnels, and other critical lifeline systems against earthquakes, winds, and other natural or manmade hazard environments.

SUMMARY AND CONCLUDING REMARKS

Over the last two decades, considerable progress has been made to improve structural safety through research in structural, earthquake, and wind engineering. The possibility of applying control systems to mitigate the structural damage against earthquakes and winds opens up a wide range of challenging research opportunities for multidisciplinary investigators in the field. Further, such challenges can be undertaken most effectively by engineering professionals through international research cooperation. In this spirit, the mechanisms which might be provided by the International Decade for Natural Disaster Reduction (IDNDR) program can efficiently serve as a launching pad for the research projects as identified. The UJNR Panel on Seismic and Wind Effects, and the US Panel on Structural Control Research and its counterpart Japanese panel are all expected to play a pivotal role in the advancement of control technology and to lead us toward a safer and more economical future for civil engineering construction.

REFERENCES

1. Chong, K.P., Liu, S.C., and Li, J.C., Editors, Intelligent Structures (1990). Elsevier Applied Science, London and New York, 460 pp.

2. Housner, G.W. and Masri, S., Editors, Proceedings of U.S. National Workshop on Structural Control Research, October, 1990, USC Publication No. CE-9013, 251 pp.

3. Roberts, R., and Li, W.P., Editors, Summary Report of Workshop on Sensors and Signal Processing for Structural Control, NSF, February, 1991, Lehigh University.

4. Shoureshi, R. and Wormley, D., Editors, Proceedings of Workshop on Research Needs in Intelligent Control Systems (1990). NSF/EPRI.

STRUCTURAL CONTROL RESEARCH NEEDS

G.W. Housner
Division of Engineering & Applied Science
California Institute of Technology
Pasadena, California 91125, U.S.A.

S.F. Masri
Department of Civil Engineering
University of Southern California
Los Angeles, California 90089, U.S.A.

ABSTRACT

An overview is presented of the recent establishment of a panel in the United States under the auspices of the National Science Foundation to deal with the field of active structural control. Information is given about the purpose of the panel, its organizational structure, future plans, and some proposed international collaborative efforts. Deliberations of the first "U.S. National Workshop on Structural Control Research" which was convened by the Panel in October 1990 to deal with research needs in the field of structural control are summarized.

INTRODUCTION

There is a growing worldwide awareness by civil engineers of the potential of active (hybrid) protective systems for earthquake hazard mitigation. These control systems have the potential for significantly reducing the dynamic forces experienced by structures under arbitrary excitations. There is also awareness of the need for organized national programs and cooperative multi-national efforts involving collaborative projects, exchange of personnel, jointly organized tests, and exchange of data and technical information.

BACKGROUND INFORMATION

With the above discussion in mind, the United States National Science Foundation approved a proposal by the authors to establish in February 1990 a "U.S. Panel on Active Structural Control." The Panel activities are funded by the Earthquake Engineering Program under the direction of Dr. S.C. Liu.

The managing organization of the Panel activities is CUREe (California Universities for Research in Earthquake Engineering) which was formed in 1988 as a not-for-profit corporation. At present, the Institutional Members of CUREe are: California Institute of Technology, University of California - Berkeley, University of California - Davis, University of California - Irvine, University of California - Los Angeles, University of California - San Diego, Stanford University, and University of Southern California.

PURPOSE OF THE PANEL

The panel is charged with the following responsibilities:

1. To develop a plan for a U.S. program in the active (hybrid) control of structures.
2. To conduct in collaboration with a counterpart panel in Japan, established by the Japan Science Council, a joint U.S.-Japan cooperative work plan for research, development, and implementation of controlled structural systems.
3. As part of the International Decade for Natural Disaster Reduction, to develop a plan of action for U.S. participation in collaborative international research in active control.
4. To monitor U.S. efforts in this area.

ORGANIZATION OF THE PANEL

The panel is composed of seven members whose names and affiliations are listed below:

1. Mr. Frank Conati, MTS Systems Corporation, Minneapolis, Minnesota
2. Professor William J. Hall, University of Illinois, Urbana, Illinois
3. Professor George W. Housner, California Institute of Technology, Pasadena, California
4. Professor Sami F. Masri, University of Southern California, Los Angeles, California
5. Professor Masanobu Shinozuka, Princeton University, Princeton, New Jersey
6. Professor Tsu T. Soong, State University of New York at Buffalo, Buffalo, New York
7. Mr. Ben K. Wada, Jet Propulsion Laboratory, Pasadena, California

An Executive Committee consisting of G.W. Housner, S.F. Masri, and T.T. Soong oversees the Panel activities. The objectives of the Panel will be accomplished through the efforts of seven Working Groups:

1. Analytical Methods
2. Experimental Methods
3. Building Applications
4. Non-Building Applications
5. Interdisciplinary Approaches
6. International Coordination

7. Information Dissemination

Figure 1 shows an overall organizational chart for the Panel.

INTERNATIONAL COLLABORATION

In keeping with the spirit and intent behind its creation, the U.S. Panel would like to pursue with all interested countries possible cooperative multi-national efforts involving conferences and workshops, collaborative projects and the exchange of technical information.

U.S. NATIONAL WORKSHOP ON STRUCTURAL CONTROL RESEARCH

In order to carry out its responsibilities, the U.S. Panel convened the first "U.S. National Workshop on Structural Control Research." The Workshop was held at the University of Southern California on 25 and 26 October 1990. Approximately one hundred participants with a variety of technical backgrounds participated in the formal presentations and discussions covering current practices, research, and future needs. The major objectives of this Workshop were:

1. To summarize the state-of-the-art of the structural control field.
2. To identify and prioritize needed research in the field.
3. To develop preliminary plans for the analytical and experimental advancement of the field and for the performance of full-scale testing.
4. To facilitate the transmission of information concerning state-of-the-art developments in this rapidly evolving field.

The Workshop program was designed for researchers, engineers, architects, and others who are interested in the theory, experimentation, and application of the vibrational control of structures, sensitive equipment, etc., under the action of dynamic loads such as earthquake and wind.

The first day of this two-day Workshop featured invited papers and summary reports by U.S. and international representatives from universities, research organizations and industrial concerns. Presentations about ongoing research activities or future research plans in the general area of adaptive structural systems were solicited. Among the specific topics addressed in the Workshop were: possible applications, reliability issues, combined active/passive approaches, large scale tests, economic issues, retrofitting problems, implementation techniques, control purpose (safety/comfort), dynamic environment (wind, earthquake, etc.), actuator development, distributed sensor technology, control energy sources, control algorithms, active parameter control methods, stability considerations, system identification procedures, computational issues, architectural considerations, educational/training issues, and international collaborative projects.

Overall Organizational Chart

U.S. Panel on Structural Control Research

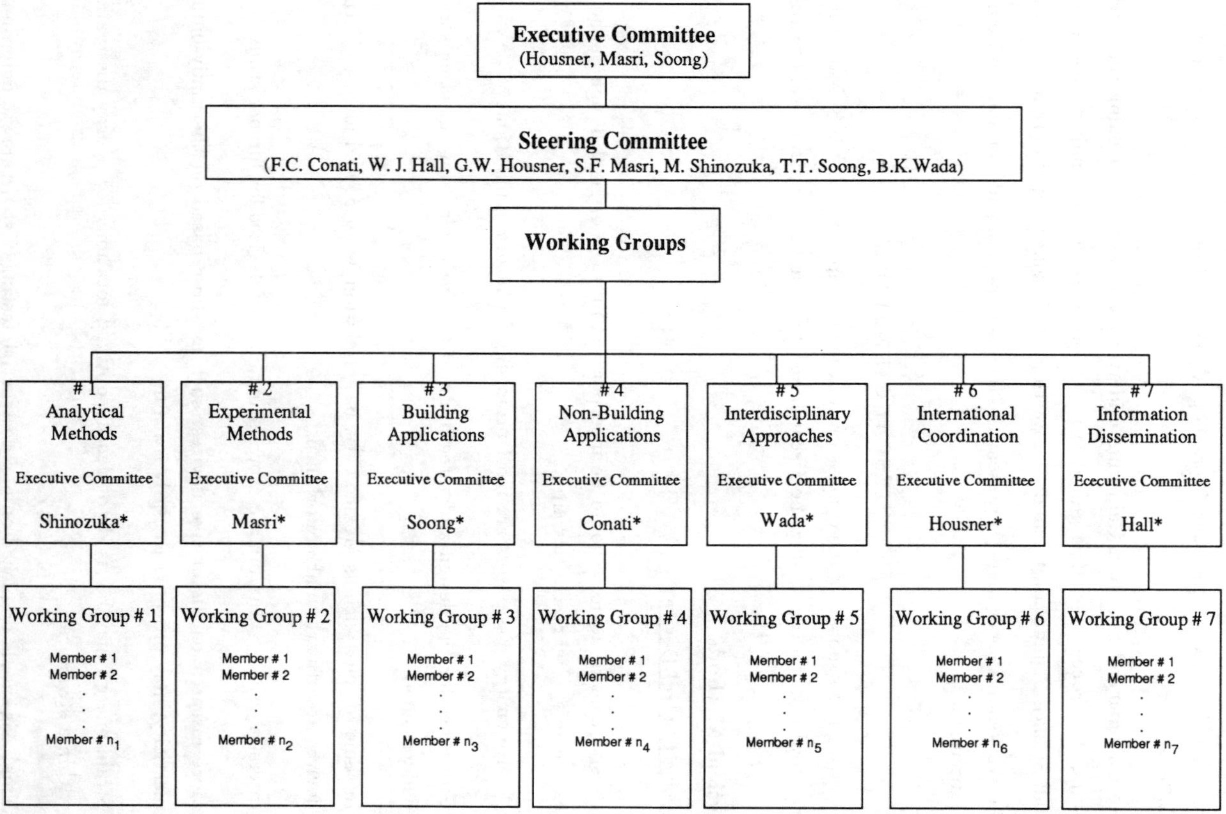

*Designates Steering Committee Member

The second day of the Workshop consisted of several workshop panels that discussed topics related to specific areas: (1) analytical methods, (2) experimental approaches, (3) building applications, and (4) non-building applications. The Workshop culminated in a prioritized research plan by each of the four Workshop Working Groups, which are included in these Proceedings.

The University of Southern California undertook the task of convening the workshop and publishing the proceedings. The guidance as well as most of the funding have come from the National Science Foundation through the Earthquake Engineering Program.

The Workshop Proceedings [1] present the technical papers and summary statements of the formal discussions.

RESEARCH NEEDS

The following is a synopsis of the conclusions of the participants in the U.S. National Workshop on Structural Control concerning the research needs in the field of structural control:

Analytical Methods

Passive Control Devices

- Passive control devices should be incorporated into the original design of the structure, except in special cases (retrofits).
- Passive energy dissipators are most promising for future applications.
- Guidelines for the design and integration of passive devices into existing structures are badly needed.
- In some regions, such as Mexico City, base isolation is not a good idea, and other passive devices are of particular importance.
- Interstory shear actuators are a promising class of devices for further study.
- More research on dissipative devices, such as those based on thermally-induced metal plasticity, may prove particularly fruitful.
- Strain energy may be used to identify optimal locations for energy dissipation devices within a structure.
- Emphasis should be placed on non-destructive testing of structures to determine accurate and realistic mathematical models of structural behavior.

- Modeling and system identification must be considered as a coupled task, and may involve decisions regarding the use of finite element, lumped parameter, modal, and continuous model descriptions.
- Model verification must be given great importance, since in reality "you don't control the model, you control the structure."
- Parameter and model uncertainty are of particular importance, and must be quantified whenever testing is performed.
- Control algorithms which are robust against uncertainty and variation in model parameters must be developed.
- Future research should focus in part on the development of a realistic representation of structural behavior, taking into account the degree of deterioration, and recognizing the continuous nature of the structure, for both initial (virgin-state) and in-service conditions.

Active Control Systems

- Actuator dynamics, including modeling and identification, is an area in need of further research.
- Performance robustness vs. stability robustness: further research on the trade-offs inherent in improving performance and improving stability for large civil structures is needed.
- While there exist robust control algorithms involving co-located sensors and actuators, more research is needed to develop algorithms which improve performance over that obtainable with co-located approaches without sacrificing stability, particularly for nonlinear hysteretic structures.
- The research community would benefit from a broader education for civil engineers: many useful control approaches exist in related disciplines, such as electrical engineering, aerospace engineering, mechanical engineering, and acoustics which are not yet familiar within the civil engineering community.
- Recent developments in nonlinear control theory may permit substantial progress in developing efficient control algorithms for hysteretic structures.
- While chaos is not likely to be of primary importance in practical problems, the quantitative use of statistical models of chaos may provide some interesting new results in structural control.
- Soil-structure interaction effects are important in that accurate definitions of base motions are necessary for many approaches within structural control.

- The development of accurate, reliable and inexpensive displacement sensors would provide a substantial benefit to many approaches in structural control.
- The torsional response of buildings is an area of continuing research need.

Hybrid Control Approaches

- Hybrid control approaches seem to offer opportunities for improving performance over either active or passive approaches taken individually.
- While passive (base isolation) systems seem to provide advantages for earthquake protection, and active devices (tuned-mass dampers, etc.) seem to offer advantages in controlling wind-induced vibrations, hybrid systems would seem an attractive approach for a wide range of excitations.
- Hybrid control approaches can lead to designs which reduce required actuator forces.
- Friction effects in actuators are always present, and are not well modeled or understood.
- More effort should be applied to the development of solutions for retrofit problems since the largest need for structural control and hazard mitigation will likely emerge from this class of problems.

Experimental Approaches

Research Directions

- Component Tests
  - Standardized tests for active control system components
  - Performance of sensors and computer control systems
  - Characterization of force-displacement-time relationships
- Component Development
  - Need for development of new sensors
  - Need for information on what structural responses can be measured (accel., vel., disp., member forces)
  - Need for information on possible sensor accuracy
  - Need to develop compact, efficient actuators
  - Need for testing of sensors used in other applications for suitability for use in active structural control

Areas Needing Further Research

- Model Testing
  - All control systems should be verified in the laboratory
  - Math models used for control systems should realistically represent nonlinear response of full-scale structures at high excitation levels
  - Promising control systems should be tested on medium-scale structures
  - Tests should include multidirectional excitations
  - Tested structures should include elements that behave in a nonlinear manner
- Full-Scale Tests
  - The most promising control systems need to be tested on a full-scale structure
  - Cooperative efforts with companies in Japan that have full-scale test facilities should be considered
  - Inexpensive, effective methods should be developed for exciting structures for testing purposes
- Development of Active Structures
  - Adaptive structures
  - Parameter control
  - Hybrid systems
- Research Collaboration
  - Experimental research can be very expensive (about $200,000 per year per PI, plus equipment costs)
  - Facilities, equipment, and models should be shared through the development of several national testbeds

Building Applications

Central Issue

Should structural control systems be considered life safety systems or merely as means of improving human comfort and function?

Further information about the effectiveness of structural control will be needed before it can finally be resolved.

Areas Needing Further Research

- Reliability
  - acceptable level of reliability
  - attainable level of reliability
  - system approach (including active control systems as well as support systems such as power)
  - reliability studies that include realistic (nonlinear) building behavior
- Nonlinear Structural Behavior
  - models used should be latest and most sophisticated models available
  - even in active control for comfort and function, nonlinear structural behavior is important
- Performance Criteria
  - establish performance goals under different loads such as wind, seismic, and man-made
  - address passive, active, and hybrid systems
  - establish which system is best for what application
- Design Criteria and Guidelines
  - design professionals as well as university researchers must be involved in the process
  - a good model to follow is process used for base-isolated structures
  - design criteria to be developed should incorporate latest design and analysis concepts, including nonlinear behavior
- Case Studies
  - at least three case study design and analysis projects should be undertaken
  - different control technologies should be highlighted
  - existing/new buildings, and eastern as well as western construction methods should be included
- Scale Model Experiments
  - more laboratory and field experiments on scale model structures need to be performed
  - special attention should be devoted to stiff (concrete) structures

- Full-Scale Testing
  - a full-scale demonstration project should be included in the latter phases of the recommended five-year program
  - there must be a set of achievable objectives
  - the structure selected should resemble a "typical" building structure
  - the site should be selected in a region where either wind or earthquakes are a problem

Non-Building Applications

Critical Function Facilities Appropriate for Structural Control

1. Emergency Response Facilities
2. Transportation Facilities
3. Communication services
4. Water and sewage facilities
5. Banking systems including their computer facilities
6. Power plants (including nuclear power and fossil fuel plants)
7. Industrial facilities including refineries; gas-oil processing and distribution systems; off-shore platforms
8. Harbor and Coastal facilities
9. Gas and liquid fuel facilities
10. Production processing lines critical to public safety

Areas Needing Further Research

1. Floor systems ranging from small to very large (single or multiple isolation and control)
2. Special equipment which includes:
   (a) Medical equipment (such as x-ray and nuclear magnetic resonance imaging facilities). Also, power sources in medical facilities which supply life-support and intensive care facilities.
   (b) Computer Systems which range from small to extended facilities and data storage.
   (c) Museum artifacts protection.

(d) Water plant facilities including cross connection between water and sewage systems.

(e) Manufacturing equipment.

(f) Food-supply sources which include warehouse and refrigeration systems.

(g) Transmission lines including towers and poles.

3. Study the nonlinearity effects on the response of long-span cable-supported bridges and equipment. Nonlinearities could be geometric, material or due to self-excited forces.

4. Provide means to control the nonlinear response of structures. Compare the use of traditional control methods such as tuned mass dampers and pulse controllers.

5. Study the relationship between structural identification and structural control considering the variability and uncertainty in estimating the structural parameters.

6. Research results should be applied to long-span bridges.

Interdisciplinary Approaches

Problems similar to the active control of structures during earthquakes and winds are studied in the fields of aerospace, aero-space structures, electrical control theory, noise control, and others. It will be important that researchers in active control of structures keep informed of similar activities that are underway in these different fields of research and, where feasible, interdisciplinary research projects should be undertaken. It may be that a Workshop devoted to clarifying the state-of-the-art in these different disciplines should be held.

International Coordination

The potential benefit of active control of structures during earthquakes and winds will be applicable to a number of countries, which either face these natural hazards or are actively engaged in engineering structures in hazardous countries. There is much to be gained by international coordination and cooperation. This can be done through international conferences, workshops, or cooperation between committees, panels, or cooperation between individual research workers.

Information Dissemination

A program on active control of structures must be supported by appropriate information dissemination both nationally and internationally. This will assist in speeding up the program and providing timely results.

RECOMMENDATIONS

The U.S. National Workshop on Structural Control Research Working Groups provided lists of research topics and other activities that were considered important. It is recommended that these topics, which are presented above, form the basis of a program in active control of structures in the United States.

In order to avoid falling hopelessly behind in research in the important area of structural control, the U.S. should allocate at least as much funding in this area as does Japan. Funding at a level of $5 million per year is recommended to cover all phases of the structural control program in the U.S., which includes analytical methods, experimental approaches, building applications, non-building applications, interdisciplinary approaches, international coordination, and information dissemination.

ACKNOWLEDGEMENTS

We are indebted to all participants whose attendance and contributions helped us focus on the important and critical issues of structural control research, and whose enthusiasm for acting on the recommendations of the Workshop will provide the motivation to complete the tasks that lie ahead of us. Special thanks are due to the Working Groups recorders.

REFERENCES

1. Housner, G.W. and Masri, S.F., (Editors), *Proceedings of the U.S. National Workshop on Structural Control Research*, Department of Civil Engineering, Publication No. CE-9013, University of Southern California, Los Angeles, California 90089-2531, October 1990.

FULL-SCALE IMPLEMENTATION OF ACTIVE STRUCTURAL CONTROL

T.T. SOONG AND A.M. REINHORN
National Center for Earthquake Engineering Research
and Department of Civil Engineering
State University of New York at Buffalo
Buffalo, NY 14260, USA

ABSTRACT

An active bracing system has been designed, fabricated, and installed in a full-scale dedicated test structure for structural response control under seismic loads. Presented in this paper are the design of the active system, control algorithm development, and simulated as well as preliminary test results. Detailed design and analysis of the active system are carried out with respect to hardware development, control force constraints, and power and energy requirements. It is shown that a full-scale efficient active structural control system can be developed within limits of current technology. Simulation and test results provide information on performance bounds that can be expected of active systems in structural control under seismic loads and under constraints imposed by practical considerations.

INTRODUCTION

The possible use of active control systems as a means of structural protection against environmental loads, such as wind and earthquakes, has received considerable attention in recent years. It has now reached the stage where active systems have been installed in full-scale structures [1,3,4]. As an example, an active mass driver system was recently installed atop of the eleven-story Kyobashi Seiwa Building in Tokyo. This system, designed primarily for comfort control of building occupants, has performed well under several hurricanes and moderate earthquakes encountered since installation.

Another active mass damper system was recently installed in a dedicated full-scale 600-ton test structure on the grounds of the Takenaka Research Laboratory. It has also performed well during a recent moderate earthquake. In addition, a full-scale active bracing system has been installed in the same structure for performance evaluation purposes. The advantage of having the performance of two active systems evaluated using the same structure is obvious. In addition to providing the same base parameters for performance comparisons, this arrangement allows the calibration of both systems by using one of the systems as motion inducer and the other as motion controller. Even without actual environmental disturbances, much of the performance characteristics of these systems can be assessed.

To provide a focus, this paper addresses design, fabrication, and performance issues related to the development of the active bracing system. These discussions provide a realistic assessment of the potential benefit which can be derived from an active system on the one hand, and system design and capability requirements on the other.

TEST STRUCTURE AND ACTIVE BRACING SYSTEM

Test Structure

As mentioned above, a dedicated full-scale structure was erected for performance verification of two active structural control systems, i.e., an active mass damper and an active bracing system. Located in Tokyo, Japan, the structure is a symmetric two-bay six-story building as shown in Fig. 1. It was constructed of rigidly connected steel frames of box columns and W-shape beams with reinforced concrete slabs at each of the floors. Weighing 600 metric tons, the structure was designed as a relatively flexible structure with a fundamental period of 1.0 sec in the strong direction and 1.5 sec in the weak direction in order to simulate a typical high-rise building. It is assumed that the floor slabs are rigid in their planes and the structure does not twist; accordingly, the space structure was modeled as a fixed-base plane frame with three degrees of freedom (horizontal, vertical and rotation) at each joint in the eigenvalue analysis. The stiffness of the concrete slabs was taken into account by considering rigid zones at ends of the members in analytical modeling. The spectral properties of the structure obtained analytically as well as experimentally are compared in Table 1, showing that their differences are within 3% for all modes.

TABLE 1
Modal frequencies of test structure

Mode	X-direction		Y-direction	
	Record (Hz)	Analysis (Hz)	Record (Hz)	Analysis (Hz)
1	0.900	0.922	0.650	0.643
2	2.700	2.703	1.800	1.750
3	4.650	4.737	2.975	2.918
4	6.850	6.973	4.125	4.117
5	9.400	9.481	5.425	5.511
6	12.75	13.03	7.250	7.623

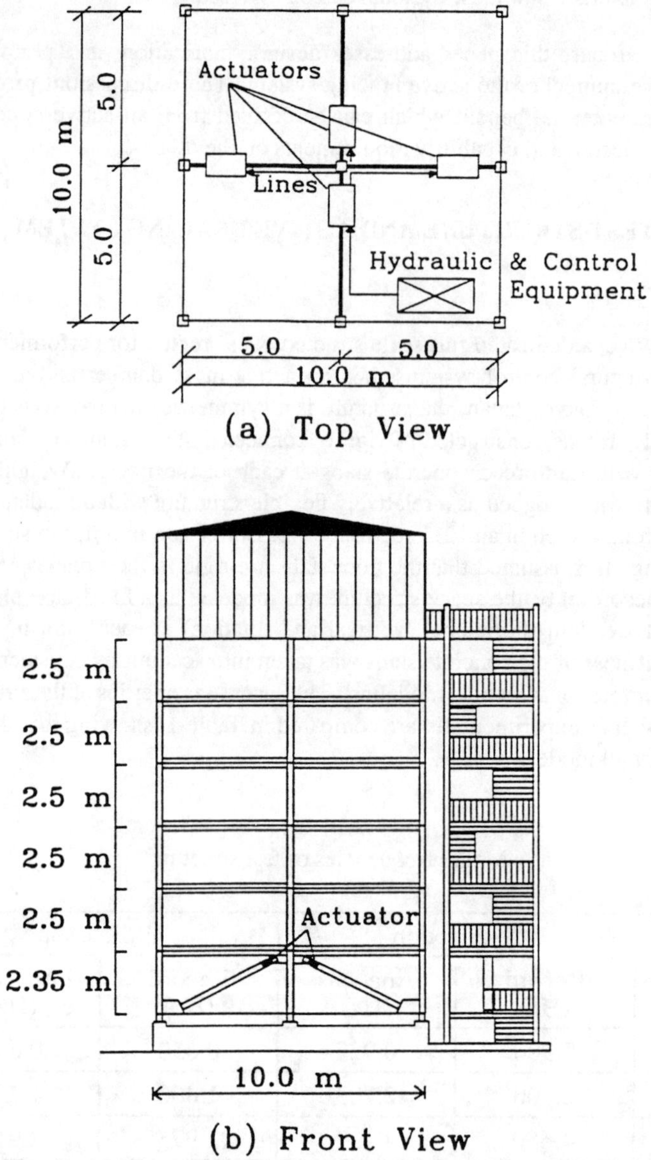

Figure 1. Test structure and active bracing system

Active Bracing System (ABS)

As shown in Fig. 1, the biaxial active bracing system consists of a pair of hydraulic actuators in each of the principal directions with the diagonal braces connected to the first floor. The actuators, monitored by two hydraulic servovalve controllers with auxiliary hydraulic accumulators, are capable of generating 685 kN of control force in each direction.

Velocity sensors are attached to the base, the first, the third and the sixth floors and an accelerometer is placed at the base. Measurements derived from these sensing devices serve as the feedback state variables in either of the two proposed control algorithms. The PC-Limited 286 microcomputer facilitated with A/D and D/A boards is used for on-line computation. The control logic is implemented on an 80286-24 MHz processor and an 80287-10 MHz co-processor.

Design Earthquakes

For design purposes, the peak velocity of the design earthquakes was taken to be 10 cm/sec based on local seismic records of the past seven years (maximum = 9.5 cm/sec). Accordingly, the scaled (32%) El Centro earthquake with 98 cm/sec^2 (0.1 g) peak acceleration was determined as the design earthquake which corresponds to the criterion of 10 cm/sec maximum velocity. Response analyses were also carried out using a series of recorded earthquake time histories to verify the adequacy of the design specifications.

Control Algorithms

The real-time computations in the microcomputer were based on a reduced order model (ROM) active control algorithm. Two basic algorithms based on the classical closed-loop linear optimal control law [2] were designed for this investigation: (a) three-velocity feedback which uses only velocities measured at three critical floors in the structure [5] and (b) velocity feedback with observer which accounts for full dimensional state feedback with aid of a state estimator [5].

Both algorithms were modified to include the compensation of delays in the force application with respect to the measured signal. The time delays are a result of phase shifts in instruments, conditioners, time required for on-line computation and time required for the actuator to respond. A phase type compensation was employed in connection with compensation of a parallel time delay in computations and actuators. All compensations were based on field measured performances of the control components [2].

Simulated Results

Simulations using several design earthquakes produced satisfactory results as shown in Table 2. The maximum power required during an earthquake reaches 20 kw for most earthquakes except for Hachinohe earthquake where 50 kw are required. The total energy consumption during an earthquake of one minute will be of the order of 0.3 kwh to 0.9 kwh which can be delivered from passive electrical power sources. Additional power should be considered for the continuous operation of changing electrical and hydraulic equipment.

TABLE 2
Summary of response analysis under design earthquakes

Earthquakes	Scale factor		Performance indices: Top fl. disp. (cm)	Top fl. acc. (g)	Base shear (kN)	Resource requirement / Actuator: Control force (kN)	Actuator disp. (cm)	Max. flow rate (gpm)	Ave. flow rate (gpm)	Vol.of accum. (gal.)	Total volume of oil (gal.)	Power (kW)
El Centro	0.32	Unctrl Ctrl Red.(%)	7.98 4.66 41.6	0.37 0.23 36.5	1018.3 618.0 39.5	313.9	0.51	19.72	2.7	0.11	0.74 in 20 s	20.02
Miyagioki	0.68	Unctrl Ctrl Red.(%)	5.30 3.27 36.4	0.43 0.21 50.7	847.6 547.4 35.8	245.3	0.40	15.24	2.0	0.12	0.71 in 30 s	16.08
Mexico (N90W)	0.65	Unctrl Ctrl Red.(%)	9.70 5.80 40.2	0.32 0.18 44.2	1347.9 877.0 35.0	255.1	0.41	6.21	1.2	0.18	1.33 in 100 s	5.65
Mexico (S00E)	1.115	Unctrl Ctrl Red.(%)	9.95 6.99 29.8	0.34 0.24 30.4	1389.1 1000.6 28.2	323.7	0.53	9.36	1.5	0.34	2.13 in 100 s	10.77
Pacomia Dam (S16E)	0.095	Unctrl Ctrl Red.(%)	5.58 3.90 30.1	0.35 0.17 50.7	741.6 500.3 32.7	255.1	0.42	10.94	2.0	0.06	0.47 in 20 s	10.06
Pacomia Dam (S74W)	0.104	Unctrl Ctrl Red.(%)	4.14 2.68 35.1	0.30 0.14 52.6	606.3 406.1 33.0	176.6	0.29	9.35	2.0	0.055	0.44 in 20 s	4.88
Taft (N21E)	0.715	Unctrl Ctrl Red.(%)	7.64 4.36 42.9	0.63 0.30 53.3	1083.0 535.6 50.5	333.5	0.54	18.29	3.0	0.12	0.96 in 20 s	21.48
Tokyo	1.48	Unctrl Ctrl Red.(%)	4.85 3.73 23.2	0.33 0.26 21.1	623.9 406.1 34.6	333.5	0.54	12.63	2.7	0.07	0.37 in 10 s	21.60
Hachinohe	0.60	Unctrl Ctrl Red.(%)	17.93 9.48 47.1	0.79 0.41 48.6	2731.1 1471.5 46.0	696.5	1.13	26.74	6.0	0.17	1.51 in 20 s	59.02
Hachinohe (limited Ctrl force)	0.60	Unctrl Ctrl Red.(%)	17.93 12.65 29.4	0.79 0.69 12.5	2731.1 1660.8 39.2	333.5	0.54	31.04	4.0	0.17	1.21 in 20 s	55.54

A typical structural response of the full-scale system is shown in Figure 2. The performance is shown for the top floor of the structure where both displacements and accelerations are reduced. As a result, the base shear is also substantially reduced. The required resources to produce such performance are shown in Figure 3 (for the strong direction of the structure for both actuators operating in this direction). The large forces are required for short period of times associated with several power peaks and a rapid initial energy demand.

DESIGN OF ACTIVE BRACING SYSTEM (ABS)

The primary parameters upon which the detail system design is based are the control force, the actuator displacement and the actuator velocity, which are related to the determination of the actuator capacity, the cylinder stroke, and the flow rate requirement of the hydraulic servovalve. In addition, the total flow of the hydraulic fluid required during a seismic event is the basis of sizing the hydraulic supply system. During the operating period of the active system, the hydraulic servo-controller supplies a constant flow of oil regardless of the actual requirement. When the demand of oil is more than what is supplied, the accumulator will pressure a subsidiary flow to the actuator cylinder; inversely, when supply is more than demand, excess oil will be discharged into the accumulator, thus no additional pumping is necessary during the short period of earthquake excitation. For economic reasons, the pumping rate of the controller is determined to be the average flow rate estimated over the time history of the cumulative flow such that the accumulator volume is minimized. These design parameters

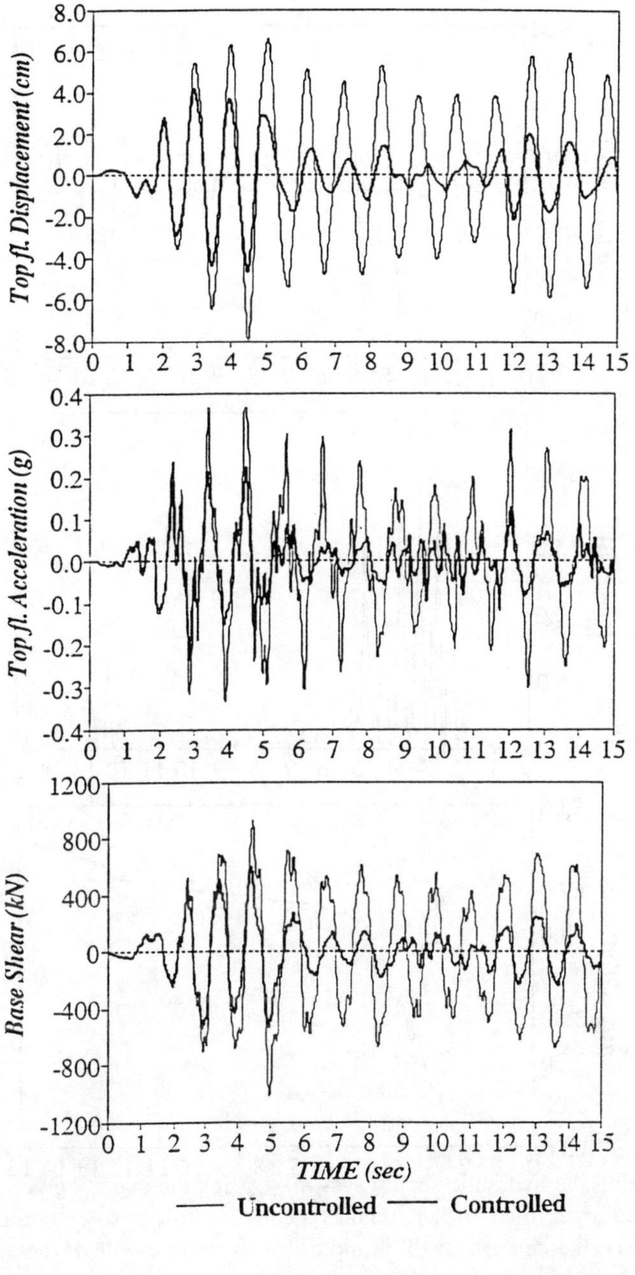

Figure 2. Structural response under 32% El Centro earthquake

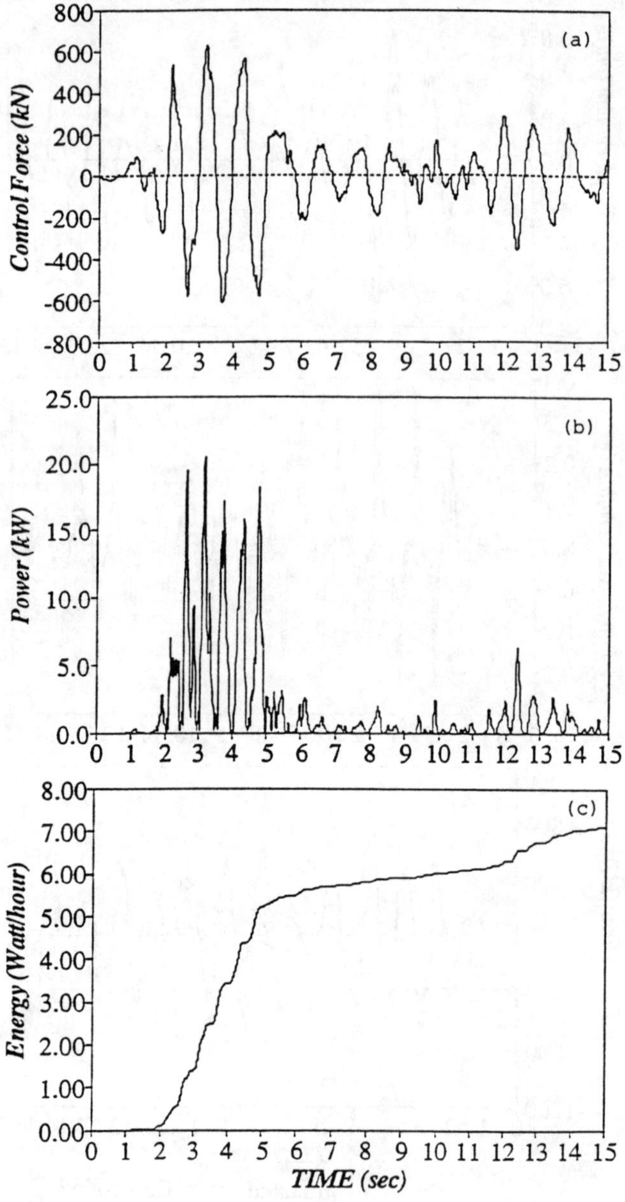

Figure 3. Control requirements under 32% El Centro earthquake

were established in accordance with the simulated results of resource demands required for a desired structural performance under the design earthquake as discussed below.

Preliminaries

The maximum control force and the specification of the hydraulic actuators were determined based upon the experimental study of a 1:4 scaled model. The fundamental frequency of the 1:4 scaled model was about 2.5 Hz and the time scale of the ground motions used in the tests was a reduction by a factor of 1/2 of the prototype, which made the model dynamically similar to the full-scale structure having a fundamental frequency of approximately 1 Hz and subjected to the design earthquake. Therefore, the expected maximum control force of the full-scale system can be estimated to be proportional to the dimensions of the structure and the intensity of the ground motion.

The maximum control force normalized with respect to the weight of the structure was 9.5% for the 1:4 scaled model associated with a 40% reduction in relative displacement under the 25% El Centro earthquake with an appropriate change in time scale. It was assumed in the scaling of the active system that, for the same degree of reduction in response, the force to weight ratio remained the same for the full-scale structure if it is subjected to earthquakes of the same intensity as used in the scaled model tests. Accordingly, the expected maximum control force for the full-scale model weighing 600 metric tons was estimated to be about 712 kN under 32% El Centro earthquake. Based on these estimates and the available manufactured units, two 35-ton hydraulic actuators were chosen in each direction. In other words, the active system can provide 685 kN maximum control force uniaxially.

Analysis and Design

Having determined the specification of the hydraulic actuators, further analysis of the control system can be performed based on the associated parameters. A stiffness of the diagonal brace (including the actuator) of 61.3 kN/mm was assumed in the simulation in order to estimate the actuator displacement and velocity. Changes in the structural eigenproperties due to implementation of additional braces are insignificant, therefore the influence of bracing on the structural dynamics was neglected in the analysis. The flow rate requirement of the actuator or the hydraulic servovalve was in turn calculated from the piston area of the cylinder and the actuator velocity based on a first-order approximation. Analytical simulations were performed only in the strong direction for design purposes and the same design specifications were used in the other direction.

Design of Passive Power Resource. The required flow rate of the hydraulic cylinders can be determined approximately in terms of the piston area and the actuator velocity. Figure 4 illustrates the cumulative flow accumulated during an earthquake, which is obtained by integrating the time history of the flow rate. The slope of this curve represents the instantaneous flow rate required to achieve the control goal. It is observed from interpreting the slope of the curve that system demand is the highest between 2 and 5 seconds and less so over the rest of the time history, a property apparently resulted from the nonstationary nature of the earthquake motion and the control effect contributed in the previous time period. The linear curve represents the cumulative volume of a constant flow which is obtained by minimizing the difference between the demand and supply of oil using the least-square criterion. The largest difference between the cumulative flow and the average flow indicates the minimum volume of the hydraulic accumulator to be considered in design.

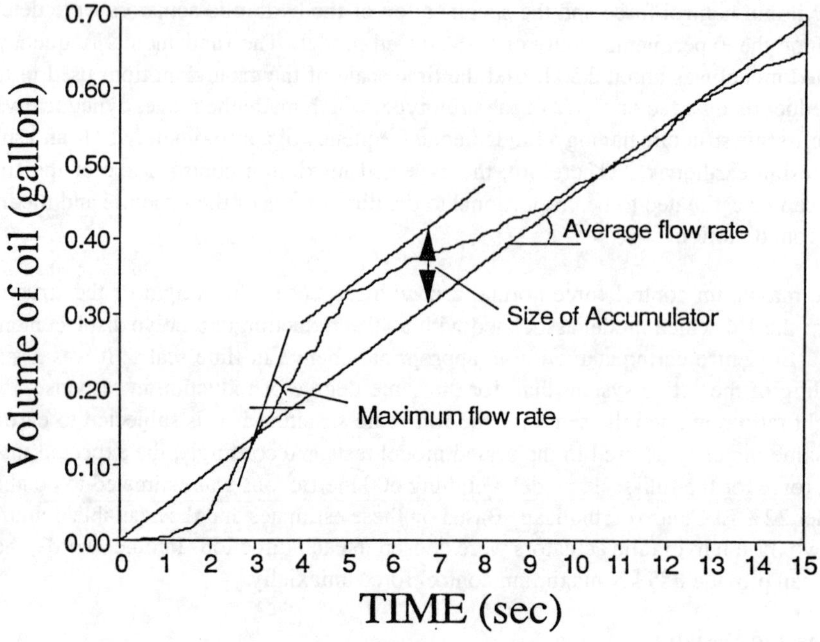

Figure 4. Cumulative oil flow during 32% El Centro earthquake

Specification

Braces. The design of the braces is based on the maximum control force and the anticipated stiffness with the assurance that buckling will not occur under actuator actions. Circular steel tubes were used as bracing members with the following specifications: length = 360.5 cm, diameter = 165.2 mm, thickness = 4.5 mm and strength = 564 kN.

Hydraulic Actuators. Four units of PARKER, heavy duty hydraulic cylinder series 2H - style TC (NFPA style Mx2) were selected as controllers with the following specifications: length = 73.5 cm, piston diameter = 152.4 mm, rod diameter = 63.5 mm, stroke = ±5 cm, and average capacity = 344 kN.

The average capacity of the actuator is calculated from the working pressure (3,000 psi) of the hydraulic oil and the average piston area, i.e., average of the piston area on one side and the same area minus the rod area on the opposite side of the piston. The capacity can be improved by increasing working pressure of the hydraulic oil.

Hydraulic Servovalve Manifold. A pair of hydraulic actuators is monitored by a unit of servovalve controllers in each direction. Hydraulic accumulators are provided to regulate oil supply without additional pumping during earthquake motions (within 1 min. in general). Only 1/3 of the accumulator volume contains hydraulic oil, the rest is filled with nitrogen. The mechanical system gains substantial pressure with only a slight volume change of the nitrogen. A margin was added in sizing the accumulator, in addition to the calculated quantity shown in Table 2, to account for energy loss due to pressure drop and error due to the first-order approximation, respectively.

Two units of MTS model 252.2x servovalves with flow rating of 15 gpm (55 lpm) at 1000 psi pressure drop were used accompanied with two five-gal. pressure accumulators and two two-gal. return accumulators.

Verification

From the analytical data shown in Table 2, it was found that the design was appropriate for almost all earthquake records used except for the case of Hachinohe earthquake. In that case, it requires a maximum control force of 696.5 kN which is far beyond the design capacity of the system if the same control strategy is to be used. Investigation was made by restricting the output control force within the design capacity of the actuator while using the same feedback gains. Results show less reduction of structural response (about 30% in displacement and 47% otherwise) when the control force is restricted to 333 kN per actuator whereas stability of the mechanical system is preserved.

This situation should also be taken into account in the on-line control practice due to erratic nature of earthquake ground motions.

EXPERIMENTAL PERFORMANCE OF CONTROL SYSTEMS

Recently, the construction was completed and initial tests were performed on the site. The tests included three stages: (a) control of microtremor excitation due to traffic and wind; (b) control of free vibration of the structure after excitation produced by the use of the active mass damper as a force exciter; and (c) control of forced vibration using the active mass damper exciter. Typical test results are presented herein as an illustration of the system performance.

Figure 5(a) shows the uncontrolled structural response in both directions. The decayed rate of the uncontrolled response is very slow because the system is only slightly damped (1% in the strong direction and 0.5% in the weak direction). Figure 5(b) shows the build-up of the structural vibration and then the influence of control applied at the peak of excitation. The increase in damping is evident, showing, for example, an increase from 0.5% to 3.5% in the weak direction. The influence of control during forced vibration is shown in Fig. 5(c). The response builds up in the transient stage and reaches its steady state, which is evidently smaller than the previous tendency as shown in Fig. 5(b).

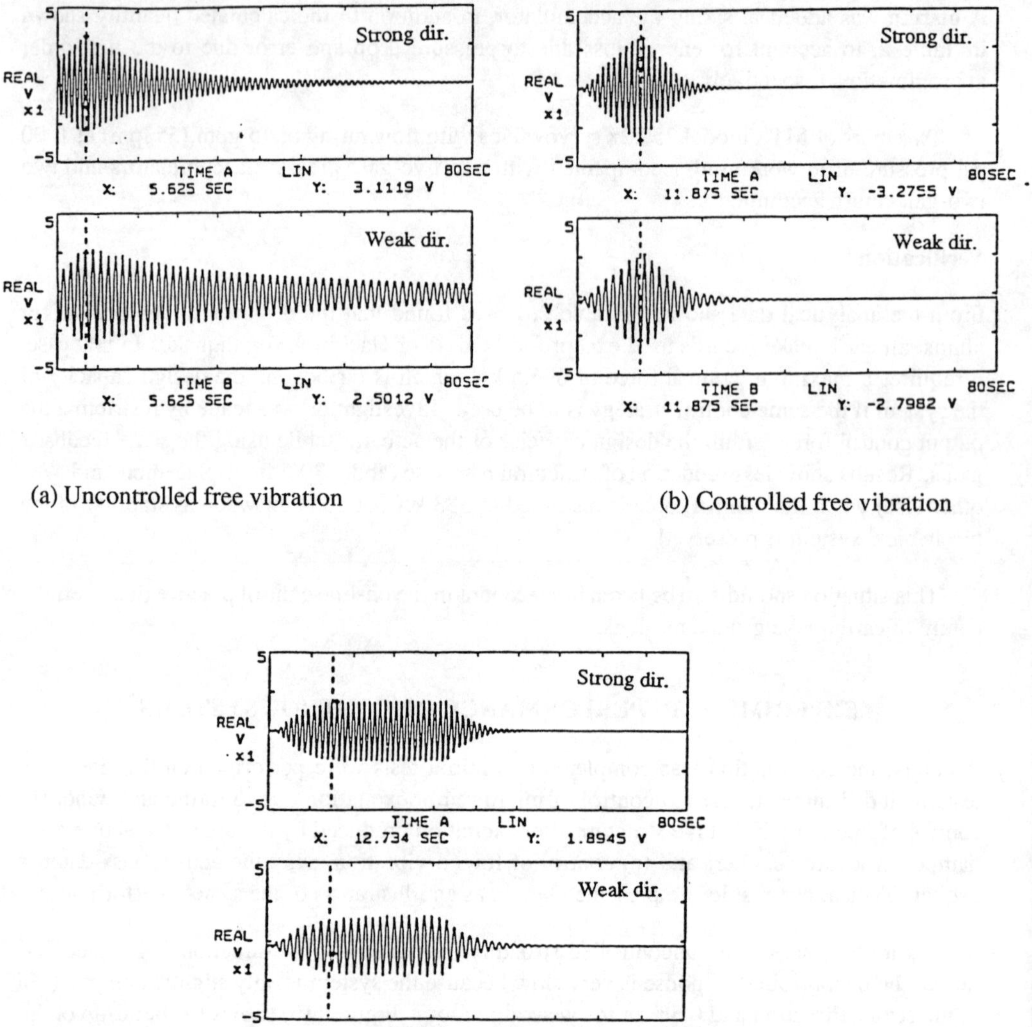

(a) Uncontrolled free vibration

(b) Controlled free vibration

(c) Controlled forced vibration

Figure 5. Preliminary test results of full-scale structure

CONCLUSIONS

Presented in this paper are design, simulation and performance of a full-scale active bracing system which was recently installed in a full-scale test structure. All components of the system were selected from currently available hardware and the power and energy resources required for this structural application were found to be within limits of current technology.

Analytical simulation and initial structural tests indicate that the system can produce significant structural response reduction. While more widespread applications of active structural control systems must await more extensive analytical and experimental verification, results obtained to date have been very encouraging and more full-scale demonstrations are expected to confirm these findings.

ACKNOWLEDGEMENT

This work was supported in part by the National Center for Earthquake Engineering Research under Grant Nos. NCEER-89-2201 and NCEER-90-2201. The authors are indebted to MTS Systems Corporation, Takenaka Corporation and Kayaba Industry Co., Ltd. for their technical assistance, consultation and financial support.

REFERENCES

1. Kobori, T. (1990), "State-of-the-art of Active Control Research in Japan," in *Structural Control Research - 1990*, eds. G.W. Housner and S.F. Masri, USC Publication, Los Angeles, CA, 1990, pp. 1-21.

2. Reinhorn, A.M., Soong, T.T., Lin, R.C. and Wang, Y.P. (1989), "1:4 Scale Model Studies of Active Tendon Systems and Active Mass Dampers for Aseismic Protection," Report NCEER-89-0026, National Center for Earthquake Engineering Research, Buffalo, NY.

3. Soong, T.T. (1990), "State-of-the-art of Active Control Research in the U.S.," in *Structural Control Research - 1990*, eds. G.W. Housner and S.F. Masri, USC Publication, Los Angeles, CA, 1990, pp. 48-65.

4. Soong, T.T. (1990), *Active Structural Control: Theory and Practice*, Longman, London and Wiley, NY.

5. Soong, T.T.,Reinhorn, A.M., Wang, Y.P. and Lin, R.C. (1991), "Full-scale Implementation of Active Control, I. Design and Simulation," *ASCE J. Structural Engineering* (in press).

HYBRID ISOLATION SYSTEM USING FRICTION-CONTROLLABLE SLIDING BEARINGS

Soichi Kawamura
Leader of Earthquake & Wind Engineering Group/Technology Research Center, Taisei Corporation, Yokohama, Japan

Masanobu Shinozuka
Director/National Center for Earthquake Engineering Research, State University of New York at Buffalo, Buffalo, NY, USA

Shunji Fujii
Research Engineer/Technology Development Department
Taisei Corporation, Tokyo, Japan

Qing Feng
Research Assistant/Department of Civil Engineering and Operation Research, Princeton University, Princeton, NJ,USA

ABSTRACT

A sliding isolation system is recognized as one of reliable base isolation technologies. Such a sliding system, however, has some limitation in its isolation capability ; its efficiency is somewhat restricted against small to medium earthquakes, and its sliding displacement during severe earthquakes can be unacceptably large.

To improve the isolation capability, a hybrid isolation system using a friction controllable sliding bearings has been proposed. By controlling the friction force with hydraulic mechanism, the sliding displacement will be confined within an acceptable range, while keeping the overall isolation preformance to reduce the response acceleration optimal under the circumstances. Friction controllable sliding bearings, hydraulic pressure control system, and control software have been developed.

A model hybrid system has been constructed and tested on a shaking table. From the test results and computer simulation analysis, the response characteristics and controllability have been identified, and the advantage of this system over the passive system has been demonstrated. Application of this system to a bridge structure has also been studied.

INTRODUCTION

In a sliding isolation system, a structure supported entirely by sliding bearings will experience the force that is always bounded by the frictional force, regardless of the level of ground acceleration or its frequency content. Such a sliding isolation system, however, has some limitation in its capability ; notably, (1) it is not necessarily efficient against weak to moderate earthquake motions, (2) it tends to suffer a large sliding displacement during severe earthquakes.

The objective of this research is to develop a hybrid isolation system that uses friction-controllable sliding bearings, wherein by actively controlling the friction force, the sliding displacement is bounded in an acceptalbe range, and yet the isolation performance is much improved against weak to moderate earthquake motions.

The friction controllable sliding bearing, which is named as Variable Friction Bearing (VFB), is a carved out steel disk with a sliding interface outside and a fluid chamber inside. The VFB consists of two parts to support the vertical load : the sliding interface and the fluid pressure. The fluid pressure and its resulting uplift force are controlled to change friction force.

A model hybrid isolation system consisting of multiple VFBs, a pressurizing system, and a controller has been developed and tested on a shaking table. The test results and computer simulations have made clear the response characteristics of the system, and also demonstrated its superiority over a passive system. Finally, an application of the hybrid isolation system to a bridge has been considered. Computer simulations with design earthquakes have demonstrated an improved isolation capability of the proposed system.

HYBRID ISOLATION SYSTEM

The composition of the hybrid isolation system by VFB as used for a buiding is schematically shown in Fig.1. The structure rests on the VFBs and a restoring force device is placed between the structure and ground. A pressurizing system, which consists of a pump, an accumulator, and a servo-valve, is connected to all VFBs. Based on the measurement of accelerations and sliding displacement, the controller computes an 'optimal' pressure according to a control algorithm and sends a corresponding control signal to the pressurizing system.

By controlling the friction force of the VFBs in this way, the isolation preformance is enhanced in reducing the response acceleration and at the same time keeping the displacement in an acceptable range.

A simplified section view of the VFB is shown in Fig. 2. The VFB is a carved out steel disk with a sliding material placed along its perimeter. Inside the VFB is a fluid chamber which is sealed at the inner edge of the sliding material with a rubber O-ring

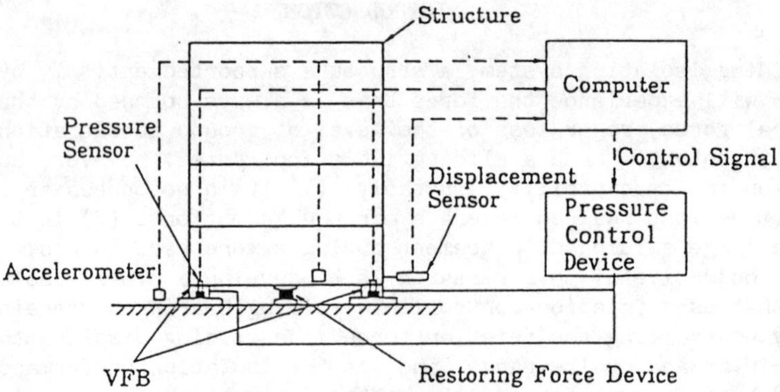

Figure 1. System Composition

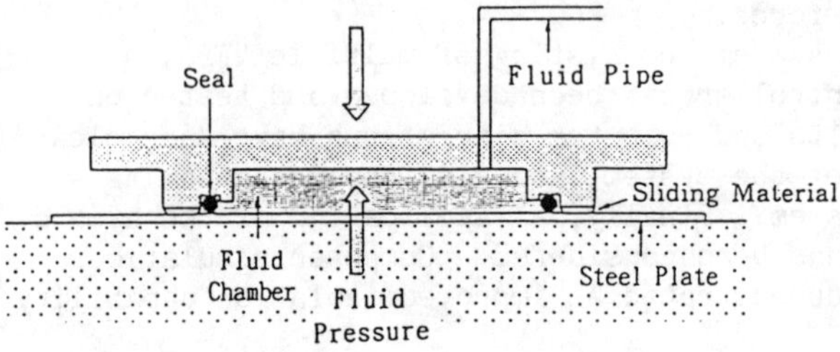

Figure 2. VFB
(Variable Friction Bearing)

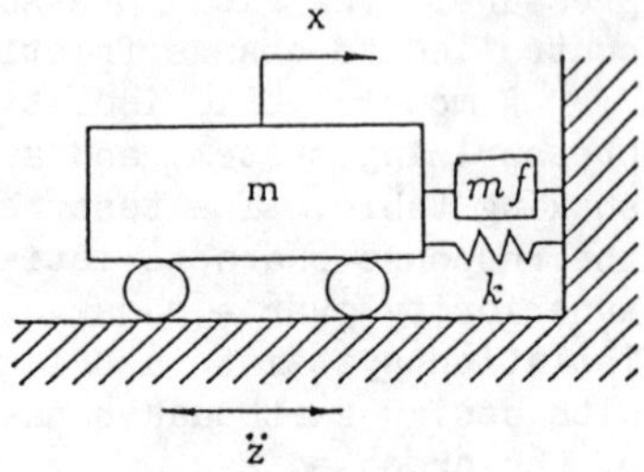

Figure .3 Analytical Model

This isolation system is designed so as to fulfil the design criteria; for example to keep the base shear coefficient

(i) less than 3% in order to maintain functional integrity of the structure against weak to moderate earthquake motions with maximum acceleration up to 100Gal,

(ii) less than 10% in order to maintain structural integrity, limit sliding displacement in an acceptable range and minimize the residual displacement against severe earthquake motions with maximum velocity 25 kine, and

(iii) less than 15% to ensure structural integrity and confinement of the sliding displacement within an allowable limit against extreme earthquake motions with maximum velocity 50 kine.

CONTROL ALGORITHMS

Analytical Model

A simple one-degree-of-freedom model shown in Fig. 3 is considered. For such a model the equation of motion during an earthquake excitation can be written as follows :

(i) No sliding - Phase I

$$\dot{x}(t)=0, \quad x(t)=\text{constant} \tag{1}$$

(ii) Sliding - Phase II

$$m\ddot{x}(t)+kx(t)=-m\ddot{z}(t)-mf\cdot \text{sgn}(\dot{x}(t)); \; f=\mu_d g \tag{2}$$

(iii) Changing condition - (Phase I → Phase II)

$$m|\ddot{z}(t)|>\mu_s mg \tag{3}$$

(iv) Changing condition - (Phase II → Phase I)

$$\dot{x}(t)=0 \; ; \; |m\ddot{x}(t)+kx(t)+m\ddot{z}(t)|\leqq\mu_s mg \tag{4}$$

where

x : displacement of the mass relative to the ground
$\ddot{z}$: ground acceleration
μ_s : static friction coefficient
μ_d : dynamic friction coefficient
m : mass
k : spring constant
g : acceleration of gravity

The dynamic characteristics of the pressure-controlling device are approximated by

$$T\dot{p}(t)+p(t)=u(t) \tag{5}$$

where

p : pressure of chamber
u : pressure control signal
T : time constant

The relationship between the pressure and the effective friction force is another important problem. This study so far assumes a linear relationship as follows ;

$$f(t)=-c_1 p(t)+c_2 \tag{6}$$

where c_1 and c_2 are constants.

Bang-Bang Control

Bang-bang control approach provides a simple and yet often quite effective algorithm to control the sliding displacement of a structure within an acceptable range against severe earthquake motion. The algorithm used in this study is as follows ; when the sliding displacement and velocity of the mass are in the same direction, the pressure control signal will be decreased to a minimal value to increase the friction force. On the other hand, when the sliding displacement and velocity are in the opposite direction, the pressure control signal will be increased to decrease the friction force [2].

$$u(t)=\begin{cases} u_{min} & \text{if } sgn(x)=sgn(\dot{x}) \\ u_{max} & \text{if } sgn(x)=-sgn(\dot{x}) \end{cases} \tag{7}$$

Instantaneous Optimal Control

In the bang-bang control algorithm, the sole control objective is to decrease the sliding displacement. In this case the absolute acceleration of the mass will exhibit at least a local maximum at each time instant at which the pressure is decreased to its minimum. Implementation of optimal control theory may, however, make it possible not only to cofine sliding displacement within an acceptable range but also to reduce the absolute acceleration.

The optimal pressure control signal is determined by minimizing the following time dependent objective function J(t) at every time instant t for the entire duration of an earthquake [2, 3].

$$J(t)=\alpha x^2(t)+\beta f^2(t) \tag{8}$$

In the case where spring constant k is negligible in Fig. 3, the sliding behavior of the system is expressed by the following equations, where the time delay of the pressure control system is ignored.

$$\ddot{x}(t)=-\ddot{z}(t)-f\ sgn(\dot{x}(t)) \tag{9}$$

$$f(t)=-c_1p(t)+c_2 \tag{10}$$

$$p(t)=u(t) \tag{11}$$

Minimizing the objective function, the optimal pressure control signal is obtained as follows ;

$$f(t)=K_f x(t)\cdot sgn(\dot{x}(t)) \tag{12}$$

$$u(t)=-\frac{1}{c_1}f(t)+\frac{c_2}{c_1} \tag{13}$$

where

$$K_f=-\frac{\alpha\Delta t^2}{6\beta} \tag{14}$$

SHAKING TABLE TEST

A model hybrid isolation system was assembled and tested on a shaking table. The 3D shaking table of Technology Research Center of Taisei Corporation Yokohama, Japan, was used. The shaking table is 4m×4m and can apply maximum ±1.0G acceleration in X, Y and Z direction with loads up to 20 tonf.

Structure

Fig. 4 shows the model structure, which is a rigid body consisting of a steel frame and steel weights. The total weight of the model was 12 tonf. The model structure was supported by four VFBs. The VFB is shown in Figs. 5 and 6. A brass sheet of 1 mm thickness was used as the sliding material, and an O-ring of 5.7 mm in diameter was used as the seal. The area of the sliding interface was 86.0 cm^2, and the area of the fluid chamber was 57.7 cm^2 when measured to the center of the O-ring width. The sliding plate was a stainless steel sheet attached to a steel plate. A servo-valve was located at the center of the structure from where the pressurized fluid was distributed to all VFBs. Accelerations of the shaking table and the structure, sliding displacements, and fluid pressure at each VFB and at the servo-valve were measured.

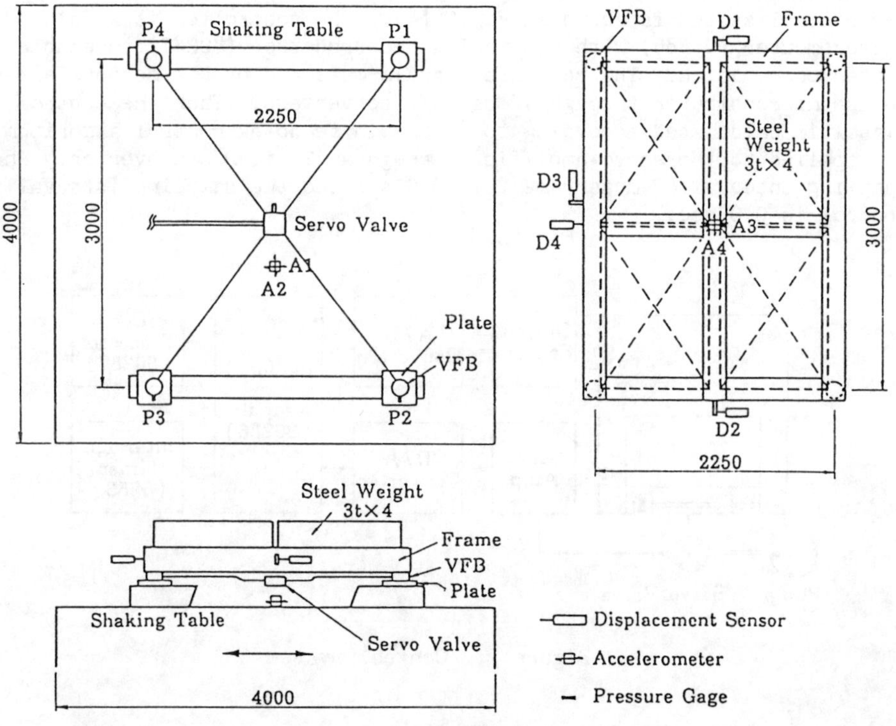

Figure 4. Model Structure

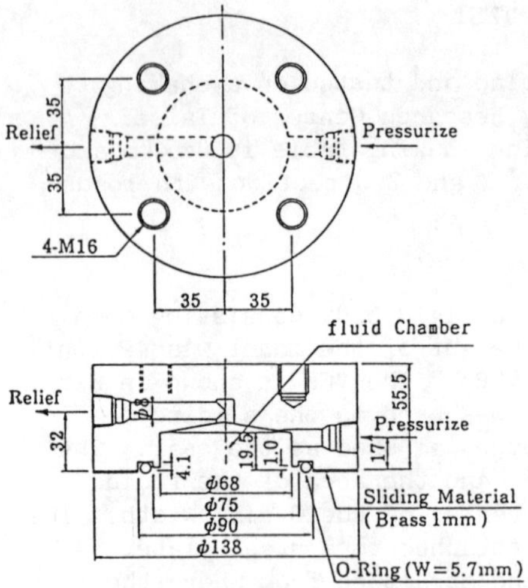

Figure 5. Model VFB

Figure 6. Model VFB

Cotrol System

The control system is shown in Fig.7 [4]. The controller is a 16 bit microcomputer (80286) with a numerical co-processor (80287) to enable faster computation. The response signals measured by the sensors are sent to the microcomputer through 12 bit A/D converters. Then the control signal is calculated according to a certain feedback control algorithm, and applied to the servo-amplifier through a 12 bit D/A converter. The sampling interval of measurement is 0.01sec and the sampling interval of control is 0.05sec.

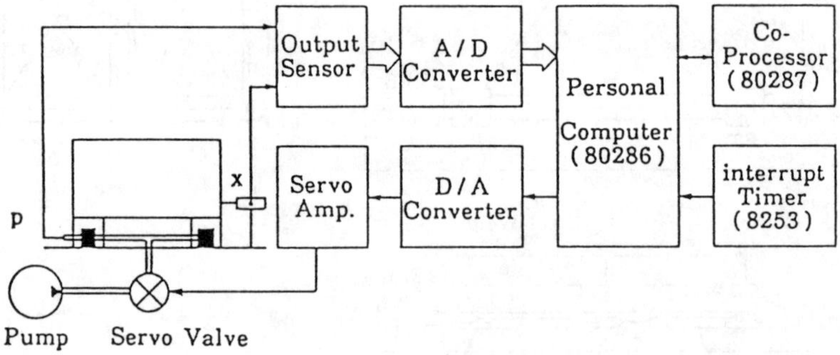

Figure 7. Control System

Identification Tests

Prior to seismic excitation tests, identification tests of the hardware response were carried out to check the relationship among control signal, pressure and friction force.

The static pressure-friction relation was obtained by a lateral loading test. The coefficient of friction was 8% at zero pressure and was reduced to 1.75% at 48 kgf/cm^2 as shown in Fig. 8.

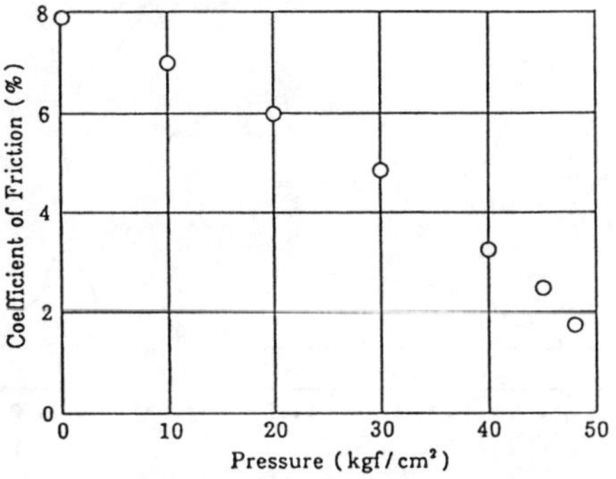

Figure 8. Static Relationship between Pressure and Friction

Fig. 9 shows the indicial response of friction force and pressure when the control signal is switched back and forth between 10kgf/cm^2 and 45kgf/cm^2, while the model structure is sliding on the shaking table moving sinusoidally at 1Hz. As pressure control signal decreases, pressure itself responds with a time delay. However, acceleration or friction force does not respond proportionally to the pressure. It increases much more slowly than the pressure. As pressure control signal increases, both of the response of pressure and friction force are faster than those when pressure control signal decreases.

The above observation indicates that the dynamic relationship between the pressure and friction force is not a simple linear relationship as suggested in Fig.8 and Eq.(6). In this study, however, linear relationship was assumed for the sake of expediences.

The relationship between pressure and control signal was identified by sending the following sinusoidal signal from the computer to the servo-valve:

$$u = A\sin(2\pi ft) + B \tag{15}$$

where A=17.5kgf/cm^2, B=27.5kgf/cm^2, f=0.5Hz~20Hz. The response of the pressure showed that in the frequency range below 10Hz the dynamic characteristics can be approximated by an model given in Eq.(5). When pressure is decreased, the time constant $T = T_d$ is larger than the time constant $T = T_i$ when pressure is increased.

From these identification tests, the parameters in the analytical model given in Eqs.(5) and (6) were determined as in Table 1.

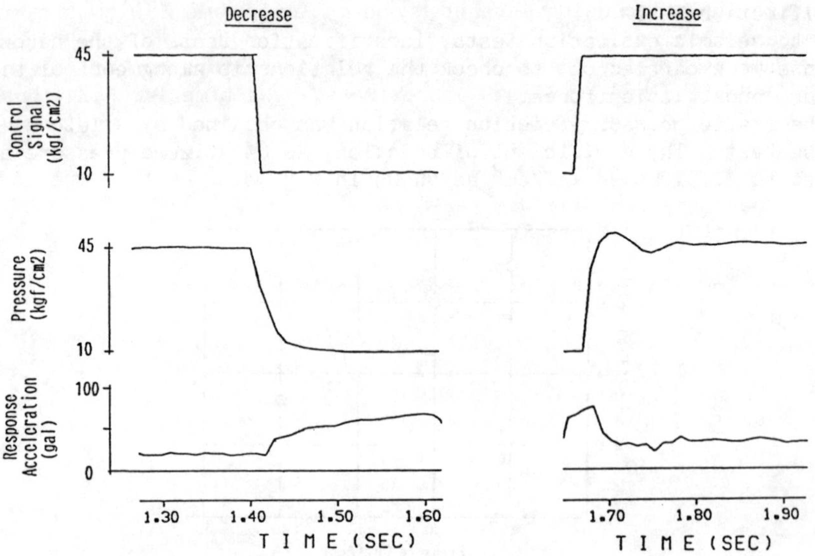

Figure 9. Step Response of Pressure and Friction to Control Signal

Table 1. Indentified Parameters

Ti	5 ms	c_1	1.29 cm³/(s²kgf)
Td	33 ms	c_2	81.49 cm/s²

Seismic Excitation Tests

In the shaking table tests, the El Centro (NS, 1940) record linearly adjusted to peak acceleration of 180 Gal was used as an input motion. The following four cases were tested.

(1) Passive isolation
The pressure control signal was kept at u=40kgf/cm^2, which corresponded to the friction coefficient of 3.3%.

(2) Active isolation using bang-bang control
The pressure control signal was switched between u_{max} = 45 kgf/cm^2 and u_{min} = 10 kgf/cm^2.

(3) Active isolation using instantaneous optimal control
The parameters in objective function (Eq.(8)) were set to

$$\alpha=1, \quad \beta=1.3*10^{-6}$$

The pressure control signal was bounded by u_{max} = 45 kgf/cm^2 and u_{min} =10 kgf/cm^2.

(4) Active isolation using instantaneous optimal control with time lag compensation

The time delay characteristics of the pressure control device as described by Eq.(5) was approximated by a time lag relationship:

$$p(t) = u(t-\tau) \qquad (16)$$

where the time lag was taken as τ = 50 ms, based on the system identification results described above.

Results of Seismic Excitation Tests and their Simulation

Figures 10, 11 and 12 show the results of three test cases excluding the time lag compensation case, with the corresponding computer simulation results. The analytical model used in the simulation is expressed by Eqs. (1) ~ (6) with the parameter values shown in Table 1.

The maximum sliding displacement of the hybrid control case shown in Figs.11 and 12 are fairly smaller than that in the passive case illustrated in Fig.10. The residual displacements in the hybrid control cases are nearly zero, which is much less than the amount observed in the passive case. The hybrid control has shown its efficiency in constraining the displacement. The computer-simulated displacement in each case shows good agreement with the test result.

For the hybrid isolation case, a decrease in the pressure is accompanied by an increase in the acceleration response (or friction force), as is anticipated. The amount of acceleration increase during the shaking table test, however, is generally larger than that expected by the numerical simulation. It is noted that this can be easily observed if such an increase in the friction force under seismic loading condition is larger than what is estimated from the static loading test as done in this study. An explanation for this behavior needs further investigation.

The test cases under instantaneous optimal control show somewhat better results in cotrolling response accerleration as well as sliding displacement than the test case under bang-bang control.

Time lag compensation produced little improvement in controlling the response beyond what was observed without compensation.

APPLICATION TO BRIDGES

To examine the feasibility of the application of this hybrid system to bridges, numerical simulation was coducted on an actual bridge shown in Fig. 13. A simple lumped mass model as described in Fig. 3 was cosidered. The weight of the mass (girder) was 501.5 tonf and the spring constant of restoring force device was 54 tonf/m.

A passive isolation and a hybrid isolation were compared. The coefficient of friction of the passive case was taken to be 0.10, while in the hybrid case, it was changed between 0.03 and 0.20 according to the other control algorithm than described previously.

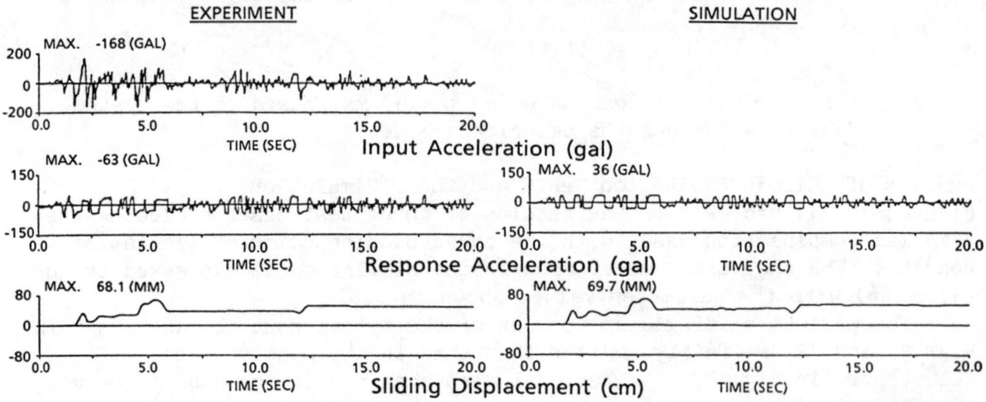

Figure 10. Experiment vs. Simulation (Passive)

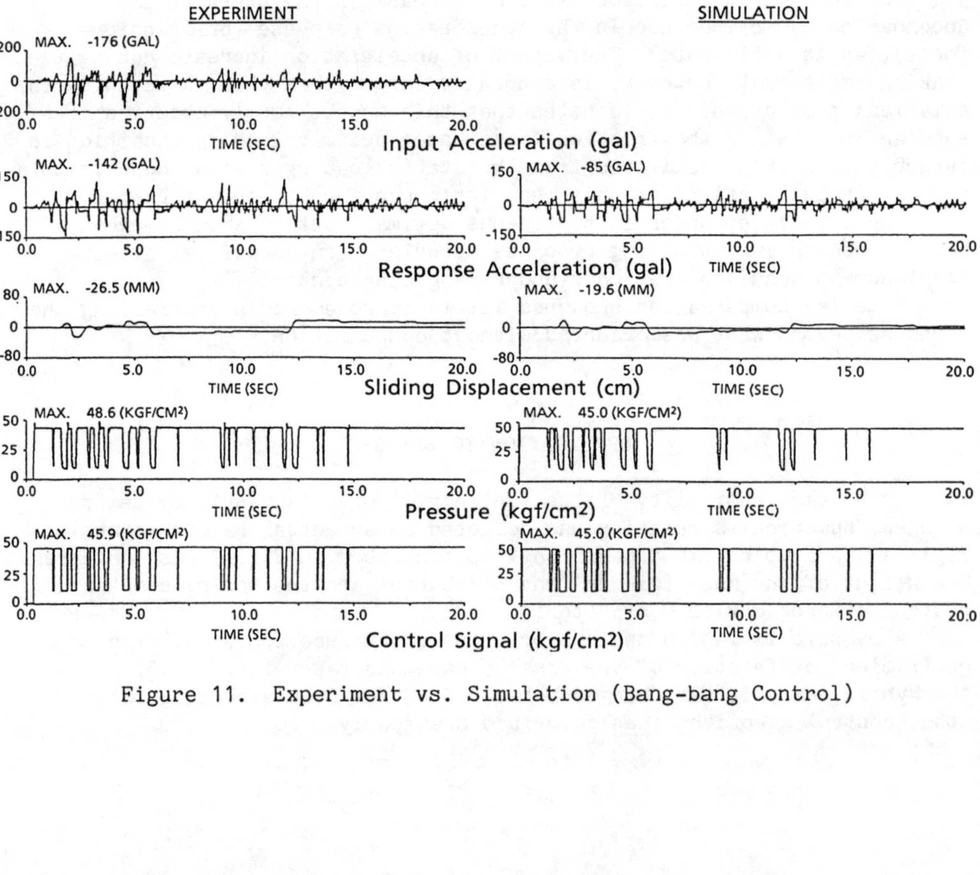

Figure 11. Experiment vs. Simulation (Bang-bang Control)

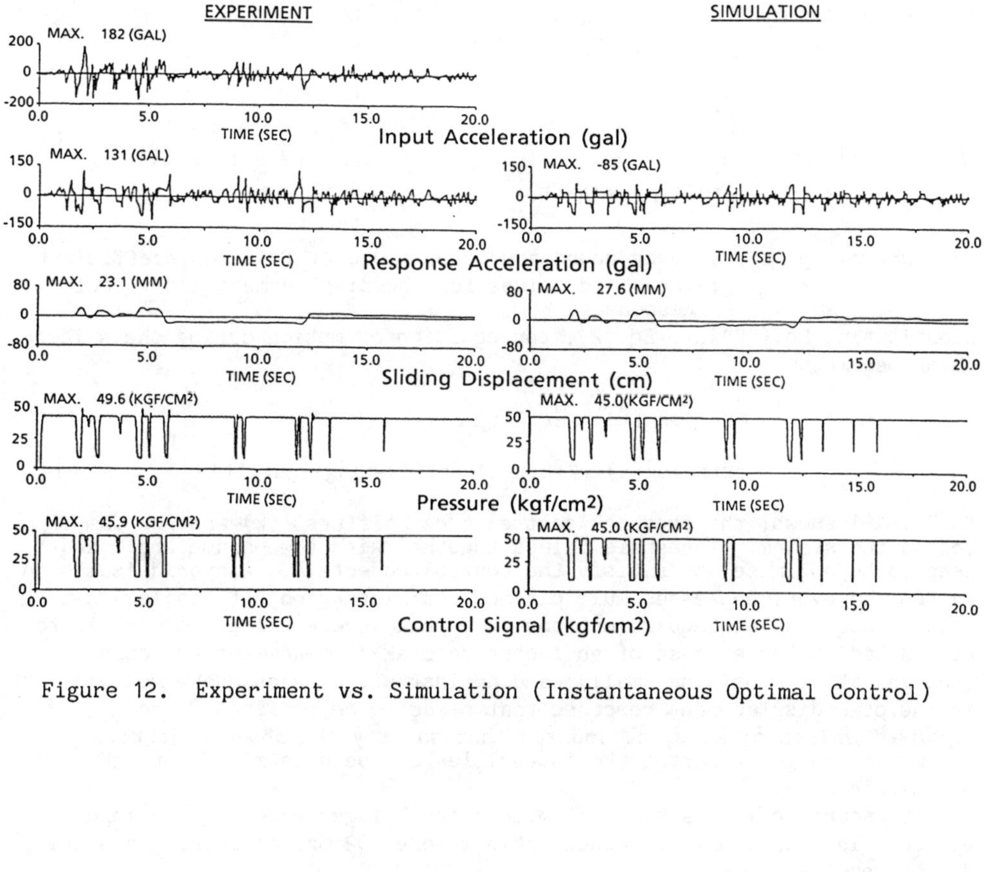

Figure 12. Experiment vs. Simulation (Instantaneous Optimal Control)

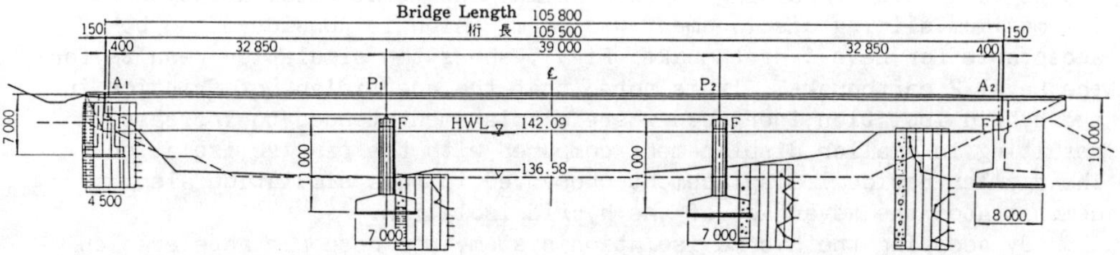

Figure 13. Analysed Bridge

The control alogorithm odopted here is based on a partial feedback linearization [7] of the equation of motion during the sliding phase as follows [6] ;

$$\mu_d(t)=\begin{cases}\mu_l & \text{if } x(t)\dot{x}(t)\leqq 0\\ \mu_l & \text{if } x_b\neq 0 \text{ and } |x(t)|\leqq x_b\\ \mu_l & \text{if } \dot{x}_b\neq 0 \text{ and } |\dot{x}(t)|\leqq \dot{x}_b\\ \min[\mu_u,\mu_l+(\bar{c}\dot{x}(t)+\bar{k}x(t))/mg] & \text{otherwise.}\end{cases} \quad (17)$$

where μ_l,μ_u : lower bound and upper bound of friction coefficient
$x_b,\dot{x}_b$: threshold values for the displacement and velocity responses

Substituting Eq.(17)into Eq.(2), the equation of motion during the sliding phase becomes

$$m\ddot{x}(t)+kx(t)=-m\ddot{z}(t)-m\mu_l g\cdot \mathrm{sgn}(\dot{x}(t)) \quad (18)$$

or

$$m\ddot{x}(t)+\bar{c}\dot{x}(t)+(k+\bar{k})x(t)=-m\ddot{z}(t)-m\mu_l g\cdot \mathrm{sgn}(\dot{x}(t)) \quad (19)$$

As Eq.(19) shows, the feedback control adds stiffness ($\bar{k}$) and /or damping ($\bar{c}$) to the system. Threshold values together with the parameters $\bar{k}$ and $\bar{c}$ need to be optimized to satisfy the control objective. A hybrid isolation system, however, possesses multiple and conflicting objectives; against weak to moderate earthquake motions, the peak acceleration response is to be reduced at the expense of an increased peak displacement response (though, still remaining small); and against severe earthquake motions, it is the peak displacement response that needs to be confined. The "optimal" values of $\bar{k}$, $\bar{c}$, x_b and $\dot{x}_b$ that satisfy the above objectives for a wide range of earthquake intensities can be determined through parametric study.

Standard design earthquake motions for bridges were used as input motions. The input maximum accelerations were 102 Gal at Level 1 and 360 Gal at Level 2.

Fig. 14 shows the simulation results for the Level 1 earthquake. The passive isolation did not work at this level and the response acceleration was equal to the input level, 102 Gal. In the hybrid isolation, on the other hand, the coefficient of friction was mostly maintained at the lowest value 0.03 resulting in the maximum acceleration of about 50 Gal. The maximum sliding displacement was 6 cm, which is considered to be acceptable for Leve 1 earthquake. Fig.15 shows the simulation results for the Level 2 earthquake. It is noted that the coefficient of friction in the hybrid isolation increased where displacement tended to increase, resulting in smaller displacement compared with the passive isolation. The smaller residual displacement obeserved in this simulation also demonstrates the advantage of the hybrid isolation.

By adopting the hybrid isolation system, the response acceleration can be reduced at the Level 1 earthquake, while reducing the maximum displacement at the Level 2 earthquake. This result shows the possibility that the hybrid isolation system can offer a better isolation capability for bridges than conventional passive isolation system.

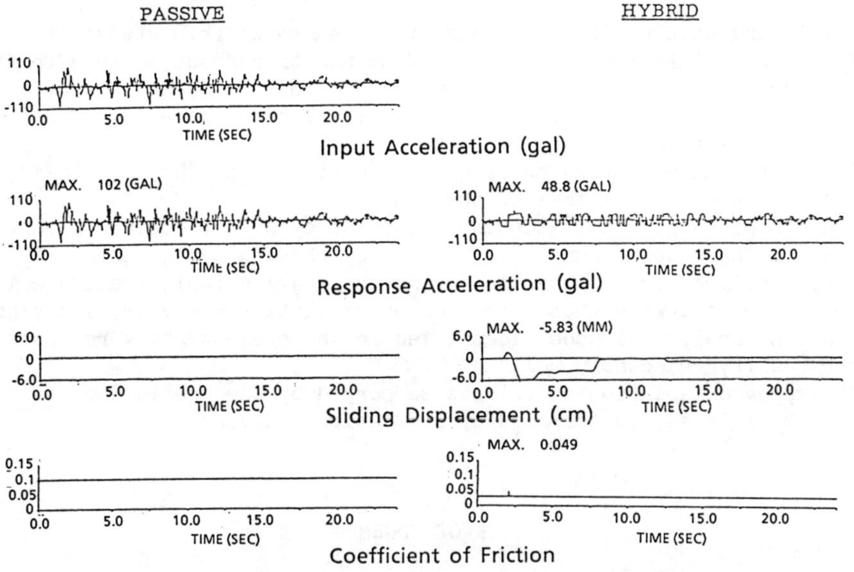

Figure 14. Response to Level 1 Design Earthquake

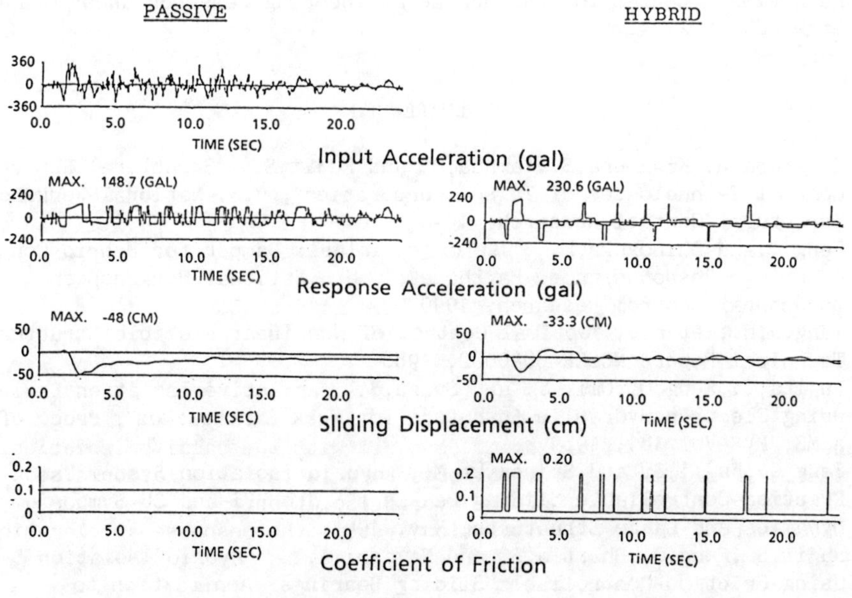

Figure 15. Response to Level 2 Design Earthquake

CONCLUSIONS

The following observations summarize the results of this study;

1) The hybrid isolation system using VFBs has turned out to be effective in reducing response acceleration against weak to moderate excitations and in eliminating the maximum and residual displacements against strong excitation.
2) The system also appears to be quite robust despite the uncertainties present in various stages of the control system.
3) The control algorithms adopted in this study are effective in confining the sliding displacement, as verified by shaking table tests.
4) Structural responses observed during the shaking table tests are in good agreement with those computed by simulation analyses, implying that the analytical model identified in the present study represents the reality reasonably well.
5) An improved isolation of bridges supported by the hybrid isolation system has been demonstrated by simulation analyses.

ACKNOWLEDGEMENT

The authors wish to acknowledge the advice and encouragement provided by professor T.Fujita of the University of Tokyo. Supervision by Dr.H.Kitamura of Taisei Corporation and co-operation by Mr.I.Nagashima, Dr.B.Bhartia and Dr. K. Hagio of Taisei are also appreciated. This work was supported partially by Taisei Corporation, Tokyo,Japan and partially by the National Center for Earthquake Engineering Research under Grant Number NCEER 90-2204.

REFERENCES

[1] Kitamura,H.,Kawamura,S.,Yamada,M. and Fujii,S., "Structural Response Control Technologies of Taisei Corporation", U.S. National Workshop on Structural Control Research, 1990

[2] Feng,Q.and Shinozuka,M., "Use of a Variable Damper for Hybrid Control of Bridge Response under Earthquake", U.S.National Workshop on Structural Control Research, 1990

[3] Yang,J.N., et al., "Optimal Control of Nonlinear Flexible Structures", Technical Report NCEER-88-0002, 1988

[4] Fujita,T.,Feng,Q.,Omi, T. and Suizu,Y., "An Active Isolation Device Using Electric-Hydraulic Actuators for Weak Earthquakes", Proc. of ASME, PVP-Vol.181, 1989

[5] Feng,Q.,Fujii,S. and Shinozuka,M., "Hybrid Isolation System Using Friction-Controllable Sliding Bearings", 8th VPI and SU Symposium on Dynamics and Large Structures, May, 1991

[6] Fujii,S.,Feng,Q.,Bhartia,B. and Nagashima,I., "Hybrid Isolation System Using Friction-Controllable Sliding Bearings -Application to Bridges-", Colloquium on Response Control, Japan Society of Civil Engineers, July, 1991

[7] Slotine,J.J. and Li,W., "Applied Nonlinear Control", Prentice-Hall, Inc. 1991

Dynamic Intelligent Buildings
-Active Seismic Response Control-

TAKUJI KOBORI
Executive Vice President of Kajima Corporation
Emeritus Professor of Kyoto University
6-5-30,KI-Building,Akasaka,Minato-ku,Tokyo,Japan

SHUICHI KAMAGATA
Senior Researcher/Kobori Research Complex
Kajima Corporation
6-5-30,KI-Building,Akasaka,Minato-ku,Tokyo,Japan

ABSTRACT

The Dynamic Intelligent Buildings (DIB) are the active seismic response controlled structures which replace the ordinary seismic structural elements with the active controllers (AC). For the feasibility study of the DIB, we develope a DIB analytical simulator, and define the DIB response spectra, the AC performance spectra, and the Performance Spectra from the analytical results. The DIB spectra of an Active Adjustable Stiffness System (A2S2) and an Active Counter Force System (ACFS) are investigated.

INTRODUCTION

The Dynamic Intelligent Buildings are explained as the advanced adaptive structures with the sensory system.
A...Kinetic Structures
This structures have a variable vibration property to escape the resonant vibration, which is diverted from the previous research of the nonlinear hysteresis property.
B...Sensory Structures [1]
This structures have a vibration measurement system. For the seismic ground motion, some industrial plants have a seismic warning system, and the elevator control system has a P-wave sensor to stop the operation.
C...Dynamic Intelligent Structures (C=A∩B) [2,3,4]
As well as the actuator system, we install the mechanical system to adjust the structural vibration property.

By the recent computer technology, the advanced control theory is applied in many kind of machines. Especially the industry of home electric machines makes the "Fuzzy Control [5,6]" popular for the Japanese in 1990. Also the active control system for the automobile is adopted for the safty and comfortable drive. Moreover the warning system for the seismic ground motion "UrEDAS" are working in the bullet train [7]. Backgrounding these technology we proposed the concept of DIB as a new research of "Mecha-st-ronics".

COMPOSITION OF DIB

The DIB is composed of four components [4].

1. Seismic Sensors: This signal is used for the decision logic of the active controllers.
2. Vibration Sensors: This signal is used to identify the current structural property.
3. Control Unit(BRAIN): An inhouse main frame or an additional computer is used as the hardware, and the decision logic is based on the recent control theory and the progressing research in the seismic engineering.
4. Active Controllers(AC): Instead of the seismic structural elements, dynamically controllable devices are installed.

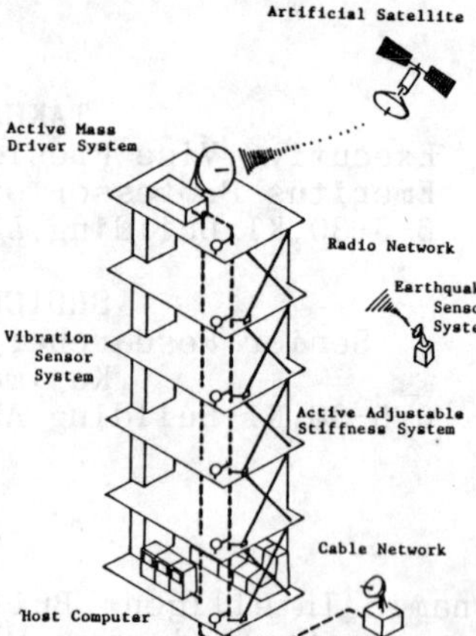

Fig.1 Concept of DIB

History of Research

	1970 1980 1990
Feasibility Study	
Theoretical Research[8,9,10,11,12]	---++++++++++++++++++
Control Effect	
Boundary of Stability	
Identification	
Experimantal Research[13,14,15,16]	+++++++++++++++++++
Active Control Systems	
Control Algorithms	
Evaluation of Performance[17,18,19]	----------+++++++++
Controllability of the AC	
Accuracy of the Sensory Systems	
Application to Structures	
Development of the AC Systems	----++++++
Reliability and Initial Cost	
Practical Application[20,21]	---++++
Evaluation of Seismic Ground Motion	
Design Criteria for Structures	
Maintenance and Running Cost	----++++

<u>PROBLEM OF DIB</u>:

Find the performance of the Active Controllers: O(t)

to minimize $J(t) = f(R(t),O(t))$ (1)

by the conditions of $R(t) = g(M,C,K,O(t),\ddot{y}(t))$ (2)

$O(t) = h(M,C,K,W,Ir(t),Is(t)$ (3)

$R(t) \leqq Rdc$ (4)

structural property matrix: M=mass, C=damping, K=stiffness

R(t): structural response O(t): operation of the AC

$\ddot{y}(t)$: seismic ground acceleration

Ir(t): information of structural response

Is(t): information of seismic ground motion

W : control condition Rdc : design criteria

DIB Analytical Simulator

To evaluate the DIB response property and the necessary capacity of the active controllers, we develope an Analytical Simulator for the DIB. As the basic structural property is numerically estimated in the design process, so we can use them as a numerical model of the structure.

$$M\ddot{x}(t)+C\dot{x}(t)+Kx(t)=-M\ddot{y}(t)+O(t) \quad (4)$$

$\ddot{x}$=Acceleration
$\dot{x}$=Velocity
x=Displacement

Based on the trapezoidal rule, we can get a integral solution for the seismic response of the controlled structure.

$$\ddot{x}(t)=E^{-1}\{Fs(t)+Fr(t-dt)+O(t)\} \quad (5)$$

$$E=M+bC+cK \quad (6)$$

$$Fs(t)=-M\ddot{y}(t) \quad (7)$$

$$Fr(t-dt)=-G_1\ddot{x}(t-dt) -G_2\dot{x}(t-dt) -G_3x(t-dt) \quad (8)$$

$G_1=cK+bC$, $G_2=aK+C$, $G_3=K$
$a=dt$, $b=0.5dt$, $c=0.25dt^2$

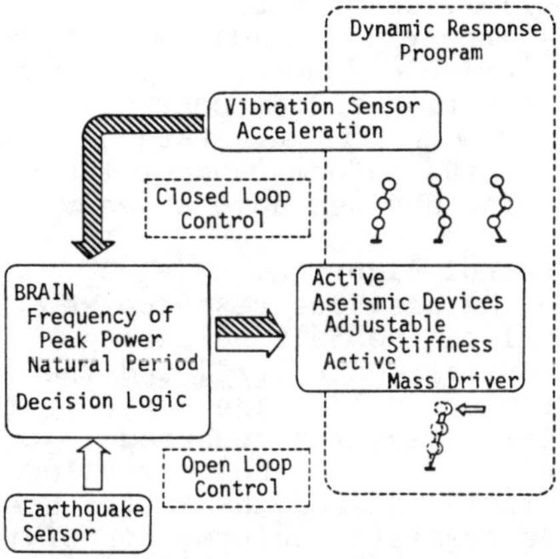

Fig.2 DIB Analytical Simulator

We adopt an iterative calculation for the nonlinear vibration when the A2S2 is installed.

Evaluation of Seismic Ground Motion

As the DIB aims to suppress the structural response in the seismic duration, so we investigate the nonstationary property of the seismic motion. For this purpose, we adopt the running power spectra ($Sr(\omega,T)$).

$$Sr(\omega,T)=\int_{-\infty}^{\infty}f(\tau,T)e^{-i\omega\tau}d\tau \quad (9)$$

ω:Frequency
T:Instant time

$$\tau=t-T+Ta \quad (10)$$

$f(\tau,T)=\ddot{y}(t)$
$T-0.5*Ta \leq t \leq T+0.5*Ta$
$f(\tau,T)=0.0$
$T+0.5*Ta \leq t \leq T-0.5*Ta+Te$
Ta:Data acquisition time
Te:FFT analytical time

For example, the NS component record of El Centro 1940 is analyzed with 5.0 seconds of the Ta-time, and the FFT calculation is conducted in the interval of 0.2 seconds with the 10.24 seconds of Te-time.

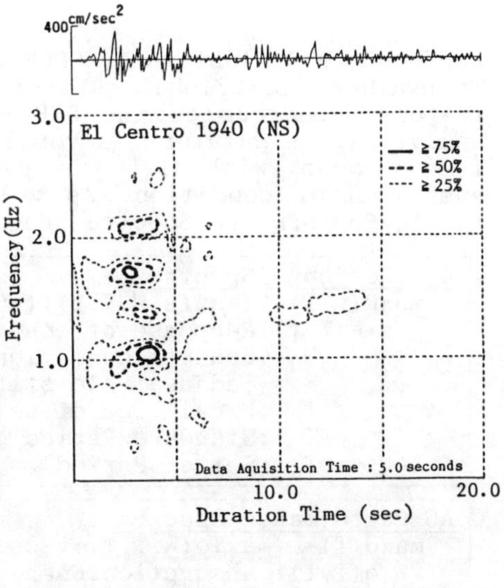

Fig.3 Running Power Spectra

ACTIVE ADJUSTABLE STIFFNESS SYSTEM

This system aims to isolate the structural period from the period of seismic dominant components actively. This detuning function is achieved by means of the mechanical adjustable stiffness devices.

Definition of A2S2

1. The Active Controller (AC) works a mechanical stiffness with a variable function. The response time of the AC is faster than the period of the controlled, structural vibration mode.
2. The stiffness state is commanded by the control unit which uses the information of the sensor system.
3. The devices have an energy absorption function.

Control Algorithm

In the previous research we derived the Self-balancing Adaptive control[22] and the Predictive Adaptive control, the former uses the measured information of the controlled structures, and the latter uses the predicted information of the seismic motion as well as the structural information.

In this paper we adopt the Self-Balancing Adaptive Control which is based on the human balance, such as straining in the state of increasing demonstration, and relaxing in the state of decreasing deformation (Fig.4).

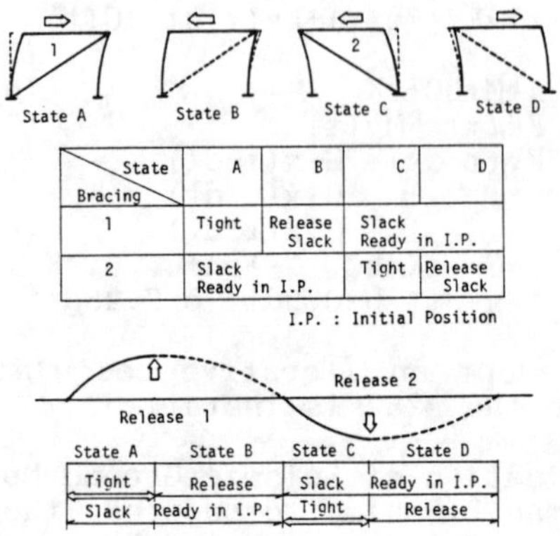

Fig.4 Self-balancing Adaptive Control

Performance Evaluation by SDOF with A2S2

To evaluate the global control effect, we apply the concept of response spectrum to the DIB structures. Namely using the DIB analytical simulator, several number of single degree of freedom model with different natural period are analyzed in the same control condition. As well as the DIB response spectra, the AC Performance Spectra are defined as follows.

<u>DIB Response Spectra</u> :

$$\max R(w,T) = (\max(s(W,T,t):t=0,\ldots,t_f) : T \subset T_G) \quad (11)$$

s(W,T,t): Response of controlled structures
W : Control condition of the AC (adjusatable stiffness range)
T_f : Final time of seismic ground motion
T : Standard Period of Controlled Structures
T_G : Set of Period

<u>AC Performance Spectra</u> :

$$\max O(T) = (\max(o(W,T,t):t=0,\ldots,T_f) : T \subset T_G) \quad (12)$$

o(T,t): Absorption Energy of Active Controllers

Analytical Conditions:

1. Controlled structures

The mass of the controlled structure is 1.0tonf*sec^2/cm, and no damping is assumed. the 20 cases of natural period are sellected from 0.2 second to 4.0 seconds.

2. Active Adjustable Stiffness System (A2S2)

Two cases of active stiffness (k_{as}) are assigned for each case of natural period.

Case-A: $k_{as}=3.0k_o$

Case-B: $k_{as}=k_o$

k_o: Fundamental Stiffness of the controlled structure

3. Seismic ground motion

The El Centro 1940 (NS) record is normalized to the maximum acceleration of 100.0 cm/sec^2.

Analytical Results:

In twenty cases of the natural period, the response of 1.0 second natural period is illustrated in Fig.5.

1. Acceleration response
2. Displacement response
3. Absorbed Energy of the A2S2
   $A.E.=\int k_{as}x(t)\dot{x}(t)dt$
4. Input Energy
   $I.E.=\int M\ddot{y}(t)\dot{x}(t)dt$
5. Stiffness state of the A2S2
   Index=1 : Hardening state
   =0 : Softening state
6. Restoring property

In these figures the solid line shows the result of the controlled system, and the dotted line shows the results of the uncontrolled system with the fixed A2S2 in the hard stiffness state.

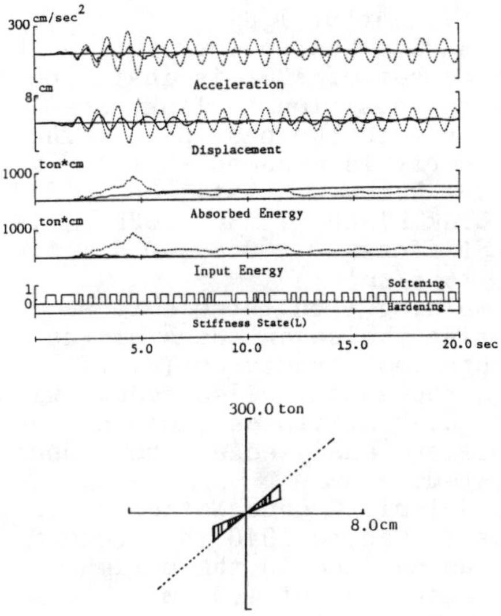

Fig.5 Response Time History

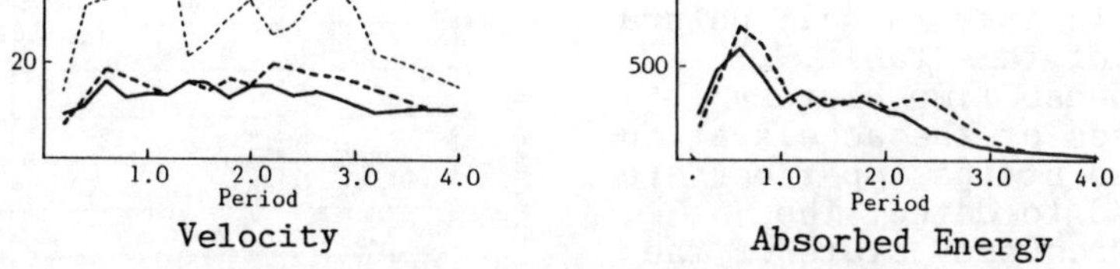

Fig.6 DIB Response Spectra and AC Performance Spectra

The DIB Response Spectra are shown with the uncontrolled linear system which is fixed in the hard stiffness state. And the absorbed energy spectra are shown as the AC Performance Spectra (Fig.6).

Performance Evaluation by MDOF with A2S2

To evaluate the control effect in the Multi-Degree of Freedom (MDOF) System with the A2S2, we developed a numerical DIB Simulator for the MDOF model.

<u>Analytical Conditions</u> :

1. Controlled Structure

Three story open frame structure is defined as the fundamental structure, which have same stiffness (1.0tonf/cm) and mass (1.0tonf*sec^2/cm) in each story. The damping property of the fundamental structure is assumed a viscous damping with the coefficent of 1.0%.

2. Active Controllers

Three set of A2S2 is installed in the position to link each floor with the base place. The property is assumed as follows.

Set Number	1st.	2nd.	3rd.
Link Place	3F	2F	1F
Stiffness (tonf/cm)	0.45	0.63	1.0

Combining with three set, we define the seven cases of the contolled structures (Fig.7). And the uncontrolled models with the high stiffness state of each case are analyzed as the linear system.

3. Seismic Ground Motion

The El Centro 1940 (NS) record is normalized to the maximum acceleration of 1.0 cm/sec^2.

<u>Analytical Results</u> :

1. Eigen Value Analysis

The natural vibration period of the fundamental and the seven uncontrolled structures are analysed (Table.1). In the control state, the case-III has a chance to take the all natural period of this table.

2. Response Time History

Two cases of the acceleration response at the top floor are analysed to detect the multi-frequency state of the vibration process (Fig.8).

(a) The controlled structure

(b) The uncontrolled structures which fix the all AC in the hard state

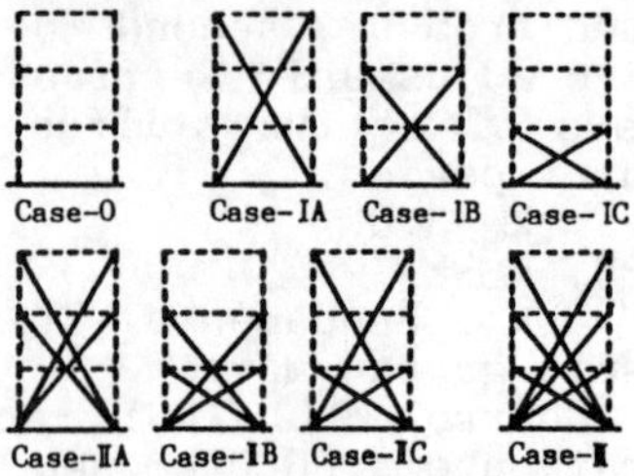

Fig.7 Placement of A2S2

Table-1 Natural Period

	1st.	2nd.	3rd.
case-0	0.451	0.161	0.111
case-IA	0.314	0.152	0.110
case-IB	0.323	0.158	0.105
case-IC	0.388	0.142	0.104
case-IIA	0.250	0.151	0.105
case-IIB	0.302	0.135	0.101
case-IIC	0.278	0.136	0.104
case-III	0.231	0.132	0.100

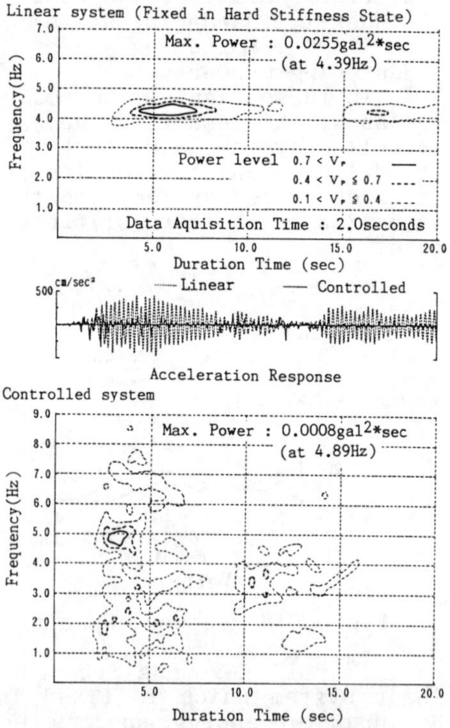

Fig.8 Property of Response

3. Maximum Response Values
Four types of the AC placement are compaired (Fig.9).
(a) Case-IA, Case-IB, and Case-IC
(b) Case-IIA, Case-IIB, and Case-IIC
(c) Case-IA, Case-IIA, and Case-IIIA
(d) Linear cases and Case-IIIA

Concluding Remarks

1. This control algorithm gives both the unresonant function by alternate two kind of frequency state in each one fourth cycles, and the energy accumulation function in each half cycles.
2. From the DIB Response Spectra and the AC Performance Spectra we can derive the design procedure of the A2S2.
(a) Find the control condition to reduce the response less than the design criteria (Rdc) at the natural period (T) of controlled structure.
(b) From the obtained control condition, find the necessary performance property from the AC Performance Spectra at the same natural period.

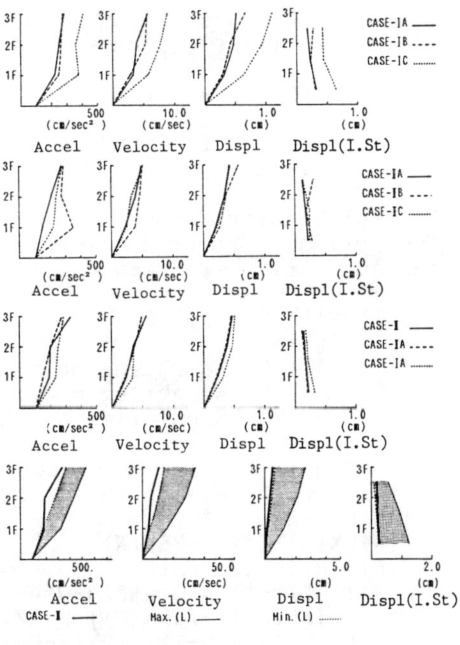

Fig.9 Maximum Response Values

ACTIVE COUNTER FORCE SYSTEM

Optimization

The counter force (Fc) is optimized by the optimality principle of the Dynamic Programming. In this optimizing process the information of the coming earthquake is required as well as the information of structural response.

State Equation

$$M\ddot{x}(t)+C\dot{x}(t)+Kx(t)= -M\ddot{y}(t)+Fc(t) \quad (13)$$

M: Mass Matrix, C: Damping Matrix, K: Stiffness Matrix
$\ddot{x}(t)$: Acceleration Response
$\dot{x}(t)$: Velocity Response
$\mathbf{x}(t)$: Displacement Response
$\ddot{y}(t)$: Seismic Ground Acceleration

$$\ddot{x}(t)=E^{-1}[\ Fs(t)+Fr(t-dt)\ +Fc(t)\ ] \quad (14)$$

$$Fs(t)= -M\ddot{y}(t) \quad (15)$$

$$Fr(t-dt)=-G_1\ddot{x}(t-dt)\ -G_2\dot{x}(t-dt)\ -G_3\mathbf{x}(t-dt) \quad (16)$$

$$E=M+bC+cK \quad (17)$$

$$G_1=bC+cK,\ G_2=C+aK,\ G_3=K$$

$$\dot{x}(t)=\dot{x}(t-dt)+b\{\ddot{x}(t-dt)+\ddot{x}(t)\} \quad (18)$$

$$\mathbf{x}(t)=\mathbf{x}(t-dt)+a\dot{x}(t-dt)+c\{\ddot{x}(t-dt)+\ddot{x}(t)\} \quad (19)$$

$$a=dt,\ b=0.5dt,\ c=0.25dt^2$$

$$\mathbf{X}(t)=A\mathbf{X}(t-dt)\ +fuFs(t)+fuFc(t) \tag{20}$$

$$A=\begin{bmatrix} A_1 & A_2 & A_3 \\ b(1+A_1) & 1+bA_2 & bA_3 \\ c(1+A_1) & b+cA_2 & 1+cA_3 \end{bmatrix} \tag{21}$$

$$A_1=-E^{-1}G_1,\ A_2=-E^{-1}G_2,\ A_3=-E^{-1}G_3$$

$$\mathbf{X}(t)=\begin{Bmatrix} \ddot{x}(t) \\ \dot{x}(t) \\ x(t) \end{Bmatrix} \quad fu=\begin{bmatrix} -E^{-1} \\ -bE^{-1} \\ -cE^{-1} \end{bmatrix}$$

Optimal Counter Force

The optimal counter force is derived as the minimum solusion of the quadratic performance function (23).

$$J(T)=\ {}^{T_c}_{o}[\mathbf{X}^T Q\mathbf{X}+\ Fc^T RFc]dt \tag{22}$$

Q: Weighting matrix for the structural response
R: Weighting matrix for the cost of the AC
T_c: Exent time of the optimization

The global optimization is derived from the final stage to the first stage in the reverse process of actual time.

(a) Nth.stage

$$J(N)=\mathbf{X}(N)^T Q\mathbf{X}(N)\ \ RFc(N)^2 \tag{23}$$

$$\begin{aligned} J(N)=&(R+fu^T Qfu)Fn(N)^2 \\ &+\{\mathbf{X}(N-1)^T A^T Qfu+2fu^T QfuFs(N)+fu^T QA\mathbf{X}(N-1)\}Fc(N) \\ &+\{A\mathbf{X}(N-1)+fuFs(N)\}^T Q\{A\mathbf{X}(N-1)+fuFs(N)\} \end{aligned} \tag{24}$$

$${}_{OP}Fc(N)=V(1)^T\mathbf{X}(N-1)+V(0)^T fuFs(N) \tag{25}$$

$$V(1)=W_{RQ}(N)fu^T W(N)A \tag{26}$$

$$V(0)=W_{RQ}(N)fu^T W(N) \tag{27}$$

$$W(N)=Q \tag{28}$$

$$W_{RQ}(N)=-(R\ \ fu^T W(N)fu)^{-1} \tag{29}$$

(b) (N-1)th.stage

$$\begin{aligned} J(N-1)=&\Delta J(N-1)+\min J(N) \\ =&\mathbf{X}(N-1)^T Q\mathbf{X}(N-1)\ \ RFc(N)^2+\min J(N) \end{aligned} \tag{30}$$

$$\begin{aligned} J(N-1)=&\mathbf{X}(N-1)^T W(N-1)\mathbf{X}(N-1)+\mathbf{X}(N-1)^T Wg(N)W(N)fuFs(N) \\ &+Fs(N)^T fu^T Wg(N)W(N)\mathbf{X}(N-1) \\ &+Fs(N)^T fu^T Wz(N)fuFs(N)+RFc(N-1)^2 \end{aligned} \tag{31}$$

$$Wf(N)=fuW_{RQ}(N)fu^T \tag{32}$$

$$W(N-1)=W(N)+A^T W(N)A+A^T W(N)Wf(N)W(N)A \tag{33}$$

$$Wg(N)=A^T\{[I]+W(N)Wf(N)\} \tag{34}$$

$$Wz(N)=W(N)\{[I]+Wf(N)W(N)\} \tag{35}$$

$${}_{OP}Fc(N-1)=V(2)^T\mathbf{X}(N-2)+V(1)^T fuFs(N-1)+V(0)^T fuFs(N) \tag{36}$$

$$V(2)=W_{RQ}(N-1)fu^T W(N-1)A \tag{37}$$

$$V(1)=W_{RQ}(N-1)fu^T W(N-1) \tag{38}$$

$$V(0)=W_{RQ}(N-1)fu^T Wg(N)W(N) \tag{39}$$

$$W_{RQ}(N-1)=-(R+fu^T W(N-1)fu)^{-1} \tag{40}$$

(c) 1st.stage

$${}_{OP}Fc(1)=V(N)^T\mathbf{X}(0)+V(N-1)^T fuFs(1)+\qquad +V(0)^T fuFs(N) \tag{41}$$

$$V(N)=W_{RQ}(1)fu^T W(1)A \tag{42}$$

$$V(N-1)=W_{RQ}(1)fu^T W(1) \tag{43}$$

$$V(N-2)=W_{RQ}(1)fu^T Wg(2)W(2) \tag{44}$$

$$V(0)=W_{RQ}(1)fu^T Wg(2)\quad Wg(N)W(N) \tag{45}$$

$$W_{RQ}(1)=-(R+fu^T W(1)fu)^{-1} \tag{46}$$

$$W(N)=Q \tag{47}$$

$$W(i)=W(N)+A^T W(i+1)A+A^T W(i+1)Wf(N)W(i+1)A \tag{48}$$

$$Wg(i+1)=A^T\{[I]+W(i+1)Wf(N)\}\ [i=N-1,\ldots,1] \tag{49}$$

We classify the optimal counter force (41) in three components.

(a) Self-balancing component : $V(N)^{T}X(0)$ (50)

(b) Instantaneous component : $V(N-1)^{T}fuFs(1)$ (51)

(c) Predictive component : $V(N-2)^{T}fuFs(2)$

+

$+V(0)^{T}fuFs(N)$ (52)

Control Algorithm

As the state vecter includes the three components, such as an acceleration, a velocity and a displacement, the weighting matrix for the structural response [Q] can be set for each components.

$$R = 1.0 \qquad (53)$$

$$[Q]=\begin{bmatrix} q(a) & 0.0 & 0.0 \\ 0.0 & q(v) & 0.0 \\ 0.0 & 0.0 & q(d) \end{bmatrix} \qquad (54)$$

From this definition three control algorithms are derived.

(a) Acceleration Control : $q(a)\neq 0.0$, $q(v)=0.0$, $q(d)=0.0$

(b) Velocity Control : $q(a)=0.0$, $q(v)\neq 0.0$, $q(d)=0.0$

(c) Displacement Control : $q(a)=0.0$, $q(v)=0.0$, $q(d)\neq 0.0$

Also we defined the other three control algorithms by restricting the extent of the predicted seismic ground motion in the optimal counter force($_{op}Fc$).

(a) Optimal Adaptive Control

Considering the structural vibration property (A) and the control condition (Q/R), the factor of optimal performance function is derived in the extent of the optimization. In this optimized counter force, the predicted information(Is) of seismic ground motion is used as well as the information(Ir)of structural vibration immediately before the instant of control.

optFc(t)=h(A,W,Ir(t- t),Is(t(i)))

t(i)=t+ t*(i-1) :i=1, ,N (55)

(b) Instantaneous Adaptive Control

The information of the predicted seismic ground motion is used in one analytical step, assuming the remaining information to be zero.

insFc(t)=h(A,W,Ir(t- t),Is(t)) (56)

(c) Self-balancing Adaptive Control

No information of the predicted seismic ground motion is used.

sbFc(t)=h(A,W,Ir(t- t)) (57)

Definition of Performance 2-Dimensional (P2D) Spectra

These counter force(Fc) is dependent on the weighting matrix, so there are numerous optimal solutions coincident with the numerous weighting matrix. On the other hand it is necessary to define the weighting matrix which makes the structural response (R(W)) to be less than the design criteria (Rdc). Therefore we define the Performance 2-Dimensional (P2D) Spectra as follows.

1. Define the two kind of set for the natural period (Tg) of controlled structure and the control condition(Wg) of the active controller.
2. Analyze each case of the combined set(Tg*Wg).
3. Find the relation between the specified counter force (maxFc,[I.Fc/I.Fs]) and the specified response (maxX(T,W),I.E.) in each natural period.

4. Draw the Performance 2-Dimensional Spectra.

$$Sp_2(T, maxFc) = [maxX(T, W) : T \subset Tg, W \subset Wg] \quad (58)$$

The X-axis : Counter Force (maxFc, [I.Fc/I.Fs])

The Y-axis : Natural Period of Controlled Structure (T)

The Z-axis : Response Values (maxX, I.E.)

$$I.Fc/I.Fs = \int abs[Fc(t)]dt / \int abs[Fs(t)]dt \quad (59)$$

$$I.E. = \int M\ddot{y}(t)\dot{x}(t)dt \quad (60)$$

Definition of Performannce 1-Dimensional (P1D) Spectra

The P2D Spectra show the global control effect, but they are unsuitable to evaluate the difference of the control efficiency among the different control algorithms. So we adopt the Performance 1-Dimensional (P1D) Spectra on the condition of the same natural period, the same maximum counter force, and the same maximum response values.

(a) Same Period Condition: $maxX = Sp_1t(maxFc, Tn)$ (61)

(b) Same Counter Force Condition: $maxX = Sp_1f(T, maxFn)$ (62)

(c) Same Response Condition: $maxFc = Sp_1r(T, maxXn)$ (63)

Analytical Examples

Analytical conditions :

(a) Controlled Structures

The controlled structure is modeled as a single degree of freedom model, and the natural period is selected from 0.2 second to 4.0 seconds with the interval of 0.1 second.

$T = 0.2 + 0.1 * (i-1) \quad i = 1, \ldots, 39$

(b) Control Conditions

The Optimal, the Instantaneous, and the Self-balancing Adaptive control are adopted to evaluate the value of information in the case of veleocity control. Twenty weighting values are assumed in the following extent.

$q(v) = 0.0,\ 0.1, \ldots, 10000.0,\ q(d) = q(a) = 0.0, \quad R = 1.0$

The optimal range (T_c) is set to 4.0 seconds.

(c) Seismic Ground Motion

The NS component of the El Centro 1940 record is normalized to the maximum acceleration of 1.0 cm/sec^2.

Analytical results : The P2D Spectra are illustrated in the case of the optimal adaptive control (Fig.10).

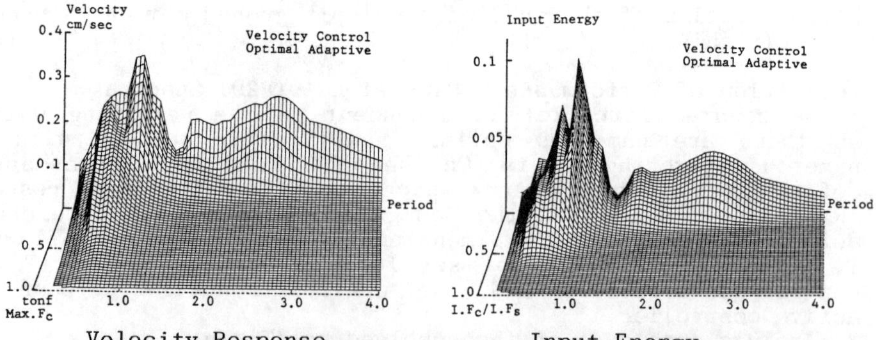

Velocity Response　　　　Input Energy

Optimal Adaptive Control/Velocity Control

Fig.10 DIB Performance 2-Dimensional Spectra

Also the contour figures of the P2D Spectra are shown (Fig.11).

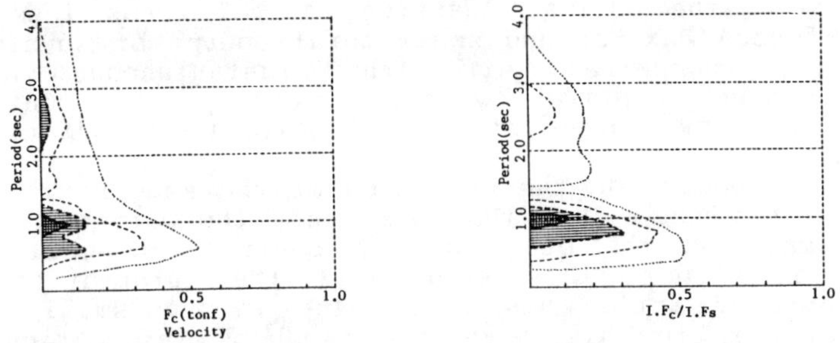

Velocity Response Input Energy

Optimal Adaptive Control/Velocity Control

Fig.11 DIB Performance 2-Dimensional Spectra (Contour)

The P1D Spectra on the condition of the same natural period and the same counter force are illustrated to evaluate the control effect of the three control algorithms (Fig.12,13).

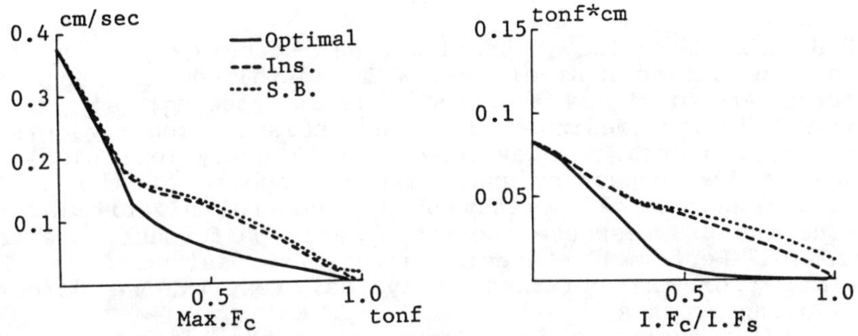

Velocity Response Input Energy

Fig.12 DIB Performance 1-Dimensional Spectra (Same Natural Period [T=1.0second] Condition)

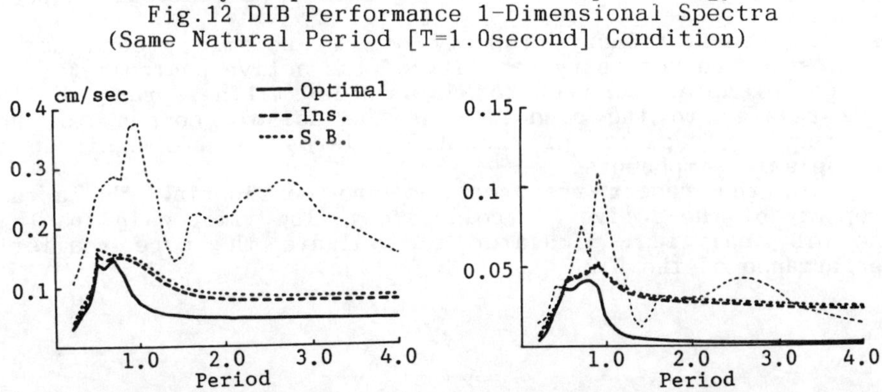

Velocity Response Input Energy

Max.Fc=0.0, 0.2, 0.4, 0.6tonf I.Fc/I.Fs=0.0, 0.2, 0.4, 0.6

Fig.13 DIB Performance 1-Dimensional Spectra (Same Counter Force Condition)

Concluding Remarks

1. Generally the Optimal Adaptive control needs less maximum counter force (Max.Fc) and small total energy of counter force (I.Fc/I.Fs) compaired with the Instantaneous and the Self-balancing Adaptive control (quasi-optimal control) to suppress the structural response. Namely the information saves the counter force.

2. The P1D Spectra on the condition of the same natural period shows, in the range less than 0.2 tonf, the difference between the Optimal and the quasi-optimal control is small and the suppression effect. This means that the information of the seismic ground motion has less value in the small range of counter force. In this range the feedback information of the structural response has enough effect to suppress the response, since the resonant vibration is mainly suppressed. On the other hand, the information of the seismic ground motion has significant effect in the large range of counter force.

3. The difference between two quasi-optimal control is small, so the information of the instant seismic ground motion has small effect to save the counter force.

CONCLUSIONS

The DIB aim for the vibration-free structures, for this purpose, the active control system is installed. In Japan the structures are designed for the seismic loading, so the AC system should be designed for the seismic loading. In a preliminary research, we develope the DIB analytical simulator to evaluate the necessary performance property of the active controllers and the efficiency of the control algorithm. Also we define the DIB response spectra, the AC performance spectra, and the DIB Performance Spectra from the results of the DIB simulator. From the obtained analytical results, we detected the following remarks.

(1) The DIB spectra show a trade-off relation between the capacity of the active controllers and the structural response values.

(2) The non-stationary property of the seismic ground motion relates to the necessary capacity of the active controllers.

(3) For example, the required adjustable stiffness range of the A2S2 relates to the bandwidth of the seismic components, and the required capacity of the ACFS relates to the magnitude of the seismic components.

In the recent research, we are considering the actual property of the active controllers(e.g. lag time, gain loss) in the DIB analytical simulator to evaluate the more realistic performance of the DIB.

REFERENCES

1. Wada,B.K., Fanson,J.L. and Crawley,E.F., Adaptive Structures, Adaptive Structures, AD-Vol.15, B.K.Wada(ed.), The winter annual meeting of the American Society of Mechanical Engineerings, San Francisco, CA, Dec.10-15,1989

2. Kobori,T., Kanayama,H., Kamagata S., New Philosophy of Aseismic Design -Approach on Dynamic Intelligent Building Systems-, ASCE, Engineering Mechanics, 6th Conference abstracts, Buffalo, New York, May 1987

3. Kobori,T., State-of-the Art Report Active Seismic Response Control, Proceedings of 9WCEE, vol.Ⅷ, pp.435-446, Aug., 1988

4. Kobori,T., Kanayama,H., Kamagata,S., A Proposal of New Anti-Seismic Structure with Active Seismic Response Control System -Dynamic Intelligent Building-, Proceedings of 9WCEE, vol.Ⅷ, pp.465-470, Aug., 1988

5. R.E.Bellman, L.A.Zadeh, Decision Making in a Fuzzy Environment, Management Sci. 17, 1970

6. L.A.Zadeh, Outline of a New Approach to the Analysis of Complex System and Decision Processes, IEEE, Transactions on System, Man, and Cybernetics, Vol.SMC-3, No.1, Jan.1973

7. Nakamura,Y., On The Urgent Earthquake Detection and Alarm System (UrEDAS), Proceedings of 9WCEE, vol.Ⅷ, pp.673-678, Aug. 1988

8. Yao,J.T.P., Concept of Structural Control, Journal of the Structural Division, ASCE Vol.98 No.ST7 July 1972

9. Yang,J.N., Application of Optimal Control Theory to Civil Engineering Structures, Journal of the Engineering Mechanics Division ASCE, Vol.101, No.EM6, Dec.1975

10. Martin,C.R., Soong,T.T., Modal Control of Multistory Structures, Journal of the Engineering Mechanics Division, ASCE, Vol.102, No.EM4, Aug.1976

11. Abdel-Rohman,M., Leipholz,H.H.E., Active Control of Flexible Structure, Journal of the Structural Division, ASCE, Vol.104, No.ST8, Aug.1978

12. Meirovitch,L., Active Control of Structures by Modal Synthesis, STRUCTURAL CONTROL H.H.E.Leipholz(ed.), North-Holland Publishing Company 1980

13. Roorda,J., Experiments in Feedback Control of Structures, STRUCTURAL CONTROL, H.H.E.Leipholz(ed.)North-Holland Publishing Company 1980

14. Masri,S.F., Bekey,G.A., Udwadia,F.E., Online Pulse Control of Tall Buildings, STRUCTURAL CONTROL, H.H.E.Leipholz(ed.) North-Holand Publishing Company 1980

15. Reinhorn,A.M., Soong,T.T., Yang,J.N., Active Control of Structures During Earthquakes, EAEE, The 8th European Conference on Earthquake Engineering, Lisbon, 1986

16. Kobori,T., et al., Dynamic Intelligent Building as Active Seismic Response Controlled Structures,
Part 1.Theoretical Control Concept
Part 2.Analytical Verification of Theoretical Control Concept
Part 3.Experimental Verification of Theoretical Control Concept
Part 4.Analytical Study on Japanese High-rise Building with Active Mass Driver System against Mexico Earthquake
Proceedings of Architectural Institute of Japan, Annual Meeting 1987

17. Yang,J.N., Control of Tall Buildings under Earthquake Excitation, Journal of the Engineering Mechanics Division, ASCE, Vol.108, No.EM5, Oct.1982

18. Wiesner,K.B., Taming Livery Building, Civil Engineering, June 1986

19. Hrovat,D., Barak,P., and Rabins,M. Semi-active versus Passive or Active Tuned Mass Dampers for Structural Control, Journal of the Engineering Mechanics Division, ASCE, Vol.109, No.EM3, June 1983

20. Kobori,T, et.al, Study on Active Mass Driver (AMD) System (Part-1,2) Active Seismic Response Controlled Structure, Tall Buildings: 2000 and Beyond, Fourth World Conference, Hong Kong, Nov. 1990

21. Kobori,T, et.al, Shaking Table Experiment and Practical Application of Active Variable Stiffness (AVS) System, Proceedings of the Second Conference on Tall Buildings in Seismic Regions, Los Angels, CA, May, 1991

22.Kobori,T., Kamagata,S., Self-Balancing Adaptive Control for Active Adjustable Stiffness System, Journal of Struct. Const. Engng, AIJ, Vol.420, pp.121-132, Feb., 1991(in Japanese)

OPTIMAL HYBRID CONTROL OF SEISMIC-EXCITED NONLINEAR AND INELASTIC STRUCTURES

J. N. YANG, Z. LI, A. DANIELIANS
Department of Civil Engineering
University of California
Irvine, California 92717

and

S. C. LIU
National Science Foundation
Washington, D.C. 20550

ABSTRACT

Instantaneous optimal control for nonlinear and inelastic systems is formulated incorporating the specific hysteretic model of the system. The resulting optimal control vector is obtained as a function of the total deformation, velocity and the hysteretic component of the structural response. The hysteretic component of the structural response can be estimated from the measured structural response and the hysteretic model used. It is shown that the optimal control vector satisfies not only the necessary conditions but also the sufficient condition of optimality. Specific applications of the optimal algorithm to two types of hybrid control systems are demonstrated. These include (i) active control of base-isolated buildings using frictional-type sliding base isolators, and (ii) active control of base-isolated buildings using lead-core rubber bearings. Numerical examples are worked out to demonstrate the applications of the proposed control algorithm. It is shown that the performance of such an optimal algorithm improves over that of the algorithm without considering the hysteretic components in the determination of the control vector.

INTRODUCTION

It has been shown recently that a combined use of active and passive control systems, referred to as the hybrid control system, is more effective, beneficial and practical, in some cases, for reducing the building response under strong earthquakes [e.g., 1-7]. However, the use of hybrid control systems involves active control of nonlinear or inelastic structural systems, since passive control systems, such as base isolation systems, usually behave nonlinearly or inelastically under dynamic loads. Unfortunately, control theories for nonlinear or inelastic structural systems are limited [e.g., 8-10]. Instantaneous optimal control algorithms for linear, nonlinear or inelastic structures were proposed in Refs. 11-14. Recently, a refined version of such optimal control algorithms was presented in Refs. 5-6 with emphasis placed on applications to aseismic hybrid control systems. In Refs. 5-6, the optimal control vector $\underline{U}(t)$ was obtained from the measured 2n state vector $\underline{Z}(t)$.

For inelastic structural systems, such as base-isolated buildings using either lead-core rubber bearing isolators or sliding-type isolation systems, the total deformation consists of an elastic component and a hysteretic component. Various mathematical models to describe the behavior of the hysteretic component have been proposed in the literature [e.g., 15-20]. To-date, the hysteretic component of the structural response has not been incorporated in the formulation of the optimal control problem, although the hysteretic model has been used in the simulation of the controlled response of buildings equipped with aseismic hybrid control systems [5-6].

Based on the concept of instantaneous optimal control [11-14,5-6], the optimal control problem is formulated incorporating the specific hysteretic model in this paper. This is accomplished by including the hysteretic component explicitly in the equations of motion, which are used as the constraints for the determination of the optimal control vector. For an n-degree-of-freedom system, the dimension of the state vector $\underline{Z}(t)$ used in Refs. 5-6 is 2n. Including the hysteretic component explicitly in the present formulation, the dimension of the new state vector $\tilde{\underline{Z}}(t)$ becomes 3n. The optimal control vector $\underline{U}(t)$ is obtained by minimizing a time dependent quadratic performance index [11-13,21] exactly in the same manner as that described in Refs. 11-14 and 5-6. In this paper we also prove that the optimal control vector satisfies not only the necessary conditions but also the sufficient condition of optimality. The optimal control vector $\underline{U}(t)$ is obtained as a linear function of the 3n state vector $\tilde{\underline{Z}}(t)$. Hence, the optimal control vector $\underline{U}(t)$ depends not only on the measured deformation and velocity but also on the hysteretic component of the structural response. The hysteretic component of the structural response can be estimated from an observer using the measured structural response and the specific hysteretic model used. A simple method of constructing the observer for the estimation of the hysteretic component is also presented.

Specific applications of the present control algorithm to two important types of aseismic hybrid control systems are demonstrated. These include (i) a frictional-type sliding isolation system connected to an actuator, and (ii) a lead-core rubber bearing isolation system connected to either an actuator or an active mass damper. These two types of aseismic hybrid systems are highly nonlinear and hysteretic in nature. Theoretically, the performance of the present optimal algorithm should be superior to that proposed in Refs. 5-6, because more information is used in the determination of the control vector. The performance of the present optimal algorithm has been investigated, evaluated and compared numerically with that of the previous algorithm [Ref. 5-6]. It is shown numerically that the performance of the present optimal algorithm is indeed better than that of the previous one for the two types of aseismic hybrid control systems investigated herein.

OPTIMAL CONTROL OF INELASTIC STRUCTURAL SYSTEMS

Nonlinear Hysteretric Model for Inelastic Systems

Various hysteretic models for the restoring force of an inelastic system have been developed in recent years [e.g., 15-20]. In this study, the differential equation model available in the literature [e.g., 15-18] will be used for both structures and passive protective systems. The stiffness restoring force, $F_{si}(t)$, of the story unit, is expressed as

$$F_{s_i}(t) = \alpha_i k_i x_i + (1 - \alpha_i) k_i D_{y_i} v_i \tag{1}$$

in which the subscript i represents the quantity associated with the ith story unit and x_i=deformation of the ith story unit, k_i=elastic stiffness, α_i=ratio of post-yielding to pre-yielding stiffness, D_{yi}=yield deformation=constant, and v_i is a nondimensional variable introduced to describe the hysteretic component of the deformation, with $|v_i| \leq 1$, where

$$D_{y_i}\dot{v}_i = A_i\dot{x}_i - \beta_i|\dot{x}_i|\,|v_i|^{n_i-1}v_i - \gamma_i\dot{x}_i|v_i|^{n_i} \quad (2)$$

In Eq. (2), parameters A_i, β_i and γ_i govern the scale and general shape of the hysteresis loop, whereas the smoothness of the force-deformation curve is determined by the parameter n_i.

The hysteretic model given by Eqs. (1) and (2) is phenomenological in nature; however, it does provide an explicit mathematical expression with many parameters flexible enough to reflect various hysteretic behaviors of inelastic systems. Likewise, Eq. (2) appears to be capable of describing the hysteretic behaviors of different base isolation systems, such as rubber bearings and sliding systems, which are the main concern of this study.

Equations of Motion

For simplicity, consider a base-isolated one-dimensional inelastic building equipped with a hybrid control system as shown in Fig. 1(a). The structure system is idealized by an n-degree-of-freedom system and subjected to a one-dimensional earthquake ground acceleration $\ddot{X}_0(t)$. The matrix equation of motion of the entire building system can be expressed as follows

$$\underline{M}\ddot{\underline{X}}(t) + \underline{C}\dot{\underline{X}}(t) + \underline{K}_e\underline{X}(t) + \underline{K}_I\underline{V}(t) = \underline{H}\underline{U}(t) + \underline{F}\ddot{X}_0(t) \quad (3)$$

in which a linear viscous damping model has been assumed, $\underline{X}(t)=[x_1,x_2,\ldots,x_n]'$ = an n vector denoting the deformation of each story unit, and $\underline{V}(t)=[v_1,v_2,\ldots,v_n]'$ = an n vector denoting the hysteretic variable of each story unit. For the notations above, an under bar denotes a vector or a matrix and a prime indicates the transpose of a vector or a matrix.

In Eq. (3), $\underline{M}$ is a (nxn) mass matrix with the i-jth element $M(i,j)=m_i$ for $i=1,2,\ldots,n$ and $j=1,2,\ldots,i$ and $M(i,j)=0$ for $j>i$, where m_i is the mass of the ith floor, $\underline{C}$ is an (nxn) band-limited damping matrix with all elements equal to zero except $C(i,i)=c_i$ for $i=1,2,\ldots,n$ and $C(i,i+1)=-c_{i+1}$ for $i=1,2,\ldots,n-1$, where c_i is the damping coefficient of the ith story unit. $\underline{K}_e$ and $\underline{K}_I$ are (nxn) band-limited elastic stiffness matrix and hysteretic stiffness matrix, respectively. All elements of $\underline{K}_e$ and $\underline{K}_I$ are zero except $K_e(i,i)=\alpha_i k_i$, $K_I(i,i)=(1-\alpha_i)k_i D_{yi}$ for $i=1,2,\ldots,n$ and $K_e(i,i+1)=-\alpha_{i+1}k_{i+1}$, $K_I(i,i+1)=-(1-\alpha_{i+1})k_{i+1}\,D_{yi+1}$ for $i=1,2,\ldots,n-1$. $\underline{H}$ is an (nxr) matrix denoting the location of r controllers and $\underline{F}$ is an n vector denoting the influence of the earthquake ground acceleration.

From Eq. (2), the vector $\underline{V}$ appearing in Eq. (3) is given by

$$\dot{\underline{V}} = f(\dot{\underline{X}},\underline{V}) \quad (4)$$

in which the ith element of $\dot{\underline{V}}$ or $\underline{f}(\dot{\underline{X}}, \underline{V})$, denoted by $\dot{v}_i$ or $f_i(\dot{x}_i, v_i)$, is given by

$$\dot{v}_i = f_i(\dot{x}_i,v_i) = D_{yi}^{-1}\left[A_i\dot{x}_i - \beta_i|\dot{x}_i|\,|v_i|^{n_i-1}v_i - \gamma_i\dot{x}_i|v_i|^{n_i}\right] \quad (5)$$

By introducing a 3n state vector $\underline{Z}(t)$ and two 3n vectors $\underline{B}$ and $\underline{W}_1$,

$$\underline{Z}(t) = \begin{bmatrix}\underline{X}\\ \underline{V}\\ \dot{\underline{X}}\end{bmatrix} \;;\; \underline{B} = \begin{bmatrix}\underline{0}\\ \underline{0}\\ \underline{M}^{-1}\underline{H}\end{bmatrix} \;;\; \underline{W}_1 = \begin{bmatrix}\underline{0}\\ \underline{0}\\ \underline{M}^{-1}\underline{F}\end{bmatrix} \quad (6)$$

the second-order nonlinear matrix equation of motion, Eq. (3), can be converted into a first order matrix equation as follows:

$$\dot{\underline{Z}}(t) = g[\underline{Z}(t)] + \underline{B}\underline{U}(t) + \underline{W}_1\ddot{X}_0(t) \quad (7)$$

in which $g[\underline{Z}(t)]$ is a 3n vector consisting of nonlinear functions of components of the vector $\underline{Z}(t)$,

$$g[\tilde{\underline{Z}}(t)] = \begin{bmatrix} \dot{\underline{X}} \\ \text{-----------} \\ \underline{f}(\dot{\underline{X}},\underline{V}) \\ \text{-----------} \\ -\underline{M}^{-1}\left(\underline{C}\,\dot{\underline{X}} + \underline{K}_e\,\underline{X} + \underline{K}_I\,\underline{V}\right) \end{bmatrix} \tag{8}$$

The first-order nonlinear matrix equation, Eq. (7), can be solved step-by-step numerically using the Fourth-Order Runge-Kutta method as follows [Refs.5-6]:

$$\tilde{\underline{Z}}(t) = \tilde{\underline{Z}}(t - 2\Delta t) + \frac{1}{6}[\underline{A}_0 + 2\underline{A}_1 + 2\underline{A}_2 + \underline{A}_3] \tag{9}$$

in which Δt is the integration time step, and $\underline{A}_0$, $\underline{A}_1$, $\underline{A}_2$, and $\underline{A}_3$ are 3n vectors

$$\begin{aligned} \underline{A}_0 &= 2\Delta t\,\{g[\tilde{\underline{Z}}(t-2\Delta t)] + \underline{B}\,\underline{U}(t-2\Delta t) + \underline{W}_1\,\ddot{X}_0(t-2\Delta t)\} \\ \underline{A}_1 &= 2\Delta t\,\{g[\tilde{\underline{Z}}(t-2\Delta t) + 0.5\underline{A}_0] + \underline{B}\,\underline{U}(t-\Delta t) + \underline{W}_1\,\ddot{X}_0(t-\Delta t)\} \\ \underline{A}_2 &= 2\Delta t\,\{g[\tilde{\underline{Z}}(t-2\Delta t) + 0.5\underline{A}_1] + \underline{B}\,\underline{U}(t-\Delta t) + \underline{W}_1\,\ddot{X}_0(t-\Delta t)\} \\ \underline{A}_3 &= 2\Delta t\,\{g[\tilde{\underline{Z}}(t-2\Delta t) + \underline{A}_2] + \underline{B}\,\underline{U}(t) + \underline{W}_1\,\ddot{X}_0(t)\} \end{aligned} \tag{10}$$

As observed from Eq. (10), $\underline{A}_0$, $\underline{A}_1$, $\underline{A}_2$, $\underline{A}_3$ are functions of t-2Δt, t-Δt and t. For simplicity of presentation, the arguments, t-2Δt, t-Δt and t, have been omitted.

To simplify the mathematical operation, it is desirable to group all vectors in Eq. (9), which are functions of $\tau < t$, into one term, denoted by $\underline{D}$(t-2Δt, t-Δt), as follows:

$$\begin{aligned} \underline{D}(t-2\Delta t,\, t-\Delta t) &= \tilde{\underline{Z}}(t-2\Delta t) + \frac{1}{3}\Delta t\,\{g[\tilde{\underline{Z}}(t-2\Delta t)] + \underline{B}\,\underline{U}(t-2\Delta t) \\ &+ \underline{W}_1\,\ddot{X}_0(t-2\Delta t) + 2g[\tilde{\underline{Z}}(t-2\Delta t) + 0.5\underline{A}_0] + 4\underline{B}\,\underline{U}(t-\Delta t) \\ &+ 4\underline{W}_1\,\ddot{X}_0(t-\Delta t) + 2g[\tilde{\underline{Z}}(t-2\Delta t) + 0.5\underline{A}_1] + g[\tilde{\underline{Z}}(t-2\Delta t) + \underline{A}_2]\} \end{aligned} \tag{11}$$

Then, Eq. (9) can be expressed as

$$\tilde{\underline{Z}}(t) = \underline{D}(t-2\Delta t,\, t-\Delta t) + \frac{\Delta t}{3}[\underline{B}\,\underline{U}(t) + \underline{W}_1\,\ddot{X}_0(t)] \tag{12}$$

It is mentioned that for the regular Fourth-Order Runge-Kutta method, the integration time step is $\Delta\tau$ that is twice of Δt used in Eqs. (9)-(12), i.e., $\Delta\tau = 2\Delta t$ or $\Delta t = \Delta\tau/2$. In other words, the integration step size Δt used in the current numerical scheme is one-half of that used in the regular Fourth-Order Runge-Kutta method.

As observed from Eqs. (4), (5) and (8), the g vector is a function of $\underline{X}$, $\underline{V}$ and $\dot{\underline{X}}$ which are elements of the vector $\tilde{\underline{Z}}$. Thus, the g vector can be expressed as a function of the response state vector $\tilde{\underline{Z}}$, i.e., $g = g(\tilde{\underline{Z}})$. To evaluate $g(\tilde{\underline{Z}})$, the argument vector $\tilde{\underline{Z}} = [\underline{X}', \underline{V}', \dot{\underline{X}}']'$ is computed first and then its components, $\underline{X}$, $\underline{V}$ and $\dot{\underline{X}}$, are substituted into the right hand side of Eq. (8). For the $g[\tilde{\underline{Z}}(t-2\Delta t) + 0.5\,\underline{A}_1]$ vector appearing in Eq. (11), the argument vector (3n-dimensional), denoted by $\underline{S}$,

$$\underline{S} = \tilde{\underline{Z}}(t-2\Delta t) + 0.5\underline{A}_1 \tag{13}$$

is computed first from Eq. (13). Then, $\underline{S}$ is partitioned in the same form as the vector $\tilde{\underline{Z}}$, i.e., $\underline{S} = [\underline{S}_1', \underline{S}_2', \underline{S}_3']'$, where $\underline{S}_1$, $\underline{S}_2$ and $\underline{S}_3$ are n-dimensional vectors. Finally, the $g[\tilde{\underline{Z}}(t-2\Delta t) + 0.5\,\underline{A}_1]$ vector is obtained from Eq. (8) by replacing $\underline{X}$, $\underline{V}$ and $\dot{\underline{X}}$, respectively, by $\underline{S}_1$, $\underline{S}_2$ and $\underline{S}_3$. Likewise, the other two vectors $g[\tilde{\underline{Z}}(t-2\Delta t) + 0.5\,\underline{A}_0]$ and $g[\tilde{\underline{Z}}(t-2\Delta t) + \underline{A}_2]$ appearing in Eq. (11) are similarily defined.

Instantaneous Optimal Control

The instantaneous optimal control algorithm developed and proposed by Yang, et al for linear structures [11-12] was extended later to nonlinear structures [13,5-6]. It is used herein for applications to control of inelastic systems.

The time dependent quadratic objective function J(t) is given as follows

$$J(t) = \underline{\tilde{Z}}'(t)\underline{Q}\underline{\tilde{Z}}(t) + \underline{U}'(t)\underline{R}\,\underline{U}(t) \tag{14}$$

in which $\underline{Q}$ and $\underline{R}$ are (3nx3n) and (rxr) weighting matrices, respectively, representing the relative importance of the response vector $\underline{\tilde{Z}}(t)$ and the control vector $\underline{U}(t)$. The optimal control vector obtained by minimizing the time-dependent objective function, J(t), is referred to as the instantaneous optimal control algorithm [e.g., 11-14].

To minimize the objective function, Eq. (14), subjected to the constraint given by Eq. (12), the Hamiltonian H is constructed by introducing a 3n-dimensional Lagrangian multiplier vector $\underline{\lambda}(t)$

$$\begin{aligned} H &= \underline{\tilde{Z}}'(t)\underline{Q}\underline{\tilde{Z}}(t) + \underline{U}'(t)\underline{R}\underline{U}(t) + \underline{\lambda}'(t)\{\underline{\tilde{Z}}(t) \\ &- \underline{D}(t-2\Delta t,\, t-\Delta t) - \frac{\Delta t}{3}[\underline{B}\underline{U}(t) + \underline{W}_1\ddot{X}_0(t)]\} \end{aligned} \tag{15}$$

The necessary conditions for minimizing J(t), subjected to the constraint of Eq. (12), are

$$\frac{\partial H}{\partial \underline{Z}} = 0, \qquad \frac{\partial H}{\partial \underline{U}} = 0, \qquad \frac{\partial H}{\partial \underline{\lambda}} = 0 \tag{16}$$

Substituting Eq. (15) into Eq. (16), one obtains the optimal closed-loop (feedback) control vector $\underline{U}(t)$ as follows

$$\underline{U}(t) = -\frac{\Delta t}{3}\underline{R}^{-1}\underline{B}'\underline{Q}\underline{\tilde{Z}}(t) \tag{17}$$

The optimal open-loop (feedforward) and closed-open-loop (feedback and feedforward) control algorithms were presented in Refs. 6 and 7.

With the optimal control vector given by Eq. (17), the response state vector $\underline{\tilde{Z}}(t)$ is obtained by substituting Eq. (17) into Eq. (12) as follows

$$\underline{\tilde{Z}}(t) = \left[\underline{I} + \left(\frac{\Delta t}{3}\right)^2 \underline{B}\underline{R}^{-1}\underline{B}'\underline{Q}\right]^{-1}\left[\underline{D}(t-2\Delta t,\, t-\Delta t) + \frac{\Delta t}{3}\underline{W}_1\ddot{X}_0(t)\right] \tag{18}$$

in which $\underline{I}$ is a (3nx3n) identity matrix.

Sufficient Condition For Optimal Control

In the original derivations for the instantaneous optimal control algorithms, [e.g., 11-13, 5-6], the optimal control vector $\underline{U}(t)$ was obtained from the necessary conditions, such as Eq. (16). In fact, the derived optimal control vector $\underline{U}(t)$ also satisfies the sufficient condition of optimality as will be proved in the following.

The sufficient condition for the optimal solution is given by [Ref. 22]

$$\left[\partial\underline{\tilde{Z}}',\ \partial\underline{U}'\right]\begin{bmatrix} \dfrac{\partial^2 H}{\partial \underline{\tilde{Z}}^2} & \dfrac{\partial^2 H}{\partial \underline{\tilde{Z}}\partial \underline{U}} \\ \dfrac{\partial^2 H}{\partial \underline{U}\partial \underline{\tilde{Z}}} & \dfrac{\partial^2 H}{\partial \underline{U}^2} \end{bmatrix}\begin{bmatrix}\partial\underline{\tilde{Z}} \\ \partial\underline{U}\end{bmatrix} > 0 \tag{19}$$

Taking derivatives of H in Eq. (15), one obtains

$$\frac{\partial H}{\partial \underline{\tilde{Z}}} = 2\underline{Q}\underline{\tilde{Z}} - \underline{\lambda} \ ; \quad \frac{\partial H}{\partial \underline{U}} = 2\underline{R}\underline{U} - \frac{\Delta t}{3} \underline{B}'\underline{\lambda}$$
$$\frac{\partial^2 H}{\partial \underline{\tilde{Z}}^2} = 2\underline{Q} \ ; \quad \frac{\partial^2 H}{\partial \underline{U}^2} = 2\underline{R} \ ; \quad \frac{\partial^2 H}{\partial \underline{\tilde{Z}} \partial \underline{U}} = \frac{\partial^2 H}{\partial \underline{U} \partial \underline{\tilde{Z}}} = \underline{0} \tag{20}$$

Substitution of Eq. (20) into the left hand side of Eq. (19) leads to the following expression

$$2\left[\partial \underline{\tilde{Z}}', \partial \underline{U}'\right]\begin{bmatrix} \underline{Q} & \underline{O} \\ \underline{O} & \underline{R} \end{bmatrix}\begin{bmatrix} \partial \underline{\tilde{Z}} \\ \partial \underline{U} \end{bmatrix} = 2\left(\partial \underline{Z}' \underline{Q} \, \partial \underline{Z} + \partial \underline{U}' \underline{R} \, \partial \underline{U}\right) > 0 \tag{21}$$

Since $\underline{Q}$ is a positive semi-definite matrix and $\underline{R}$ is a positive definite matrix, Eq. (21) is greater than zero. Thus, the sufficient condition for the optimal solution, Eq. (19), is satisfied.

Estimation of Hysteretic Component Vector $\underline{V}$

The optimal control vector $\underline{U}(t)$ given by Eq. (17) is regulated by the 3n response state vector $\underline{\tilde{Z}}(t)=[\underline{X}'(t), \underline{V}'(t), \underline{\dot{X}}'(t)]'$. The displacement and velocity response vectors, $\underline{X}(t)$ and $\underline{\dot{X}}(t)$, are available from measurements using sensors. However, the hysteretic component vector $\underline{V}(t)=[v_1(t),v_2(t),\ldots,v_n(t)]'$ is not measurable and hence it should be estimated. An approach to estimate $v_i(t)$ for i=1,2,...,n is by use of a reduced order nonlinear state observer. In the present case, the observer can be constructed easily, since $v_i(t)$ is only a function of the interstory velocity $\dot{x}_i(t)$ as shown in Eq. (5). As a result, the estimate of the hysteretic component $v_i(t)$, denoted by $\hat{v}_i(t)$, follows from Eq. (5) as

$$\dot{\hat{v}}_i(t) = f_i(\dot{x}_i, \hat{v}_i) = D_{yi}^{-1}\left[A_i\dot{x}_i - \beta_i \mid \dot{x}_i \mid \mid \hat{v}_i \mid^{n_i - 1}\hat{v}_i - \gamma_i \dot{x}_i \mid \hat{v}_i \mid^{n_i}\right] \tag{22}$$

in which the argument t for $\dot{x}_i(t)$ and $\hat{v}_i(t)$ has been omitted for simplicity and $\dot{x}_i(t)$ is obtained from measurements.

Given the measured $\dot{x}_i(t)$, Eq. (22) can be integrated on-line to obtain the estimate $\hat{v}_i(t)$ for $v_i(t)$ without any difficulty. It is mentioned that when the response level is within the elastic limit, the hysteretic component v_i is very close to the interstory deformation x_i, Eq. (1). As in the case of any observer, the accuracy of the estimates $\hat{v}_i(t)$ for i=1,2,...,n, depends on the accuracy of the hysteretic model representing the behavior of the system as well as the corresponding parameters, such as A_i, β_i, γ_i and n_i. This subject, however, is beyond the scope of this paper.

NUMERICAL EXAMPLE

The applications of the optimal control algorithm developed in this paper to aseismic hybrid control systems are illustrated in this section. The same eight-story building that exhibits binlinear elasto-plastic behavior used in Refs. 5 and 6 is considered. The stiffness of each story unit is designed such that yielding occurs simultaneously for each story unit. The properties of the structure are as follows: (i) the mass of each floor is identical with m_i = m = 345.6 metric tons; (ii) the preyielding stiffnesses of the eight-story units are k_{i1} (i=1.2,...,8) = 3.4x10^5, 3.26x10^5, 2.85x10^5, 2.69x10^5, 2.43x10^5, 2.07x10^5, 1.69x10^5 and 1.37x10^5 kN/m, respectively, and the postyielding stiffnesses are k_{i2} = 0.1 k_{i1} for i=1,2,....,8; and (iii) the viscous damping coefficients for each story unit are c_i = 490, 467, 410, 386, 348, 298, 243 and 196 kN.sec/m, respectively. The damping coefficients given above result in a classically damped structure with a damping ratio of 0.38% for the first vibrational mode. The natural frequencies of the unyielded structure are 5.24, 14.0, 22.55, 30.22, 36.89, 43.06, 49.54 and 55.96 rad./sec. The yielding level for each story unit varies with respect to the stiffness; with the results, D_{yi} = 2.4, 2.3, 2.2, 2.1, 2.0, 1.9, 1.7, and 1.5 cm. The bilinear elasto-plastic behavior can be described by the hysteretic model, Eqs. (1) and (2), with $A_i=1.0$, $\beta_i=0.5$, $n_i=95$ and $\gamma_i=0.5$ for $i=1.2,...,8$. A simulated earthquake with a maximum ground acceleration of 0.3g as shown in Fig. 2 is used as the input excitation.

With the eight-story bilinear elasto-plastic building structure described above and the earthquake ground acceleration shown in Fig. 2, time histories of all the response quantities have been computed. Within 30 seconds of the earthquake episode, the maximum interstory deformation, x_{mi}, and the maximum absolute acceleration of each floor, a_i, are shown in Table 1. As observed from Table 1, the deformation of the unprotected building is excessive and that yielding takes place in each story unit.

Example 1: Structure Equipped With Rubber-Bearing Base Isolation System and Active Mass Damper

To reduce the structural response, a lead-core rubber-bearing isolation system connected to an active mass damper is implemented as shown in Fig. 1(a). The restoring force of the lead-core rubber-bearing system is modeled as

$$F_{sb} = \alpha_b k_b x_b + (1-\alpha_b) k_b D_{yb} \nu_b \tag{23}$$

in which ν_b is given by Eq. (2) with i=b. The mass of the base isolation system is $m_b=450$ metric tons and the viscous damping coefficient is assumed to be linear with $c_b=26.17$ kN.sec/m. The restoring force of the base isolation system given by Eq. (23) is not bilinear elasto-plastic and the parameter values are given as follows: $k_b=18050$ kN/m, $\alpha_b=0.6$ and $D_{yb}=4$cm. The parameters governing the scale and general shape of the hysteresis loop of the isolation system, Eq. (2), were assumed to be $A_b=1.0$, $\beta_b=0.5$, $n_b=3$ and $\gamma_b=0.5$. The hysteretic characteristics of such a base isolation system, i.e., x_b versus v_b, are shown in Fig. 3(a). The response vector $\underline{X}(t)$ is given by $\underline{X}=[x_b,x_1,...,x_8]'$.

Without an active mass damper, time histories of all the response quantities were computed and presented in Ref. 6. The maximum response quantities of the structure in 30 seconds of the earthquake episode are shown in the column of Table 1 designated as "With BIS". As observed from Table 1, the interstory deformation and the floor acceleration are drastically reduced. However, the deformation of the base isolation system shown in row B of Table 1 may be excessive.

To protect the safety and integrity of the base isolation system, an active mass damper was proposed in Refs. 3-4 to be connected to the base isolation system as shown in Fig. 1(a). The properties of the mass damper are as follows. The mass of the mass damper, m_d, is equal to 50% of the floor mass, i.e., $m_d=0.5m_i$ and the natural frequency of the mass damper is the same as the first natural frequency of the base isolated building, i.e. 2.21 rad/sec. The damping ratio of the mass damper is 10%. With the active mass damper, the response vector is given by $\underline{X}(t)=[x_d,x_b,x_1,...,x_8]'$ and the structural response depends on the weighting matrices $\underline{R}$ and $\underline{Q}$. For this example, the weighting matrix $\underline{R}$ consists of only one element, denoted by R_0, whereas the dimension of the $\underline{Q}$ matrix is (30x30). R_0 is chosen to be 10^{-3} for simplicity. The $\underline{Q}$ matrix is partitioned as follows

$$Q = \alpha \left[\begin{array}{ccc} \underline{0} & \underline{0} & \underline{0} \\ \hline Q_{21} & Q_{22} & Q_{23} \end{array} \right] \tag{24}$$

in which $\underline{0}$ is a (20x10) matrix with all elements equal to zero, and $\underline{Q}_{21}$, $\underline{Q}_{22}$ and $\underline{Q}_{23}$ are (10x10) matrices.

Because of the fact that there is only one actuator installed on the basement floor as shown in Fig. (1), only the elements in one row of the $\underline{Q}$ matrix are relevant to the control vector $\underline{U}(t)$, see Eq. (17). In the present case, only the elements in the first row of $\underline{Q}_{21}$, $\underline{Q}_{22}$ and $\underline{Q}_{23}$ matrices are related to the control vector $\underline{U}(t)$. As a result, all elements of the $\underline{Q}$ matrix which have no effect on the control vector $\underline{U}(t)$ are set to be zero for convenience.

It is mentioned that matrices $\underline{Q}_{21}$, $\underline{Q}_{22}$ and $\underline{Q}_{23}$ are associated with the feedbacks of $\underline{X}(t)$, $\underline{V}(t)$ and $\dot{\underline{X}}(t)$, respectively. For instance, setting all the elements of $\underline{Q}_{21}$ equal to zero, the control vector $\underline{U}(t)$ given by Eq. (17) is no longer a function of $\underline{X}(t)$. Hence, the main difference between the current optimal algorithm and that presented in Refs. 5 and 6 lies in the matrix $\underline{Q}_{22}$. If all elements of $\underline{Q}_{22}$ are zero, there is no feedback of $\underline{V}(t)$ and the results will be identical to those presented in Refs. 5 and 6.

The building equipped with a base isolation system tends to behave like a rigid body and the interstory deformation is small. Hence, sensors are not needed for the building as demonstrated in Refs. 5-7. Consequently, displacement and velocity sensors are installed on the active mass damper and the base isolation system only, i.e., no sensor is installed on the building. Consider the first case in which there is no feedback of the hysteretic component $\underline{V}(t)$ in the determination of $\underline{U}(t)$, i.e., all elements of the $\underline{Q}_{22}$ matrix are zero. Further, since there is no sensor installed on the building, all elements of $\underline{Q}_{21}$ and $\underline{Q}_{23}$ are zero, except $Q_{21}(1,1)$, $Q_{21}(1,2)$, $Q_{23}(1,1)$ and $Q_{23}(1,2)$. For simplicity, the following values are assigned; $Q_{21}(1,1)=102$, $Q_{21}(1,2)=-2363$, $Q_{23}(1,1)=67$ and $Q_{23}(1,2)=731$.

Time histories of all the structural response quantities have been computed for $\alpha/R_0=50{,}000$ and $\Delta t=0.375\text{x}10^{-2}$. Within 30 seconds of the earthquake episode, the maximum response quantities including the maximum control force, U_m, are summarized in column (A) of Table 1 designated as "BIS&AMD". Time histories of many response quantities were plotted and presented in Ref. 7.

We next consider the case in which the hysteretic component $\underline{V}(t)$ is used for the determination of $\underline{U}(t)$. In this case, the following elements are assigned: $Q_{21}(1,1)=102$, $Q_{21}(1,2)=-3363$, $Q_{22}(1,1)=0$, $Q_{22}(1,2)=-100$, $Q_{23}(1,1)=67$, and $Q_{23}(1,2)=931$. All other elements of the $\underline{Q}$ matrix are zero. The maximum response quantities for $\alpha/R_0=50{,}000$ and $\Delta t=0.375\text{x}10^{-2}$ sec. are summarized in column (B) of Table 1 designated as "BIS&AMD".

It is observed from Table 1 that (i) the performance of the present optimal algorithm with a feedback of the hysteretic component is better than that of the optimal algorithm without a feedback of the hysteretic component, and (ii) the aseismic hybrid control system using a lead-core rubber bearing isolation system and an active mass damper is very effective in protecting building structures against strong earthquakes.

Example 2: Building Equipped with Rubber Bearing Isolation System and Actuator

Instead of using an active mass damper described above, the base isolation system can be connected directly to an actuator. With such a hybrid control system, the (27x27) $\underline{Q}$ matrix is partitioned as shown in Eq. (24), where $\underline{Q}_{21}$, $\underline{Q}_{22}$ and $\underline{Q}_{23}$ are (9x9) matrices. For convenience of instrumentation, displacement and velocity sensors are installed on the base isolation system only, i.e., no sensor is installed on the building. It was demonstrated in Refs. 6-7 that sensors are not necessary to be installed on the building. Further, we consider the first case in which $\underline{V}(t)$ is not used as a feedback for the determination of $\underline{U}(t)$. Hence, all elements of $\underline{Q}_{21}$, $\underline{Q}_{22}$ and $\underline{Q}_{23}$ are zero, except $Q_{21}(1,1)$ and $Q_{23}(1,1)$. For simplicity, the following values are assigned: $Q_{21}(1,1)=0.9$ and $Q_{23}(1,1)=3.4$. The maximum response quantities for $\alpha/R=1.55\text{x}10^{8}$ and $\Delta t=0.375\text{x}10^{-2}$ sec. are summarized in column (A) of Table 1 designated as "BIS&AF".

When $\underline{V}(t)$ is used in the determination of $\underline{U}(t)$, one additional non-zero element from the $\underline{Q}_{22}$ matrix is assigned. For illustrative purposes, the following values are assigned: $Q_{21}(1,1)=0.1$, $Q_{22}(1,1)=0.5$ and $Q_{23}(1,1)=3.5$ and all other elements of the $\underline{Q}$ matrix are zero. The maximum response quantities in 30 seconds of the earthquake episode are presented in column (B) of Table 1 designated as "BIS&AF" for comparison. It is observed from Table 1 that the performance of the present control algorithm is slightly better than that of the previous algorithm without considering the feedback of $\underline{V}(t)$.

Example 3: Building Equipped With Sliding Base Isolator and Actuator

Instead of using a rubber-bearing base isolation system, a frictional-type sliding base isolation system, as shown schematically in Fig. 1(b), is considered for the same eight-story bilinear elasto-plastic building described in Example 1. This type of isolators allows greater resistance to damage by permitting the building to slide on its foundation during severe earthquakes. This isolation system decouples the structure from its foundation with nearly frictionless teflon on stainless steel sliding plates that have very low frictional resistance.

When the sliding system slides, the frictional force, F_{sb}, developed in the sliding system is given by $F_{sb} = \mu \bar{m} g v_b$, in which $\bar{m}g=w$ is the weight of the building system above the sliding bearings and μ is the coefficient of friction. When the sliding system sticks, the frictional force F_{sb} is smaller than the one given above. In the dynamic analysis and active control of the sliding system, the highly nonlinear frictional force is represented in this paper by the following analytical function

$$F_{sb} = \mu \bar{m} g v_b \qquad \textbf{(25)}$$

in which v_b is a nondimensional hysteretic quantity described by Eq. (2) with i=b. It is mentioned that during the sliding phase, v_b takes a value of either 1 or -1. During the sticking phase, the absolute value of v_b is less than unity, i.e., $|v_b| \leq 1$. The conditions of sticking and sliding are accounted for by Eq. (2) automatically. The parameters governing the scale and general shape of the hysteresis loop of the sliding system are $A_b=1.0$, $\beta_b=0.5$, $n_b=2$, $\gamma_b=0.5$ and $D_{yb}=0.012$cm. The hysteresis loop of the hysteretic component, v_b, of such a sliding system is schematically shown in Fig. 3(b). The mass of the sliding system is $m_b=450$ metric tons and the coefficient of friction is $\mu=10\%$. The response vector is given by $\underline{X}=[x_b,x_1,\ldots,x_8]'$.

For the building equipped with the base sliding system alone and the earthquake ground excitation given in Fig. 2, time histories of all the response quantities were computed [6-7]. The time history of the first story deformation, $x_1(t)$, of the building without the base isolation system is shown in Fig. 4(a). The corresponding result when the building is equipped with the sliding isolation system is shown in Fig. 4(b). Other results were presented in Refs. 6-7. The maximum response quantities of the structure within 30 seconds of the earthquake episode are shown in Table 2 in the column designated as "With SIS". As observed from Table 2, the interstory deformations and floor accelerations are drastically reduced using a sliding type base isolation system. However, the deformation of the eighth story unit is still in the inelastic range.

To further reduce the structural response and to bring the deformation of the 8th story unit into the elastic range, an actuator is connected to the sliding system as shown in Fig. 1(b). The actuator will apply the active control force directly on the sliding system. The basic idea of the active force, in this case, is to counteract the frictional force in order to maintain the system in the sliding condition as much as possible, so that the transmission of the earthquake ground motion to the building can be kept to a minimum.

With the active control force, the structural response depends on the weighting matrices $\underline{R}$ and $\underline{Q}$. Again, the weighting matrix $\underline{R}$ consists of only one element, denoted by R_0, whereas the dimension of the $\underline{Q}$ matrix is (27x27). R_0 is chosen to be 10^{-3} for simplicity and the $\underline{Q}$ matrix is partitioned as shown in Eq. (24), where $\underline{Q}_{21}$, $\underline{Q}_{22}$ and $\underline{Q}_{23}$ are (9x9) matrices and $\underline{O}$ is a (18x9) zero matrix.

First, consider the case in which the hysteretic component vector $\underline{V}$ is not used as a feedback quantity for the determination of the control vector $\underline{U}(t)$. In this case, all the elements of the $\underline{Q}_{22}$ matrix are zero, i.e., $\underline{Q}_{22}=\underline{O}$. Note that the deformation of the building is in-phase with the frictional force and the velocity of the building is out of phase of the frictional force. As a result, we set $\underline{Q}_{23} = \underline{O}$ and only the elements in the $\underline{Q}_{21}$ matrix which correspond to the deformation of the building are needed. These elements are assigned in the following for illustrative purposes: $Q_{21}(1,j)=[0, -30{,}000, -3{,}000, -300, -30, -3, -0.3, -0.03, -0.003]$. All other elements of the $\underline{Q}_{21}$ matrix are set to be zero.

Time histories of all the structural response quantities were computed for $\alpha/R=2.25x10^8$ and $\Delta t=0.03x10^{-2}$ sec. The time history of the deformations of the first story unit is shown in Fig. 4(c). Other response quantities and hysteresis loops for the sliding system were also plotted in Ref. 6. Within 30 seconds of the earthquake episode, the maximum response quantities including the maximum control force, U_m, are summarized in column (A) of Table 2.

Secondly, the hysteretic component vector $\underline{V}$ is used for the feedback to determine the control vector $\underline{U}(t)$ and hence $\underline{Q}_{22} \neq \underline{0}$. Recall that the basic idea of installing an actuator is to overcome the frictional force in the sliding system in order to minimize the sticking phase of the motion during earthquakes. In other words, the active control force is intended to keep the sliding phase of the motion as much as possible. It is observed from Eq. (25) that the frictional force is proportional to v_b of the sliding system. Hence, a feedback of v_b will be sufficient to reduce the frictional force. Consequently, all elements of the $\underline{Q}_{21}$ and $\underline{Q}_{23}$ matrices are set to be zero. Further, since v_b of the sliding system is reflected only by the element $Q_{22}(1,1)$, all elements of the $\underline{Q}_{22}$ matrix are also set to be zero except $Q_{22}(1,1)$ that is chosen to be -10.258, i.e. $Q_{22}(1,1) = -10.258$.

The time history of the deformation of the first story unit, $x_1(t)$, is presented in Fig. 4(d) for $\alpha/R_0=3.39x10^9$ and $\Delta t=0.03x10^{-2}$ sec. Within 30 seconds of the earthquake episode, the maximum response quantities are summarized in column (B) of Table 2 for comparison. It is observed from Table 2 and Fig. 4 that (i) the performance of the present optimal algorithm with $\underline{V}$ feedback is superior to that of the optimal algorithm without considering $\underline{V}$ feedback, and (ii) the hybrid control system consisting of a frictional-type sliding system and an actuator is quite effective.

CONCLUSIONS

Optimal control for nonlinear and inelastic structural systems has been formulated incorporating the specific hysteretic model of the system. This is accomplished by including the hysteretic components of the structural system in the equations of motion which serve as constraints in the optimization process. The resulting control vector is obtained as a function of the deformations, velocities and hysteretic components of the structural system. The hysteretic components of the structural response can be estimated easily from a simple observer using the measured structural response quantities.

In this paper, the optimal control vector has been shown to satisfy not only the necessary conditions but also the sufficient condition of optimality. Specific applications of the optimal algorithm to two types of aseismic hybrid control systems have been demonstrated. These include (i) a lead-core rubber bearing isolation system connected to either an actuator or an active mass damper, and (ii) a frictional-type base sliding isolation system connected to an actuator. Numerical results indicate that (i) the performance of the present optimal algorithm is better than that of the algorithm without considering the feedback of hysteretic components of the structural response, (ii) the present optimal algorithm is simple for on-line operations and practical applications, (iii) the two types of aseismic hybrid control systems considered are effective for practical applications, and (iv) for active control of base-isolated buildings, the installation of sensors on the building is not necessary.

ACKNOWLEDGEMENTS

This research is partially supported by the National Science Foundation Grant Nos. NSF-BCS-89-04524 and BCS-91-96089 and National Center for Earthquake Engineering Research Grant No. NCEER-90-2202. Partial support by the University of California, Irvine through an allocation of computer time is acknowledged.

REFERENCES

1. Chong, K.P., Liu, S.C., and Li, J.C., Editors, Intel. Struc., Elsevier Applied Science Publishers, London and New York, 1990.
2. Housner, G.W. and Masri, S., Editors, Proc. U.S. Natl. Workshop on Struc. Control Research (in Press).
3. Yang, J.N, Danielians, A. and Liu, S.C., Aseismic Hybrid Control Systems For Building Structures Under Strong Earthquakes, Intel. Struc., ed. K.P. Chong, S.C. Liu and J.C. Li, Elsevier Applied Science Publshers, London, 1990, pp. 179-195, Proc. of International Workshop on Intelligent Structures, Taipei, Taiwan, July 23-26. Also J. of Intel. Mater. Sys. and Struc., 1, 4, October, 1990, pp. 432-446.
4. Yang, J.N., Danielians, A., and Liu, S.C., Aseismic Hybrid Control Systems for Building Structures, J. of Eng. Mech., ASCE, April 1991, 117, 4, pp. 836-853.
5. Yang, J.N., Li, Z., Danielians, A., and Liu, S.C., Instantaneous Optimal Control of Nonlinear Hybrid Systems, paper presented at the Eighth VPI&SU Symposium on Dynamics and Control of Large Structures, May 6-8, 1991, Blacksburg, VA; to appear in the Conference Proceedings.
6. Yang, J.N., Li, Z., and Danielians, A., Hybrid Control of Seismic-Excited Nonlinear and Inelastic Structural Systems I, to appear as NCEER technical report, 1991.
7. Yang, J.N., Li, Z., and Danielians, A., Hybrid Control of Seismic-Excited Nonlinear and Inelastic Structural Systems II, to appear as NCEER technical report, 1991.
8. Kelly, J.M., Leitmann, G., and Soldators, A.G., Robust Control of Base-Isolated Structures Under Earthquake Excitation, J. Optim. Theory and Appl., 1987, 53, 2, pp. 159-180.
9. Reinhorn, A.M., Manolis, G.D., and Wen, C.Y., Active Control of Inelastic Structures, J. of Eng. Mech., ASCE, 1987, 113, 3, pp. 315-332.
10. Masri, S.F., Bekey, G.A., and Caughey, T.K., On-Line Control of Nonlinear Flexible Structures, J. of Appl. Mech., ASME, 1981, 49, 4, pp. 871-884.
11. Yang, J.N., Akbarpour, A., and Ghaemmaghami, P., New Optimal Control Algorithms for Structural Control, J. of Eng. Mech., ASCE, June 1987, 113, 9, pp. 1369-1386.
12. Yang, J.N., Akbarpour, A., and Ghaemmaghami, P., Instantaneous Optimal Control Laws For Building Structures Under Earthquake Excitations, National Center for Earthquake Engineering Research, Technical Report NCEER-TR-87-0007, SUNY, Buffalo, 1987.
13. Yang, J.N., Long, F.X., and Wong, D.,"Optimal Control of Nonlinear Structures, J. of Appl. Mech., ASME, December 1988, 55, pp. 931-938.
14. Yang, J.N., and Soong, T.T., Recent Advancement in Active Control of Civil Engineering Structures, J. of Prob. Eng. Mech., December 1988, 3, 4, pp. 179-188.
15. Baber, T.T., and Wen, Y.K., Random Vibration of Hysteretic Degrading Systems, J. of Eng. Mech., ASCE, 1981, 107, EM6, pp. 1069-1987.
16. Park, Y.J., Wen, Y.K., and Ang, A.H-S., Two Dimensional Random Vibration of Hysteretic Structures, J. of Earthq. Eng. and Struc. Dyn., 1986, 14, pp. 543-557.
17. Wen, Y.K., Methods of Random Vibration for Inelastic Structures, J. of Appl. Mech. Reviews, February 1989, 42, 2, pp. 39-52.
18. Li, Z., Katukura, H., and Izumi, M., Synthesis and Extension of One-Dimensional Nonlinear Hysteretic Models, J. Eng. Mech., ASCE, 1991, 117, 1, pp. 100-109.
19. Kelly, J.M., Aseismic Base Isolation: Review and Bibliography, J. of Soil Dyn. and Earthq. Eng., 1986, 5, 3, pp. 202-216.
20. Contantinou, M.C., Mokha, A., and Reinhorn, A., Teflon Bearings in Base Isolation II: Modeling, J. of Struc. Eng., ASCE, February 1990, 116, 2, pp. 455-474.
21. Soong, T.T., Active Structural Control: Theory and Practice, Technical Report NCEER-87-0023, SUNY, August, 1989.
22. Papalambros, P.Y. and Wilde, D.J., Principles of Optimal Design, Chapter 3, Cambridge University Press, Cambridge, 1988.

Table 1: Maximum Response Quantities of Building Equipped with Rubber-Bearing Isolation System and Active Mass Damper (BIS & AMD) or Actuator (BIS & AF) : (A) = without $\underline{V}$ feedback; (B) = with $\underline{V}$ feedback

FLOOR NO	Yield Displ D_{yi}	without control		with BIS		BIS & AMD (A) U_m=234.4kN		BIS & AMD (B) U_m=196kN		BIS & AF (A) U_m=808.6kN		BIS & AF (B) U_m=799.2kN	
i	(cm)	x_{mi} (cm)	a_i (cm/s^2)	x_{mi} (cm)	a_i (cm/s^2)	x_{mi} (cm)	a_i (cm/s^2)	x_{mi} (cm)	a_i (cm/s^2)	x_{mi} (cm)	a_i (cm/s^2)	x_{mi} (cm)	a_i (cm/s^2)
B	-	-	-	42.70	158	24.21	118	23.3	118	26.11	117	25.83	116
1	2.4	3.05	486	1.31	155	0.69	112	0.66	111	0.86	118	0.85	116
2	2.3	2.80	456	1.25	152	0.69	109	0.65	108	0.82	123	0.81	121
3	2.2	3.14	571	1.28	157	0.72	105	0.68	106	0.83	122	0.82	120
4	2.1	2.95	529	1.17	164	0.68	94	0.64	97	0.74	116	0.72	114
5	2.0	2.70	558	1.07	177	0.63	101	0.59	98	0.66	125	0.65	123
6	1.9	2.91	720	0.97	185	0.59	111	0.54	106	0.58	117	0.57	115
7	1.7	2.90	580	0.82	198	0.53	119	0.49	117	0.48	131	0.47	129
8	1.5	2.08	617	0.52	206	0.36	142	0.33	130	0.33	129	0.32	127

Table 2 : Maximum Response Quantities of Building Equipped with Sliding Isolation System and Actuator (SIS & AF) : (A) = without $\underline{V}$ feedback; (B) = with $\underline{V}$ feedback

FLOOR NO	Yield Displ	without control		with SIS		(A) U_m=3074.7kN		(B) U_m=2321.3kN	
	D_{yi} (cm)	x_{mi} (cm)	a_i (cm/s^2)	x_{mi} (cm)	a_i (cm/s^2)	x_{mi} (cm)	a_i (cm/s^2)	x_{mi} (cm)	a_i (cm/s^2)
B	-	-	-	11.54	658	18.13	688	44.86	380.
1	2.4	3.05	486	1.30	370	0.56	325	0.50	214
2	2.3	2.80	456	1.23	547	0.76	342	0.50	239
3	2.2	3.14	571	1.38	531	0.91	379	0.56	230
4	2.1	2.95	529	1.40	404	0.83	300	0.68	217
5	2.0	2.70	558	1.50	424	0.89	379	0.78	242
6	1.9	2.91	720	1.72	462	1.00	412	0.74	298
7	1.7	2.90	580	1.82	541	1.09	341	0.93	218
8	1.5	2.08	617	1.64	601	0.99	391	0.87	344

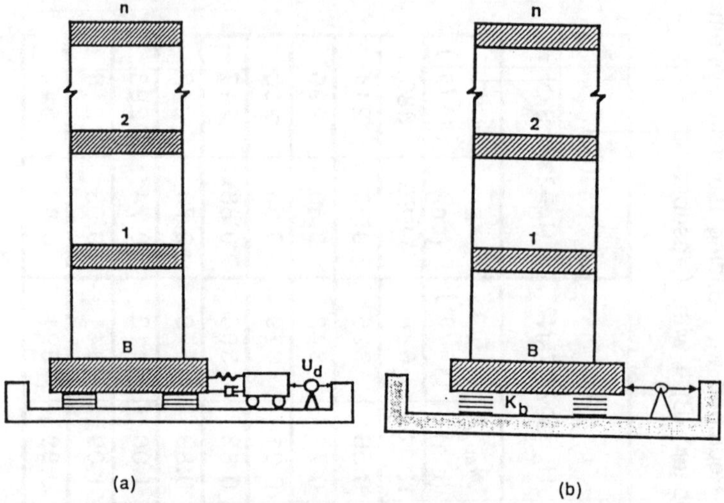

Fig.1 : Structural Model of a Multi-Story Building Equipped with Aseismic Hybrid Control Systems : (a) Rubber Bearing Isolation System and Active Mass Damper; (b) Base Sliding Isolation System and Actuator

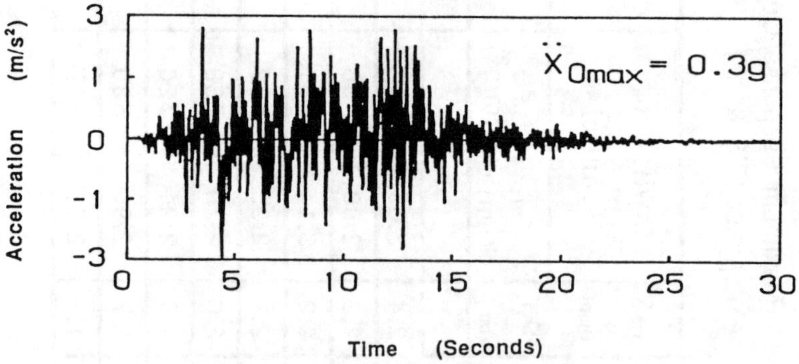

Fig.2 : A Simulated Earthquake Ground Acceleration

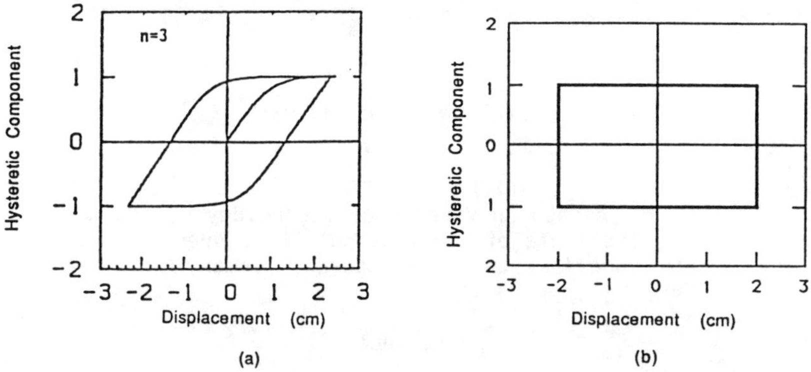

Fig.3 : Hysteretic Characteristics of Base Isolation System : (a) Rubber Bearing Isolation System; (b) Sliding Isolation System

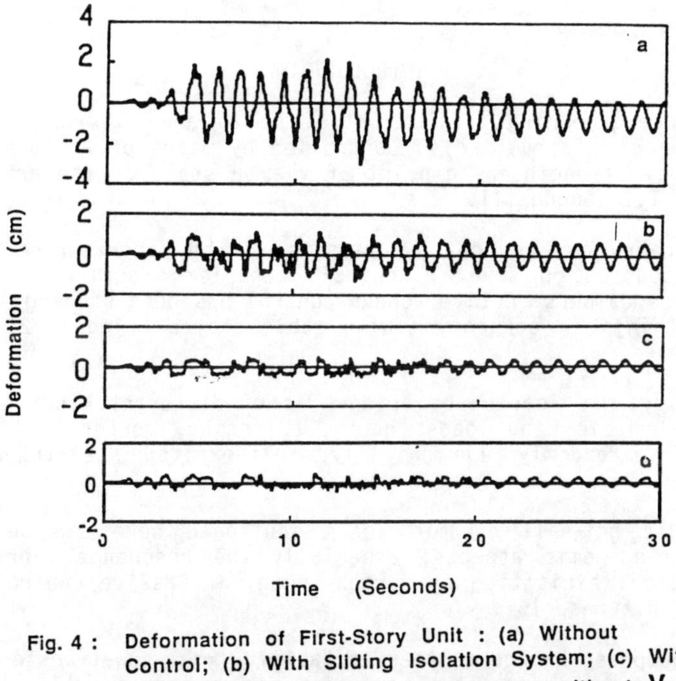

Fig. 4 : Deformation of First-Story Unit : (a) Without Control; (b) With Sliding Isolation System; (c) With Sliding Isolation System and Actuator without $\underline{V}$ feedback; (d) With Sliding Isolation System and Actuator with $\underline{V}$ feedback

ACTIVE CONTROL OF BRIDGES

GERHARD H. HIRSCH
Aachen University of Technology
Institute of Light Weight Structures
Wuellnerstrasse 7, Aachen, Germany

ABSTRACT

The objective of this paper is to present experiences of passive and active control research work. Activities of the Aachen University Working Group in control system design and practical aspects of their application are summarized. Know-how-transfer from other engineering disciplines and feature development work will be outlined.

INTRODUCTION

In structural engineering,active control is an area of research in which the deflection of a structure is controlled by means of actions of control system. From civil engineering point of view a state of the art review has been given by T.T. Soong |1|.

In order to carry out control experiments in the laboratories, investigations have focussed on several control mechanisms: active tendon control and active mass dampers. Active tendon control has been studied analytically among other things in connection with cable stayed bridges |2|, |3|, |4|, |5|.

However, the dynamic loading of bridges is of different kind: Traffic load (randomly) - Dynamic wind loads (gusts, buffeting, vortex-shedding, galloping, flutter), randomly, harmonically, self-excited - Earthquake (transient).

From the first not all dynamic loads mentioned here can be controlled actively from economic aspects, especially the resonance vibrations due harmonic critical excitation by vortex-shedding. Passive control has to be preferred in that case |6|.

To avoid or suppress bridge-flutter instability know-how-transfer from other disciplines of engineering (ship- and aircraft-engineering |7|, |8|) is possible and suitable in order to understand the control application.

More recently a comprehensive experimental study of tendon control war carried out by an Aachen University Working Group |9|. Active deformation control (ADC) of a 10 m girder was developped analytically and experimentally. Pneumatic cushions were supported on steel cables which were connected to both ends of the girder. Pneumatic pressures in the cushions were adjusted according to the deflection of the girder until the cable forces counteracted the sum of dead weight and live load.

This paper is concerned with the comparison between passive and active control of bridges. A project is presented which avoids difficulties in the field of reliability and liability. The total supervision justifies the additional means for the application of active elements in a pilot project of a 50 m bridge. Weak points in that structure will be detected automatically long before irreparable damage arises |10|.

DYNAMIC LOADS OF BRIDGES

The two most important dynamic loads for which long span bridges must be designed are moving loads due traffic and wind loads. Moreover, in some regions also seismic loads are to be considered.

Moving load

The most obvious examples of structures subjected to moving loads are highway and railway bridges. Analysis is complicated by the fact that the mass and the dynamic behaviour of moving vehicles are to compare with that of the bridge itself. Furthermore, there is a problem of interaction between motion of the bridge and the suspension of vehicles. Stresses under moving loads are much greater than under the same loads applied statically, especially if the riding surface is uneven. This is an important consideration in the design of suspension of cable-stayed bridges, as those problems of the Bangkok Rama IX cable bridge are shown in |13|. The main girder and the cables of the bridge have been subjected to intensive vibrations starting from the beginning, the brige was opened to traffic in December 1987. At that time the Bangkok cable bridge was the largest world-wide. The persistent vibrations of the cables were mainly the results of dynamic movements of the cable supports caused by heavy trucks travelling and to a lesser extent the result of wind actions. In order to reduce the vibration level it had been proposed to install cable dampers, and to carefully maintain the smoothness of the bridge pavements.

From structural control point of view it is to distinguish between the crawl deflection and the oscillation of the main girder about that. On a passive way, for example by means of tuned mass dampers, the vibration control is possible, but not the crawl deflection. Thoroughly, the working group of Aachen University has developped an active deformation control system to compensate the quasi-static load |9|.

Another important load is that of a force travelling at constant speed, but with a pulsating amplitude. This is an historical interest in connection with the vibration produced by the unbalanced forces of the driving wheels of steam locomotives. Modern railway locomotives do not generate dynamic loads of this kind.

Finally, footbridges may be analysed by assuming that pedestrian loading is equivalent to a moving pulsation force (BS 5400), and against vandalism, for example, passive control by means of tuned mass dampers is very economically |6|.

Wind load

Due to the potential susceptibility of the flexible long span bridges to wind, it has been the main focus of both experimental and theoretical researches |14|. The main effects of wind from dynamic point of view are the forced vibrations due to buffeting and vortex-shedding and the self-excited vibrations: galloping and flutter. Also the vortex-shedding resonance including lock-in effects are of self-excited type.

In case of forced vibrations with steady state dynamic response the structural damping is of great influence, and that reason passive structural control is available to alleviate the dynamic deflections of main girder and pylons. In case of bridge-flutter passive control is ineffective.Moreover, also the transient gust load can not alleviate remarkable by means of passive structural control.

Earthquake load

Cable structures as cable-stayed bridges have often been favoured for long span construction due to their leight weight: this is particularly advantageous for the design of structures subjected to earthquake. The observed seismic energy is typically concentrated between 3 Hz and 10 Hz, while the main girder of long span bridges have lower frequencies, and the supporting structures are having much higher natural frequencies. The result is that high frequency components are dynamically amplified by the support structure, and the low frequency components are attenuated. From that reason the base-isolation of main girder of long span bridges is to discuss, however, in passive isolation system the relative displacement between the bridge girder and piers tends to be excessive. Increasing damping of isolation system can reduce relative displacement, but will decrease the isolation effectiveness. To alleviate this situation the use of hybrid control system will be contemplated |15|.

STRUCTURAL CONTROL

As an additional redundancy in safety consideration, the possible application of passive and active control to civil engineering structures has been studied by many investigators during these past two decades |1|. Various applications and techniques have been introduced including the use of active tendons or active tuned mass dampers to control the response of slender structures, flexible bridges and flexible framed structures to unexpected excitations. Such disturbances included wind forces, moving loads, earthquakes or uneveness in bridge decks.

Passive control systems have been used in practice to-date |6|, |11|. Combining them with active systems may decrease the control energy consumption. Moreover, passive systems may be used to compensate such active control related problems as time delay and power failures.

Passive control in comparison to active control

In case of critical excitation of bridges due to vortex-shedding resonance the passive vibration control, for example, is more economically than the active control. Disadvantageous is that the tuning of each auxiliary tuned mass damper system is optimum for one of the critical frequencies. Nevertheless, with a restricted number of damping systems the passive control of the most important frequencies is possible. Here, for example, the Bangkok cable-stayed bridge is to be mentioned.

The total length of the bridge - figure 1 shows the elevation - is 782 m, composed of a 450 m main span and two 166 m back spans. The width of the bridge varies between 31 m and 33 m with three traffic lanes in each direction. There were three external dynamic loads which played an important role in the initial design: wind, traffic and seismic loads. Among these wind loads were assumed to be dominant. Wind tunnel tests with section models showed unacceptable vortex-induced flexural and torsional oscillations can be expected at wind speeds of 23 m/s and 32 m/s, respectively.

In order to reduce the amplitude of vortex-shedding excited amplitudes of the bridge deck and to remove the risk of galloping of the pylons, tuned mass dampers have been installed both in the deck structure and also in the pylons, as shown in figures 2 and 3. Tuned mass dampers have been tuned to first bending (0.32 Hz) and first torsion mode (0.71 Hz), the mass ratio is $\mu = 0.015$. Vortex-shedding vibrations have not been observed. However, the higher modes are not influenced by the tuned mass dampers, and therefore the heavy truck travelling moves the cable supports. Relating fatigue problems additional damping of the cables - hydraulic shock absorbers between bridge and cables near supports - had to be installed.

Moreover, the passive systems are not very effective in case of earthquake response alleviation. In earthquake (transient) excitation the maximum response is not influenced in the initial periods of oscillations. The random excitation results more influence of the control system and therefore some authors have proposed passive control also in case of earthquakes.

Finally, passive control devices are very effective in reducing footbrige vibrations. Many footbridges were made as all-welded steel box-girders, and the inherent vibration of such constructions is very low. The problem of vibration control is of classical nature and results in a provision of additional damping.

Figure 4 shows a cable-stayed footbridge |6| as an example of vibration control. The natural frequencies (1st and 2nd mode) of the 140 m span bridge were found to be 0.86 and 1.72 Hz and the log. decrement of damping was measured to 0.096 and 0.029, respectively. In case of the 2nd mode the longest cables are vibrating, too.

The mass of the auxiliary system was 2700 kg (1.84 % of reduced mass of bridge in case of fundamental vibration mode). The effective damping of the bridge with passive control was investigated to be 0.38 (log. decrement) in case of fundamental mode. Also the damping of 2nd mode increased at a factor of 2 approximately.

In |6| some more examples of passive control of bridges in case of wind induced vibrations are handled.

Active control

A research group of Aachen University 1981 has started interdisciplinary concerning active deformation control of civil engineering structures. First results have been presented on International Structural Conference at Waterloo University, Canada, 1985 |9|.

The aim of the theoretical and experimental investigations can be demonstrated by the hand of the test girder shown in figure 5. The 10 m girder functions as a normal prestressed concrete girder, when coupling links are inactive. The high cable tension results a small sag, when, subjected to increased loads (slowly moved force), the coupling links. They are pneumatic cushions in that active control concept and push the steel cables down until the resulting counterforces are equal to the sum of dead weight and life load. Figure 5 also shows the beam with and without active deformation control (ADC) with different load positions. The block diagram of control is depicted in figure 6.

This active control concept does not compare with the active tendon control, because the control force is not a direct control of cable tension. An electrical control system measures the fluctuation in the distance between girder and a constant reference - in the test example the floor - and computes the appropriate orders to set the pneumatic valves. This directly shows the need of a regulator for that ADC-girder. The cushion pressure, i.e. the volume flow into the cushions, is to be controlled in such a manner that, independent of varying load, the girder stays free of deflection without deviating from the above mentioned reference. Ideally, only the cables bear the load, while the concrete merely returns the compressive stress, and that fact is remarked in the ADC-concept presented here.

More details are described in |17|: Figure 7 obviously shows the combined efforts of cushion height and girder deflection with regulator, as load of 16 kN passes (slowly) over the girder.

Increasing load passing velocity results instability of control system. In |18| stability limits will be shown. Moreover, the surface roughness of test girder - and also of real bridge structures - induces oscillations of the girder. Suppression of most important frequency by means of additional passive control device (tuned mass damper) is possible, and has been successfully realized at test beam.

The further development of the ADC-concept at Aachen University will be the realization of full-scale application at 50 span bridge in cooperation with industry |10|.

Active bridge-flutter suppression

Experiences in active flutter suppression of wings in aeronautical engineering bring about a basis relating to active bridge-flutter suppression. Accordingly, in |19| results of model tests have been reported. Figure 8 shows the wind-tunnel test results of the bridge-section model. Two control wings are attached above the both ends of the bridge deck. Each control wing is fixed to the deck with a hinge at the mid-chord of the wing. Thus the control wing allows pitching motion. When flutter motion of the deck is detected, the wings are driven (by miniature DC motors installed inside of the bridge deck) in proportion to the deck. The aerodynamic forces acting on the wings can prevent the flutter instability of the deck. Model sizes are: 232 x 14 mm deck and 20 x 2 mm wings, the length is 340 mm. As figure 8 shows, the critical reduced flutter velocity has been increased from 2 (without) to 4 with control.

FINAL REMARKS

Finally, some remarks are necessary. Concerning the application of active structural control concept in civil engineering it is to distinguish between:

- Strength and stability
- Serviceability
- Comfort

Long time full scale test experiences will be needed to realize safe active control concepts in the most important relation to strength and stability of controlled structure. The realization concerning seviceability and comfort is easier, because the energy loss does not result catastrophic destabilization of the controlled structure.

Unfortunately, the realization of active controlled civil engineering structures and the full use of their remarkable advantages is not yet possible owing to the actual lack of official regulations. The new Aachen University project |10| avoids these difficulties in the field of reliability and liability. By abdication, the increase of load-bearing capacity allows the riskless test of the whole active system together with a complete analysis of the structure performance, its actual load and subsequent stress distribution released hereby. Such total supervision justifies the additional means for the application of active elements in that pilot project. Weak points in that structure will be detected automatically a long time before irreparable damage arises.

In critical comparison between active and passive control, in cases of steady state more or less harmonical oscillations at distinguished frequencies and at low structural damping, the passive control is most economically, and without problems due to energy loss. Other dynamic responses, especially due to transient loading (e.g. earthquake) can not be controlled passively with respect to success.

Experimental work in feature and international cooperation in this interesting field of civil engineering will be most important.

REFERENCES

1. Soong, T.T., State of-the-art review - Active structural control in civil engineering. Eng. Struct., 1988, 10, 74-84.

2. Murata, M. and Ito, N., Suppression of wind-induced vibrations of a suspension bridge by means of gyroscope. Proc. 3rd Int. Conf. of Wind Effects on Buildings and Structures. Tokyo, 1971, 1057-1066.

3. Yang, J.N. and Giannopoulos, F., Active control of two-cable stayed bridge. ASCE Journ. Eng. Mech. Div., 1979, 105, 50-68.

4. Yang, J.N. and Giannopoulos, F., Active control and stability of cable-stayed bridges. ASCE Journ. Eng. Mech. Div., 1979, 105, 677-694.

5. Meirovitch, L. and Ghosh, D., Control of flutter in bridges. ASCE Journ. Eng. Mech. Div., 1987, 113, 720-736

6. Aschrafi, M. and Hirsch, G., Control of wind-induced vibrations of cable-stayed bridges. Journ. Wind. Eng. and Ind. Aer., 1983, 14, 235-246

7. Ducane, P., The reduction of roll in ships. Transact. I.N.A., 76, 1964,

8. Sensburg, O., Becker, J. and Hönlinger, H., Active control of flutter and vibration of an aircraft. Structural Control (ed. H.H.E. Leipholz), IUTAM 1980, North-Holland Publ. Comp. & SM Publ.

9. Domke, H., Survey, advantages and restraints of ADC in civil engineering. Structural Control (ed. H.H. Leiphol. Martinus Nijhoff, Amsterdam, 1985, 172-184.

10. Domke, H., Active control of structures (in German). Bauingenieur, 1991, 66, 205-213.

11. Hüffmann, G., Freeman, R.A. and Brown, C.A., Dao Khanong cable stayed bridge tuned mass dampers in deck and pylon. Cable stayed bridges. Proc. of Int. Conf. Cable-stayed Bridges, Bangkok, 1987, 2, 1485-1493.

12. Malhorta, P.K. and Wieland, M., Tuned mass damper for suppressing wind effects in a cable-stayed bridge. Cable stayed bridges. Proc. of Int. Conf. Cable-stayed Bridges, Bangkok, 1987, 1, 557-568.

13. Wieland, M. and Lukkunaprasit, P., Dynamic strain measurements in three stay cables of main span of Rama IX cable-stayed bridge in Bangkok. AIT report, (AIT and Chulalongkorn University), January 1990.

14. Walther, R., Cable stayed bridges. Thomas Telford, London, 1985.

15. Feng. Q. and Shinozuka, M., Use of a variable damper for hybrid control of bridge response under earthquake. Proceedings U.S. National Workshop on Structural Control Research. USC-Publ. No. CE-9013. Los Angeles, October 1990.

16. Housner, G.W. and Masri, S.F. (ed.), Proceedings of the U.S. National Workshop on Structural Control Research. USC-Publ. No. CE-9013. Los Angeles, October 1990.

17. Bouten, H. and Meyr, H., Control design for ADC-girder. Structural Control (ed. H.H. Leipholz), M. Nijhoff, Amsterdam, 1985, 199-214.

18. Streck, D., Instabilitäten des geregelten Verhaltens aktiver unterspannter Biegetragwerke bei wachsender Lastüberfahrtgeschwindigkeit. Dr.-Thesis, RWTH Aachen, 1989.

19. Kobayashi, H. and Nagaoka, H., Active control of flutter of a suspension bridge. Proc. 8th Int. Conf. on Wind Engineering, London, Ontario, Canada, 1991 (in preparation)

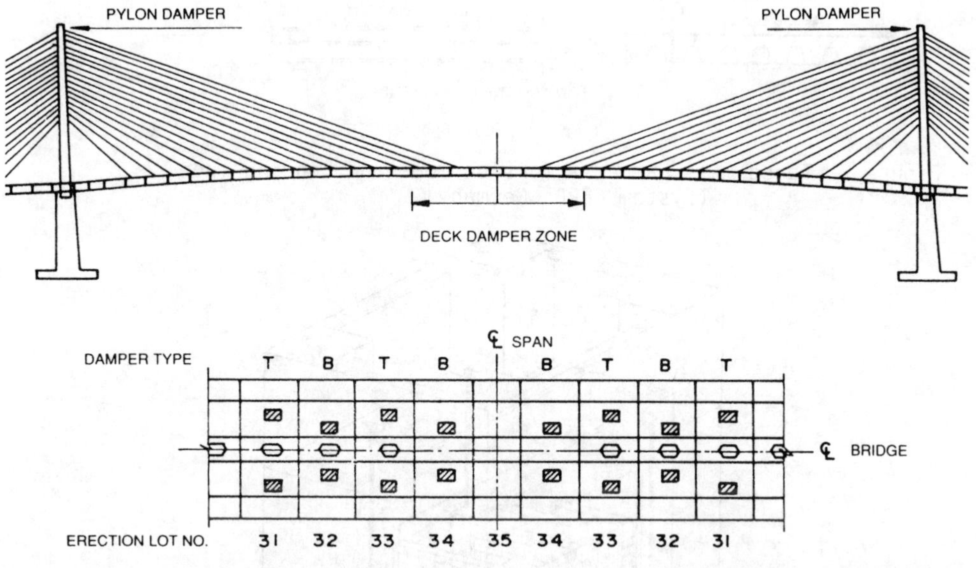

Figure 1. Elevation of Bangkok Rama IX cable-stayed bridge and location of tuned mass dampers

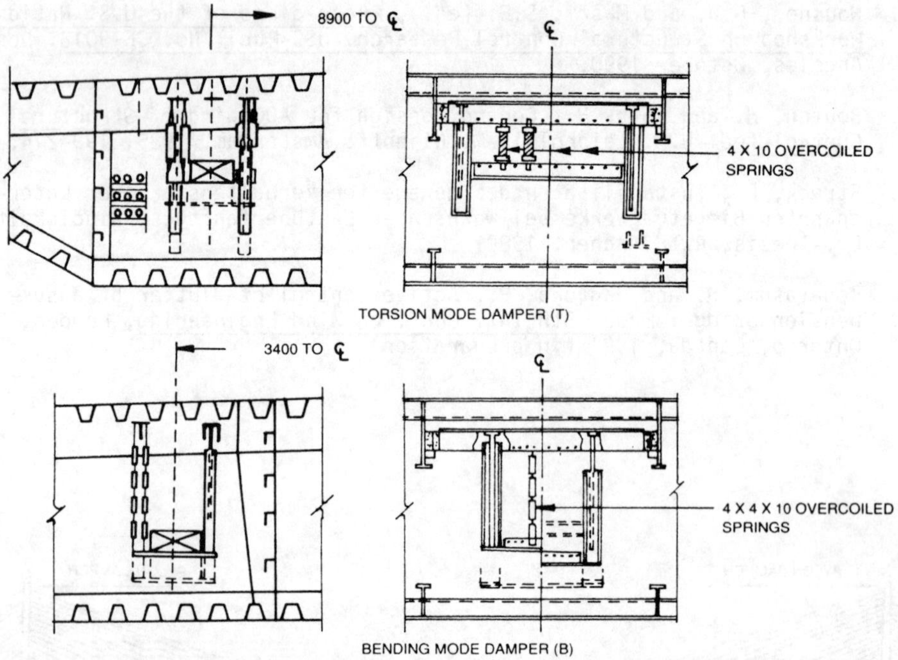

Figure 2. Tuned mass damper - Torsion mode and bending mode (System GERB, Germany)

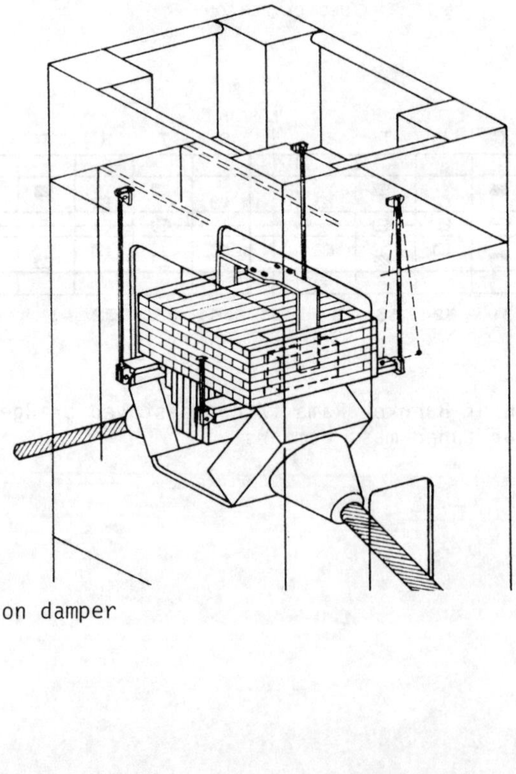

Figure 3. Pylon damper

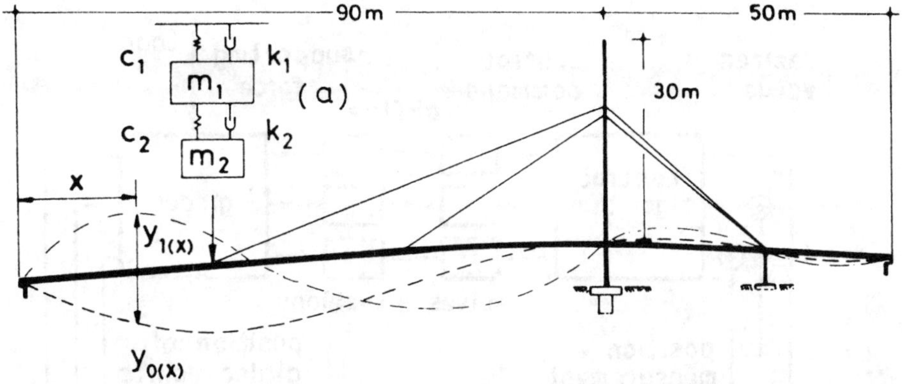

Figure 4. Cable-stayed footbridge
a) Bridge elevation and mathematical model of normalized system (fundamental bending mode)

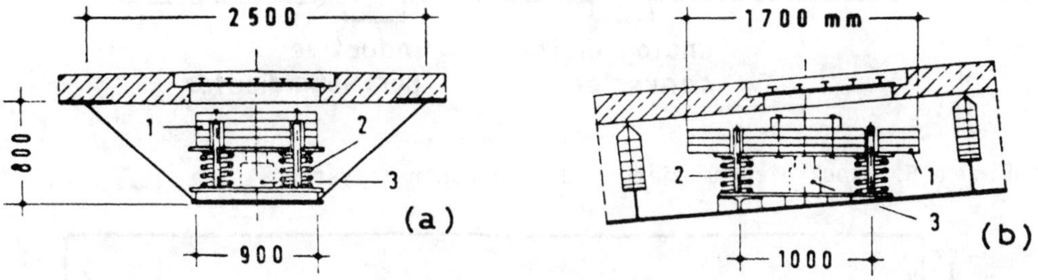

b) Tuned mass damper drawing, cross-section (a) longitudinal section (b)

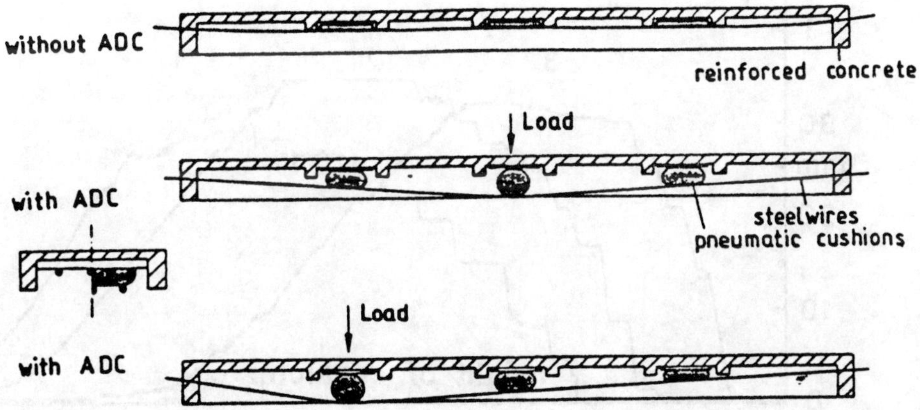

Figure 5. Test girder of Aachen University with and without active control device

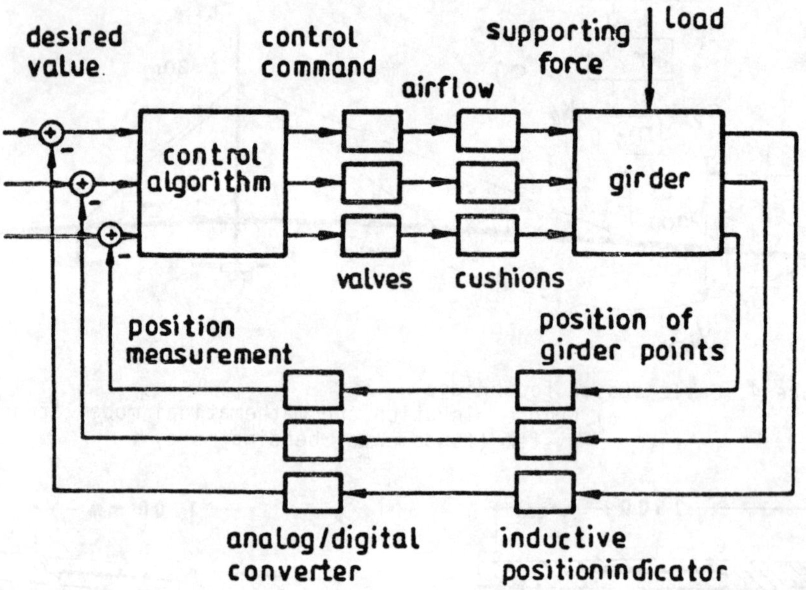

Figure 6. Block diagram of closed loop control system

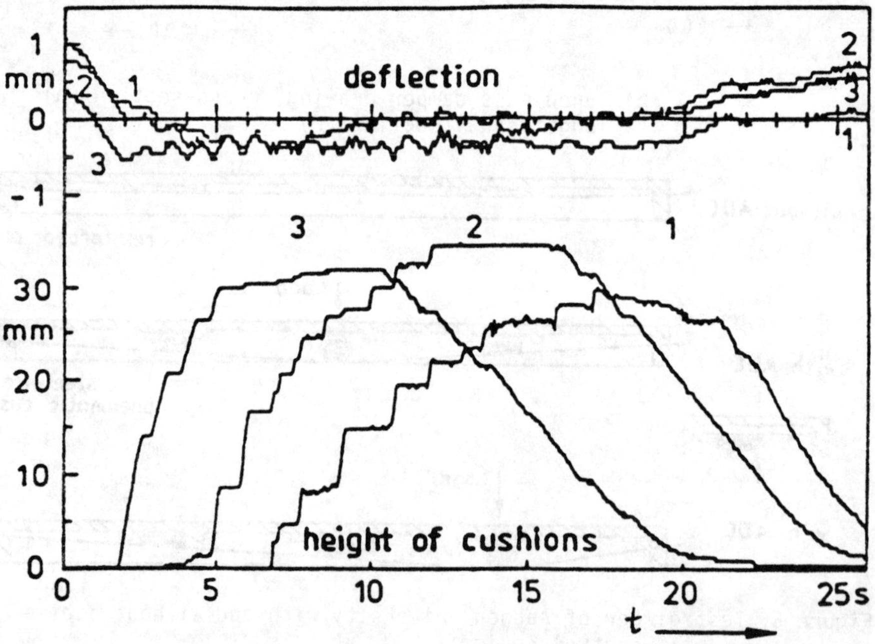

Figure 7. Cushion height and girder deflection with control

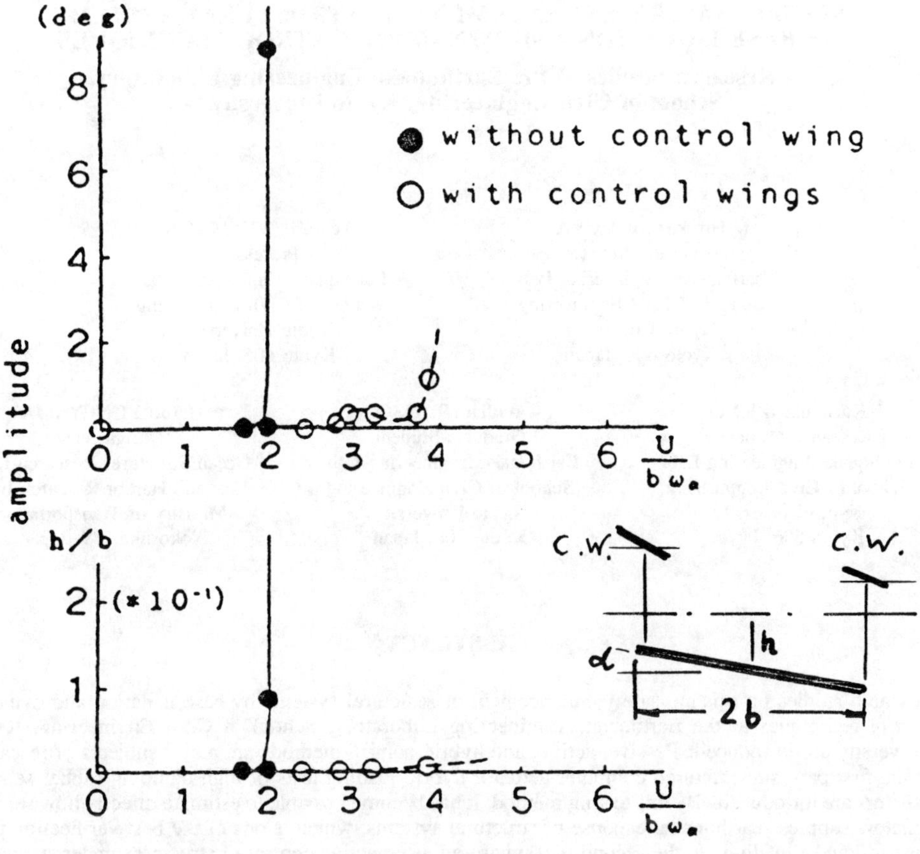

Figure 8. Response of bridge-deck with and without active flutter suppression

SEISMIC SAFETY ENHANCEMENT OF STRUCTURAL SYSTEMS BY BASE ISOLATION AND DYNAMIC CONTROL TECHNIQUES

— Research Studies at the Earthquake Engineering Laboratory, School of Civil Engineering, Kyoto University —

Hirokazu IEMURA
Associate Professor
Earthquake Engineering Lab.
School of Civil Engineering
Kyoto University
Kyoto 606, Japan

Yoshikazu YAMADA
Professor
Earthquake Engineering Lab.
School of Civil Engineering
Kyoto University
Kyoto 606, Japan

Kazuyuki IZUNO
Research Associate
Earthquake Engineering Lab.
School of Civil Engineering
Kyoto University
Kyoto 606, Japan

William TANZO
Graduate Student
Earthquake Engineering Lab.
School of Civil Engineering
Kyoto University
Kyoto 606, Japan

Haruo YONEYAMA
Engineer
Ocean Structures Research Lab.
Port and Harbor Research Inst.
Ministry of Transportation
Yokosuka 239, Japan

ABSTRACT

Research studies on seismic safety enhancement of structural systems by base isolation and dynamic control techniques at the Earthquake Engineering Laboratory, School of Civil Engineering, Kyoto University are introduced. Passive, active, and hybrid control methods are major subjects of research. In the first part, substructured computer-actuator hybrid loading tests of high-damping rubber seismic isolators are introduced. By this testing method, it has become possible to estimate directly how seismic isolators suppress earthquake response of structural systems, which is one of the best verification tests for earthquake loading. In the second part, phase-adjusted active control of structures under stochastic base excitation is investigated. The effectiveness of the phase-adjusted active control law including identification of ground motion is demonstrated with a numerical simulation. The results show that the proposed control law brings considerable control force reduction compared with the feedback control for the same level of response reduction in both cases.

INTRODUCTION

With vast improvement in recent construction techniques and materials, large-size and flexible structures such as high-rise buildings and long-span bridges have been designed and constructed in Japan. It is becoming critically important to suppress dynamic response of structures due to earthquake excitation, not only for their safety but also their serviceability.

There has been a lot of interests in developing active and passive control methods to reduce earthquake response. While active control methods are effective in controlling structural response under a wide range of situations, the amount of energy required to generate the control forces within a certain small time interval may be too great to be economical and practical. On the other hand, passive control methods such as mass dampers and base isolators are 'tuned' to be effective only in certain range of earthquake ground input. In addition, the displacement in a base-isolated structure may be excessive so as to cause girder fall-off in highway bridges. Recently, hybrid control has been proposed by combining active and passive control methods. The required energy to generate control forces is smaller because of the suppressed structural response due to passive control devices such as seismic isolators. Moreover, excessive displacement due to seismic isolation may be reduced by active control mechanism.

At the Earthquake Engineering Laboratory, School of Civil Engineering, Kyoto University, active and passive control as well as hybrid control methods are major subjects of analytical and experimental studies. A summary of the major researches on structural control are summarized in the following:

- Phase-adjusted active control of structures with identification of random earthquake ground motion [Yamada, Iemura et al (1990)];
- Analytical simulation of active, passive, and hybrid control methods [Iemura, Yamada, et al (1991)];
- Experimental verification of the proposed active control law [ongoing];
- Experimental testing on high-damping rubber seismic isolators [Yamada, Iemura, Tanzo, et al (1990)];
- Analytical studies on earthquake energy partitioning of bridge structures with seismic isolators [Iemura, Nakata, Ohshima (1990)];
- Analytical study on tuned mass dampers for site-dependent earthquakes [Quinn, Yamada, Iemura (1987)].

Current results obtained on the experimental testing on the high-damping rubber seismic isolators and phase-adjusted active control are presented in this paper.

TESTS ON HIGH-DAMPING RUBBER SEISMIC ISOLATORS

Extensive tests were conducted on high-damping rubber (HDR) seismic isolators. Seismic performance of HDR-supported systems can be evaluated under different earthquake ground motions. Resonant response under frequency-sweeping sine input excitations can also be tested. In addition, the normal procedure of repeated cyclic loading test can be conducted using the same testing facility. Most recently, a substructured hybrid loading system for MDOF bridge structures with seismic isolators has been developed. Thus, the integrated testing system can test for: (a) fundamental mechanical properties, (b) resonant response, and (c) earthquake response. This offers an ideal testing system in the early development stage, as well as for final verification and proof tests.

A series of high-damping rubber (HDR) seismic isolators with equivalent damping rated at about 15% had been tested. The specimens are of 25-cm × 25-cm square in plan. A 3-layer configuration, a 4-layer configuration, and a 9-layer configuration were fabricated to determine differences in stiffness and damping properties due to different heights. Each rubber layer has a thickness of 12mm and 1mm thin steel plates are interleaved between the rubber layers to form laminated elastomers with high vertical stiffness but are laterally flexible.

The setup shown in Fig. 1 was used in the test. A specimen is bolted to the underside of the rigid load-transfer beam and to a rigid platform that is attached to the rails of the strong reaction floor. Different heights of isolator specimens could be accommodated by inserting spacer plates between the specimen and the test platform. In this series of tests, the load-transfer beam was controlled to impose horizontal displacements while maintaining a constant axial load on the isolator specimen.

Basic Formulation of Hybrid Loading Test

On-line hybrid test (pseudo-dynamic test) is a computer-controlled experimental technique in which direct numerical time integration is used to solve the equations of motion. The computed displacement at each step is statically imposed on a specimen through a computer-controlled load actuators in order to measure its restoring forces at the current deformation state. The measured restoring forces are then fed into the equations of motion to compute the next set of displacements.

The equations of motion of a nonlinear structural system may be expressed as follows:

$$[M]\{\ddot{u}\}_t + [C]\{\dot{u}\}_t + \{R_f\}_t = \{P\}_t \tag{1}$$

where $\{u\}_t$, $\{\dot{u}\}_t$, $\{\ddot{u}\}_t$: displacement, velocity, and acceleration vector at time t;
$[M]$, $[C]$: mass and viscous damping matrices;
$\{R_f\}_t$: restoring forces to be measured experimentally from a specimen loaded based on its current deformation state $\{u\}_t$;
$\{P\}_t$: load vector at time t.

In an on-line hybrid test, the restoring force vector $\{R_f\}_t$ is measured from a specimen loaded based on its current deformation state. Using an explicit integration scheme such as the central difference scheme, the equations of motion can be solved to obtain the displacement vector at the current step $\{u\}_{t+\Delta t}$ depending only on quantities computed in the previous steps.

$$\{u\}_{t+\Delta t} = \frac{\{P\}_t - \{R_f\}_t + \left(\frac{2}{\Delta t^2}[M]\right)\{u\}_t - \left(\frac{1}{\Delta t^2}[M] + \frac{1}{2\Delta t}[C]\right)\{u\}_{t-\Delta t}}{\frac{1}{\Delta t^2}[M] - \frac{1}{2\Delta t}[C]} \tag{2}$$

Substructured Hybrid Method

In the basic formulation of on-line hybrid test, specimen modeling the total structural system is fabricated and the total restoring force vector is obtained experimentally. However, inelastic deformations in structures during strong earthquake are often confined to some localized regions. By incorporating substructuring concepts into the basic form of on-line hybrid test, only the critical regions need to be fabricated and tested.

A complete structural system is considered to be divided into analytical substructures and experimental substructures. Portions of the structure that are as yet difficult to model mathematically are taken as experimental substructures. The remaining portions are taken as analytical substructures in which presently available analytical models are used to describe their restoring-force characteristics. For instance, the critical seismic isolators may be taken as the experimental substructures. The equations of motion may be similarly expressed as in Eqn. 1. However, the contributions into the restoring-force vector $\{R_f\}_t$ come from both analytical models and measured restoring forces from the loaded specimen.

Using the conditionally stable explicit integration method in substructured on-line hybrid tests of a MDOF system would necessitate extremely small integration time intervals in order to satisfy the required numerical stability criterion. This requirement is especially stringent in systems such as structural frame where the axial and rotation D.O.F.s are very much stiffer than the lateral ones. A class of so-called mixed integration methods originally developed for finite-element analysis of structure-fluid interaction has been found to be suitable for substructured hybrid tests.

To implement the OS algorithm, Eqn. (1) is expressed as follows:

$$[M]\{\ddot{u}\}_{t+\Delta t} + [C]\{\dot{u}\}_{t+\Delta t} + \{R_f\}^{linear}_{t+\Delta t} + \{R_f\}^{nonlinear}_{t+\Delta t} = \{P\}_{t+\Delta t} \tag{3}$$
$$[M]\{\ddot{u}\}_{t+\Delta t} + [C]\{\dot{u}\}_{t+\Delta t} + [K]\{u\}_{t+\Delta t} + (\{R_f\}_{t+\Delta t} - [K]\{\tilde{u}\}_{t+\Delta t}) = \{P\}_{t+\Delta t}$$

An explicit trapezoidal rule for the predictive displacement vector is given by:

$$\{\tilde{u}\}_{t+\Delta t} = \{u\}_t + \Delta t\{\dot{u}\}_t + \left(\Delta t^2/4\right)\{\ddot{u}\}_t \tag{4}$$

On the other hand, an implicit trapezoidal rule (Newmark method with $\gamma = 0.5$ and $\beta = 0.25$)

$$\{u\}_{t+\Delta t} = \{\tilde{u}\}_{t+\Delta t} + \left(\Delta t^2/4\right)\{\ddot{u}\}_{t+\Delta t} \tag{5}$$

$$\{\dot{u}\}_{t+\Delta t} = \{\dot{u}\}_t + (\Delta t/2)(\{\ddot{u}\}_t + \{\ddot{u}\}_{t+\Delta t}) \tag{6}$$

Substituting the displacement and velocity expressions given above into Eqn. (2),

$$\{\ddot{u}\}_{t+\Delta t} = \frac{\{P\}_{t+\Delta t} - \{R_f\}_{t+\Delta t} - [C]\{\dot{u}\}_t - [C](\Delta t/2)\{\ddot{u}\}_t}{([M] + [C]\Delta t/2 + [K]\Delta t^2/4)} \tag{7}$$

It has been shown [Plesha and Belytschko, 1985] that the OS integration method is unconditionally stable for materials with decreasing stiffness (hysteresis is of 'softening' type).

Three DOF General In-Plane Loading System

A 3-DOF general in-plane loading system capable of subjecting a specimen under combined axial, shear, and bending loads has been developed and implemented. The general set-up of the test bed and test rig is shown in Fig. 1. One actuator is attached horizontally to the reaction wall, while two others are hanged vertically from the overhanging reaction girder. A rigid load-transfer beam was fabricated to transfer displacements and forces between the actuators and the specimen. Universal joints consisting of swivel heads and swivel bases are attached to both ends of each actuator. The same loading system has been used to test RC and steel box-girder columns, and high-damping rubber seismic isolators.

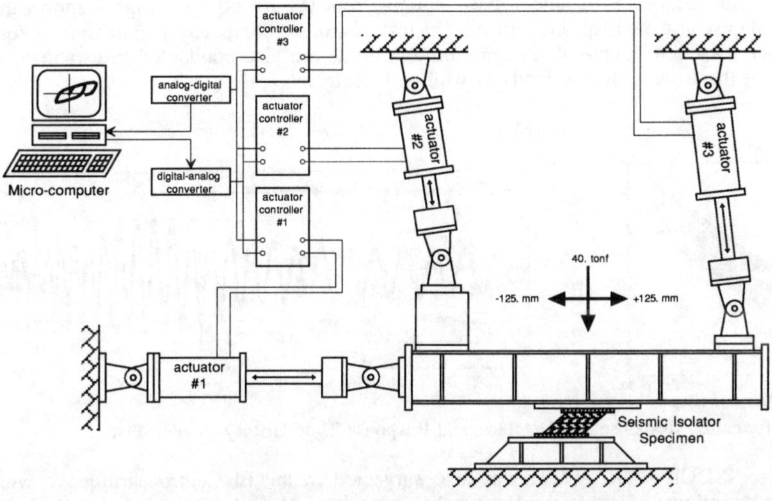

Fig. 1 Set-up for Testing Seismic Isolator

The control computer used in these sets of experiments is a TEAC PS-9000 series 300 technical computer (Yokogawa Hewlett-Packard) that is equipped with sophisticated interface programming control. An internal GPIB (IEEE 488 standard) interface bus is used to connect the analog-digital (A/D) converter, hard disks and other storage media, printers, pen plotters, and other peripherals. Digital strain reading device is also attached through the GPIB bus. Three channels of 12-bit digital-analog (D/A) converters are multiplexed through a Serial Output I/F module to the GPIO (General-Purpose Input/Output has 16 input lines and 16 output lines) interface in the control computer. In substructured hybrid tests, a Sony NEWS workstation or a PC microcomputer is connected by RS-232C serial interface to handle all the computational tasks, while the TEAC computer is solely devoted to controlling the actuators and other instruments.

Test Results

Fig. 2 shows the static load-deformation behavior of the HDR isolator subjected to cyclic displacements of ±25%, ±50%, ..., ±150% of the specimen height with two repetitions of each cycle. The specimen was finally loaded to ±200% to check its behavior at extreme overload situations. Other tests under this mode include different levels of axial loads and effects of loading rate.

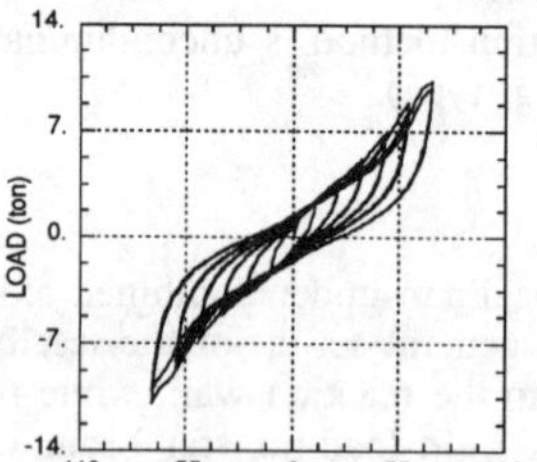

Fig. 2 Cyclic Load-Deformation

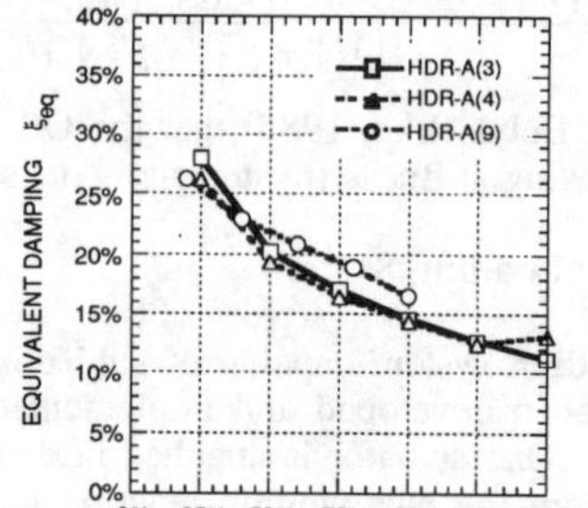

Fig. 3 Equivalent Damping

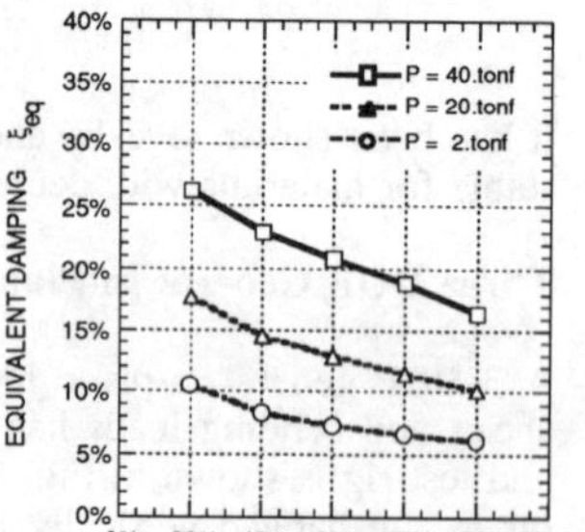

Fig. 4 Effect of Axial Load

A HDR-isolated structure is then modeled as a 2-sec SDOF system and subjected to sweeping sine excitation with frequency of sine waves ranging from 0.2 to 2.0 cps. Fig. 5 shows the load-deformation response and the displacement time history response (compared with analytical equivalent linear response). Different levels of acceleration amplitude are also conducted to establish resonant characteristics of the isolator at low loads as well as high loads.

(a) (b)

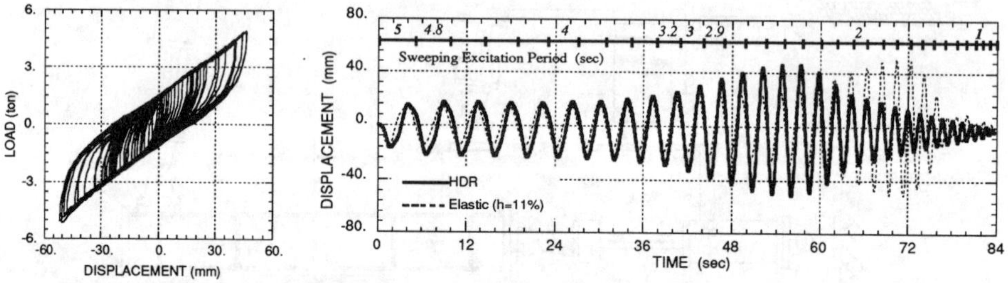

Fig. 5 Load-Deformation Response and Displacement Response Time History (Sweep Test)

The 2-sec SDOF HDR system was also subjected to the El Centro earthquake which has relatively short-period waves and to the Hachinohe earthquake which has relatively long-priod waves. The load-deformation response and displacement response of the HDR isolator under 0.1g El Centro earthquake are shown in Fig. 6. The load-deformation response and displacement response of the HDR isolator under 0.1g Hachinohe earthquake are shown in Fig. 7.

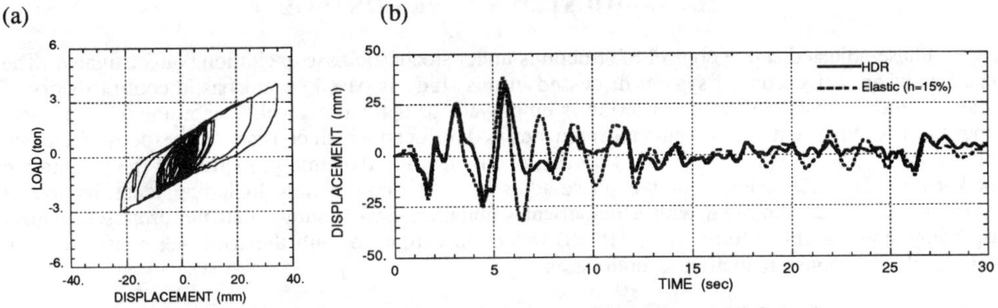

Fig. 6 Load-Deformation Response and Displacement Response Time History (El Centro)

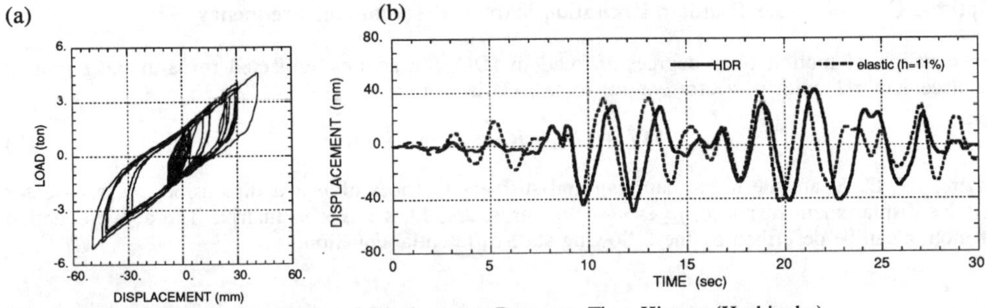

Fig. 7 Load-Deformation Response and Displacement Response Time History (Hachinohe)

Most recently, a substructured hybrid loading system for structures with seismic isolators has been developed. Fig. 8 shows two examples of a base-isolated building and a seismic isolated bridge pier. In the substructured loading test, the isolator elements are taken as experimental substructures and are tested. This test provides a proof test of the effectivenes of the isolator for a specific application. After the verification tests of the HDR isolators presented above, the second phase of the testing on the HDR isolators includes substructured hybrid loading test of the isolator for inelastic earthquake response of a MDOF bridge structure model.

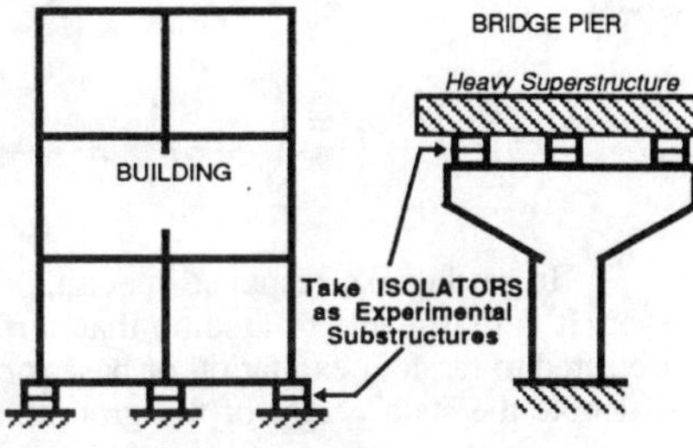

Fig. 8 Substructured Test of Structures with Seismic Isolators

The following main results have been obtained from this study:

- From cyclic loading tests, it has been confirmed that the HDR isolators bahave as bilinear elements without strain hardening when strain is below 100%. Equivalent stiffness and damping decreases with increasing strain. Also, lateral stiffness decreases but damping increases with increasing axial bearing pressure.
- From resonant tests using sweeping sine excitation, the resonant response of a 2.sec SDOF HDR system was obtained and has been found to possess sufficient damping and resistance (especially with increasing strain and strain hardening) to suppress large amplification.
- Under El Centro earthquake, equivalent damping of about 15% has been observed for strain of about 100%, the same level obtained from static tests. However, equivalent damping of only 11% was observed when subjected to the Hachinohe earthquake which contains long-period waves.
- Finally, the performance of HDR isolators as part of a MDOF bridge structures can be determined and verified by the newly developed substructured hybrid loading system.

PHASE-ADJUSTED ACTIVE CONTROL

Phase-adjusted active control of structures under stochastic base excitation is investigated. The synthesis of an active control system discussed in this study is based on stochastic control theory in which an energy-based stochastic criterion is employed for optimal control. Accordingly, an optimal control law for input with a predominant frequency is developed which consists of a response feedback portion and a ground-input feedforward portion even when the dynamic properties of the ground are not known. The effectiveness of the phase-adjusted active control law including identification of ground motion is demontrated with a numerical simulation, which shows that the proposed control law brings considerable control force (RMS) reduction compared with the feedback control for the same level of response reduction in both cases.

Optimal Control Under Random Excitation With A Predominant Frequency

An equation of motion of n degrees-of-freedom (DOF) structures subjected to earthquake ground displacement $z(t)$ with m-vector control force $\boldsymbol{u}(t)$ is written as

$$\boldsymbol{M}\ddot{\boldsymbol{y}} + \boldsymbol{C}\dot{\boldsymbol{y}} + \boldsymbol{K}\boldsymbol{y} = -\underline{m}\ddot{z} + \boldsymbol{D}\boldsymbol{u} \tag{8}$$

where, $\boldsymbol{M}, \boldsymbol{C}, \boldsymbol{K}$ are the mass, damping and stiffness matrices of $n \times n$ dimensions, $\boldsymbol{y}$ is n-vector relative displacement response, $\underline{m}$ is n-vector mass, and $\boldsymbol{D}$ is a $n \times m$ matrix. The above equation of motion can be described by the following state differential equation:

$$\dot{\boldsymbol{x}} = \boldsymbol{A}\boldsymbol{x} + \boldsymbol{B}\boldsymbol{u} + \boldsymbol{G}\ddot{z} \tag{9}$$

where

$$\boldsymbol{x} = \begin{pmatrix} \boldsymbol{y} \\ \dot{\boldsymbol{y}} \end{pmatrix} \quad \boldsymbol{A} = \begin{pmatrix} \mathbf{o} & \boldsymbol{I} \\ -\boldsymbol{M}^{-1}\boldsymbol{K} & -\boldsymbol{M}^{-1}\boldsymbol{C} \end{pmatrix} \quad \boldsymbol{B} = \begin{pmatrix} 0 \\ \boldsymbol{M}^{-1}\boldsymbol{D} \end{pmatrix} \quad \boldsymbol{G} = \begin{pmatrix} 0 \\ -\boldsymbol{M}^{-1}\underline{m} \end{pmatrix}$$

In earthquake response spectra, it is often found that ground motion is affected by soft surface layer. It is reasonable to assume that surface ground motion is the dynamic response of surface layer subjected to random excitation of base rock, from which the Kanai-Tajimi model is developed. Letting $\boldsymbol{Z}$ denote the state vector of the ground, an equation of motion is written as

$$\dot{\boldsymbol{Z}} = \boldsymbol{A}_z\boldsymbol{Z} + \boldsymbol{G}_z w \tag{10}$$

where, w : a white-noise process with intensity V_0

$$\boldsymbol{Z}^T = (z, \dot{z}) \qquad \boldsymbol{A}_z = \begin{pmatrix} 0 & 1 \\ -\omega_g^2 & -2\zeta_g\omega_g \end{pmatrix} \qquad \boldsymbol{G}_z^T = (0, 1)$$

Then, the power spectrum of the ground acceleration is obtained as

$$P(\omega) = |H(i\omega)|^2 = \frac{\left(\dfrac{\omega}{\omega_g}\right)^4}{\left\{\left(\dfrac{\omega}{\omega_g}\right)^2 - 1\right\}^2 + 4\zeta_g^2\left(\dfrac{\omega}{\omega_g}\right)^2} V_0 \tag{11}$$

Combining Eq. (9) for structures and (10) for surface ground layers, an equation of motion of total system is described as

$$\dot{\tilde{\boldsymbol{x}}} = \tilde{\boldsymbol{A}}\tilde{\boldsymbol{x}} + \tilde{\boldsymbol{B}}\boldsymbol{u} + \tilde{\boldsymbol{G}}w(t) \tag{12}$$

where

$$\tilde{\boldsymbol{x}} = \begin{pmatrix} \boldsymbol{x} \\ \boldsymbol{Z} \end{pmatrix}_{(n+2)\times 1} \qquad \tilde{\boldsymbol{A}} = \begin{pmatrix} \boldsymbol{A} & \boldsymbol{G}\boldsymbol{D}_z \\ \boldsymbol{O} & \boldsymbol{A}_z \end{pmatrix}_{(n+2)\times(n+2)}$$

$$\tilde{\boldsymbol{B}} = \begin{pmatrix} \boldsymbol{B} \\ \boldsymbol{O} \end{pmatrix}_{(n+2)\times m} \qquad \tilde{\boldsymbol{G}} = \begin{pmatrix} \boldsymbol{G} \\ \boldsymbol{G}_z \end{pmatrix}_{(n+2)\times 1} \qquad D_z = \begin{pmatrix} 0 & 1 \end{pmatrix} \boldsymbol{A}_z$$

As an optimal control strategy, minimization of the objective function with quadratic form is employed. Assuming that $\ddot{z}(t)$ and $\boldsymbol{y}(t)$ are stationary random processes, the expected total energy of the structural system is expressed as,

$$J = E\left[\tilde{\boldsymbol{x}}^T \tilde{\boldsymbol{R}}_1 \tilde{\boldsymbol{x}} + \boldsymbol{u}^T \boldsymbol{R}_2 \boldsymbol{u}\right] \longrightarrow \min \tag{13}$$

With the Riccati matrix $\tilde{\boldsymbol{P}}$ calculated from next equation,

$$\tilde{\boldsymbol{R}}_1 - \tilde{\boldsymbol{P}}\tilde{\boldsymbol{B}}\boldsymbol{R}_2^{-1}\tilde{\boldsymbol{B}}^T\tilde{\boldsymbol{P}} + \tilde{\boldsymbol{A}}^T\tilde{\boldsymbol{P}} + \tilde{\boldsymbol{P}}\tilde{\boldsymbol{A}} = \boldsymbol{O} \tag{14}$$

the optimal control force $\boldsymbol{u}(t)$ is obtained as

$$\boldsymbol{u}(t) = -\boldsymbol{R}_2^{-1}\tilde{\boldsymbol{B}}^T\tilde{\boldsymbol{P}}\tilde{\boldsymbol{x}}(t) = \tilde{\boldsymbol{F}}\tilde{\boldsymbol{x}}(t) = \boldsymbol{F}_x\boldsymbol{x} + \boldsymbol{F}_z\boldsymbol{Z} \tag{15}$$

From the above equation, it is easily understood that the optimal control force is a function of the structural response $\boldsymbol{x}$ and the input ground motion $\boldsymbol{Z}$. This is schematically illustrated in Fig. 9, where $\boldsymbol{F}_x\boldsymbol{x}$ and $\boldsymbol{F}_z\boldsymbol{Z}$ are called closed-loop and open-loop control, respectively. $\boldsymbol{F}_x$ and $\boldsymbol{F}_z$ are called feedback and feedforward gain, respectively.

Identification of Ground Parameters

In using Eq. (10), we have to know the natural frequency of the ground ω_g and the damping constant ζ_g. To get the optimal feedforward gain even when these parameters are not known, identification theory is introduced.

Using the linear acceleration method, Eq. (10) can be expressed in the discretized form;

$$z_{t+1} = A_z z_t + G_z w_{t+1} \tag{16}$$

where

$$z_t = \begin{pmatrix} z_t \\ \dot{z}_t \\ \ddot{z}_t \\ \zeta_{gt} \\ \omega_{gt} \end{pmatrix} \qquad G_z = \frac{1}{1 + \Delta t \zeta_g \omega_g + \frac{\Delta t^2}{6}\omega_g^2} \begin{pmatrix} \frac{\Delta t^2}{6} \\ \frac{\Delta t}{2} \\ 1 \\ 0 \\ 0 \end{pmatrix} \qquad \Delta t\text{: interval of calculation}$$

$$A_z = \frac{1}{1+\Delta t\zeta_g\omega_g+\frac{\Delta t^2}{6}\omega_g^2} \times \begin{pmatrix} 1+\Delta t\zeta_g\omega_g & \Delta t+\frac{2}{3}\Delta t^2\zeta_g\omega_g & \frac{\Delta t^2}{3}+\frac{\Delta t^3}{6}\zeta_g\omega_g & 0 & 0 \\ -\frac{\Delta t}{2}\omega_g^2 & 1-\frac{\Delta t^2}{3}\omega_g^2 & \frac{\Delta t}{2}-\frac{\Delta t^3}{12}\omega_g^2 & 0 & 0 \\ -\omega_g^2 & -2\zeta_g\omega_g-\Delta t\omega_g^2 & -\Delta t\zeta_g\omega_g-\frac{\Delta t^2}{3}\omega_g^2 & 0 & 0 \\ 0 & 0 & 0 & 1+\Delta t\zeta_g\omega_g+\frac{\Delta t^2}{6}\omega_g^2 & 0 \\ 0 & 0 & 0 & 0 & 1+\Delta t\zeta_g\omega_g+\frac{\Delta t^2}{6}\omega_g^2 \end{pmatrix}$$

As the equation above is nonlinear for ω_g and ζ_g, extended Kalman filtering technique is used to identify these parameters. The filtering algorithm is shown in Table 1.

Table 1 Filter variables and Kalman filtering algorithm

Variable	Definition	Dimension
$\hat{x}(k/k)$	State estimate at t_k given y_k	$n \times 1$
$P(k/k)$	Covariance matrix of the error in $\hat{x}(k/k)$	$n \times n$
$\Phi(k+1/k)$	State transition matrix (from t_k to t_{k+1})	$n \times n$
$\Gamma(k+1/k)$	System noise coefficient matrix	$n \times r$
$Q(k)$	System noise covariance matrix	$r \times r$
$\hat{x}(k+1/k)$	State estimate at t_{k+1} given y_k	$n \times 1$
$P(k+1/k)$	Covariance matrix of the error in $\hat{x}(k+1/k)$	$n \times n$
$M(k+1)$	Measurement matrix	$m \times n$
$R(k+1)$	Measurement noise covariance matrix	$m \times m$
$K(k+1)$	Filter (Kalman) gain matrix at t_{k+1}	$n \times m$
y_{k+1}	Measurement (observation) at t_{k+1}	$m \times 1$
1.	Store the filter state $[\hat{x}(k/k),\ P(k/k)]$;	
2.	Compute the predicted state $\hat{x}(k+1/k) = \Phi(k+1/k)\cdot\hat{x}(k/k)$;	
3.	Compute the predicted error covariance matrix $P(k+1/k) = \Phi(k+1/k)\cdot P(k/k)\cdot\Phi^T(k+1/k)+\Gamma(k+1/k)\cdot Q(k)\cdot\Gamma^T(k+1/k)$;	
4.	Compute the filter gain matrix $K(k+1) = P(k+1/k)\cdot M^T(k+1)[M(k+1)\cdot P(k+1/k)\cdot M^T(k+1)+R(k+1)]^{-1}$;	
5.	Process the observarion y_{k+1} $\hat{x}(k+1/k+1) = \hat{x}(k+1/k)+K(k+1)[y_{k+1}-M(k+1)\cdot\hat{x}(k+1/k)]$;	
6.	Compute the new error covariance matrix by $P(k+1/k+1) = [I-K(k+1)\cdot M(k+1)]P(k+1/k)$;	
7.	Set $k=k+1$, and return to step 1.	

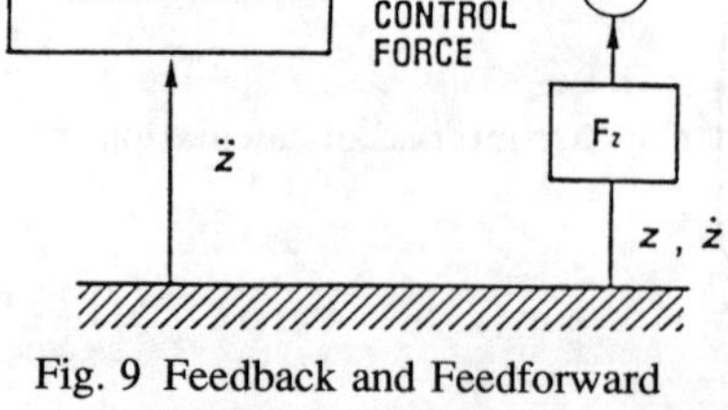

Fig. 9 Feedback and Feedforward Control System

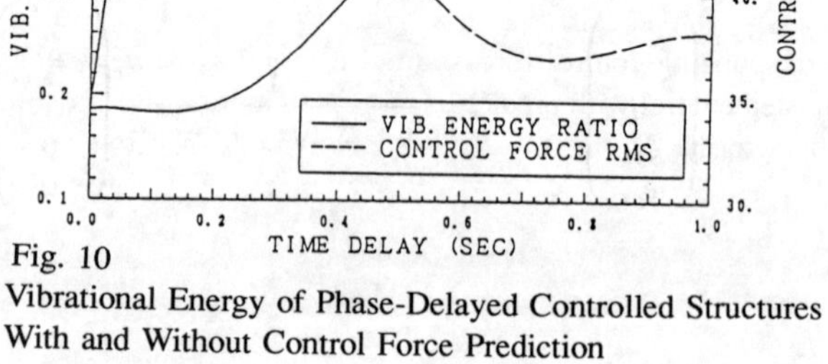

Fig. 10 Vibrational Energy of Phase-Delayed Controlled Structures With and Without Control Force Prediction

Phase-Adjusted Active Control

In order to overcome the effects of phase lag between measurement and activation of control force, earthquake response is predicted using Kalman filtering technique; and from the predicted response, optimal control force is applied.

Using the linear acceleration method, equation of motion of a SDOF structure with control force u is expressed in the following discretized form.

$$\mathrm{x}_{t+1} = A\mathrm{x}_t + Bu_{t+1} + Gz_{t+1} \tag{17}$$

where

$$\mathrm{x}_t = \begin{pmatrix} x_t \\ \dot{x}_t \\ \ddot{x}_t \end{pmatrix} \qquad A = \frac{1}{m + \frac{\Delta t}{2}c + \frac{\Delta t^2}{6}k} \begin{pmatrix} m + \frac{\Delta t}{2}c & \Delta tm + 2\Delta t^2 c & \frac{\Delta t^2}{3}m + \frac{\Delta t^3}{12}c \\ -\frac{\Delta t}{2}k & m - \frac{\Delta t^2}{3}k & \frac{\Delta t}{2}m - \frac{\Delta t^3}{12}k \\ -k & -c - \Delta tk & -\frac{\Delta t}{2}c - \frac{\Delta t^2}{3}k \end{pmatrix}$$

$$B = \frac{1}{m + \frac{\Delta t}{2}c + \frac{\Delta t^2}{6}k} \begin{pmatrix} \frac{\Delta t^2}{6} \\ \frac{\Delta t}{2} \\ 1 \end{pmatrix} \qquad G = \frac{-m}{m + \frac{\Delta t}{2}c + \frac{\Delta t^2}{6}k} \begin{pmatrix} 0 & 0 & \frac{\Delta t^2}{6} \\ 0 & 0 & \frac{\Delta t}{2} \\ 0 & 0 & 1 \end{pmatrix}$$

However, phase lag between measurement and activation cannot be eliminated. In this study, x_{t+1} is predicted from x_t as follows.

$$\hat{\mathrm{x}}_{t+1/t} = (I - BF)^{-1}\{A\mathrm{x}_t + (G + BF_z)\hat{\mathrm{z}}_{t+1/t}\} \tag{18}$$

Hence optimal control force for the random excitation with a predominant frequency is obtained as

$$u_{t+1} = F_x\hat{\mathrm{x}}_{t+1/t} + F_z\hat{\mathrm{z}}_{t+1/t} \tag{19}$$

Numerical Examples

As a model of flexible and lightly-damped structures, a single-DOF structure with damping ratio of 0.02 and natural period of 2 second ($m = 4\ kg \cdot sec^2/cm$) is adopted for numerical simulations.

From Fig. 10, it is found that prediction of ground motion and structural response in phase-adjusted control suppresses vibration energy of a structure effectively. If no prediction is performed, vibrational energy goes up very sharply with time delay. This is due to the predominance of frequency component in random excitation.

Time history displacement response of structures and control force is shown in Fig. 11, for the case of 0.4 seconds delay time. It is found that the response of the phase-adjusted controlled structure with prediction and the optimally controlled structure without time delay have little difference. Corresponding control forces are also found to be very similar. These results verify that the phase-adjusted control with prediction of ground motion and structural response proposed in this study works satisfactorily.

Then, the proposed algorithm is applied using SCT record of 1985 Mexico earthquake, which has a predominant frequency of 0.5 Hz. Fig. 12 shows the estimated ground parameters ζ_g (the broken line) and ω_g (the solid line) at each time step. These values converge to the constant values after 5 seconds. From Fig. 13, the maximum displacement response is 70.4 cm when the structure is

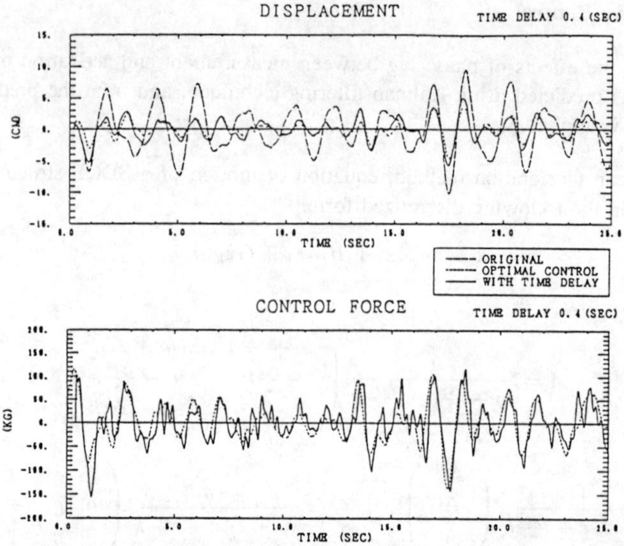

Fig. 11 Simulated Time History of Displacement Response and Control Force (Filtered White-Noise Excitation, Feedback and Feedbackward Control)

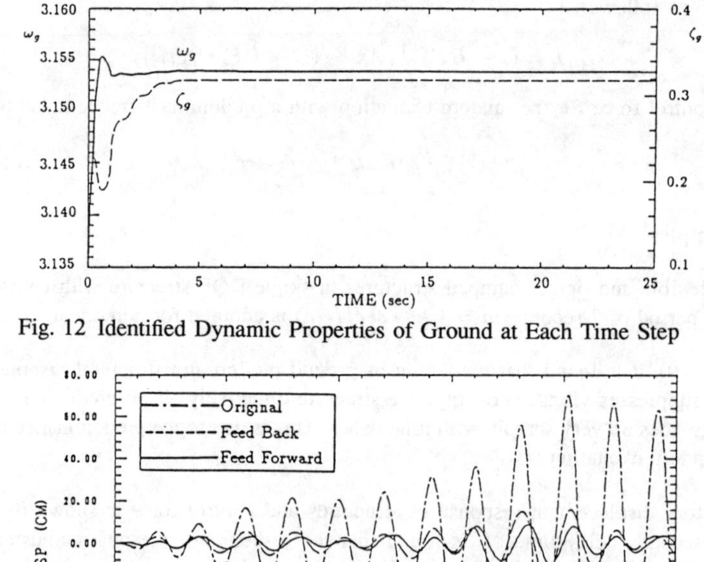

Fig. 12 Identified Dynamic Properties of Ground at Each Time Step

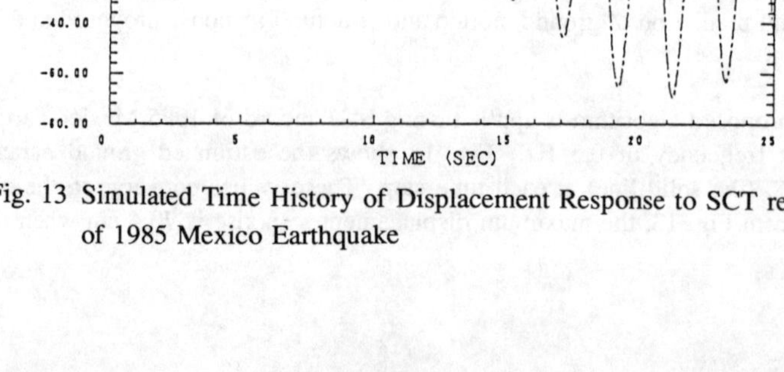

Fig. 13 Simulated Time History of Displacement Response to SCT record of 1985 Mexico Earthquake

uncontrolled; 18.1 cm (25.7% of the uncontrolled case) when it is controlled with feedback gain; and 9.92 cm (14.1% of the uncontrolled case) when it is controlled with both feedback and feedforward gain. Using this algorithm, the structure can be well controlled even if the predominant frequency of the ground motion changes during an earthquake.

Major conclusions from this study are:

- When the earthquake ground motion have a predominant frequency, future input ground motion can be predicted using Kalman filtering technique and then structural response is predicted. Feedforward gain is multiplied to the predicted ground motion and feedback gain is multiplied to the predicted response to obtain the optimal control force.
- Phase-lag effect is reduced with prediction of input and response, as illustrated by a numerical simulation. However, if phase lag becomes more than one-third of the natural period of structure, efficiency of control force decreases rapidly.
- The predominant frequency can be estimated during an earthquake using extended Kalman filtering technique. This active control algorithm with identification of ground parameters can be applicable when the predominant frequency changes during an earthquake.

REFERENCES

1. D. Quinn, Y. Yamada, H. Iemura, "Tuned Mass Dampers for Local Site Dependent Earthquake Ground Motion," *Research Report No. 87-ST-01,* School of Civil Engineering, Kyoto Univ., March 1987.
2. S. Mahin, P.B. Shing C. Thewalt, R. Hanson: "Pseudodynamic Test Method—Current Status and Future Directions," *Journal of Structural Engineering, ASCE,* 115:8; 1989.
3. M. Nakashima, T. Kaminosono, M. Ishida, K. Ando: "Integration Techniques for Substructure Pseudo Dynamic Test," *4th U.S. National Conference on Earthquake Engineering,* 1990.
4. H. Iemura, K. Ohshima, T. Nakata: "Earthquake Energy Partitioning and Reliability of Bridge Structures with Seismic Isolators," *Proc. of the 8th Japan Earthquake Engineering Symposium,* 1713–1718, 1990.
5. Y. Yamada, H. Iemura, W. Tanzo, Y. Uno, S. Nakamura: "Hybrid Loading Test of the High Damping Rubber Bearings Subjected to Sinusoidal and Random Earthquake Ground Motion," *Proc. of the Structural Engineering, JSCE,* 1991.
6. H. Kwakernaak, R. Sivan: *Linear Optimal Control Systems,* Wiley Interscience, 1972.
7. J.N. Yang, M.J. Lin: "Optimal Critical-Mode Control of Building under Seismic Load," *Proc., ASCE,* Vol. 108 (EM6), pp.1167–1185, 1982.
8. L. Meirovitch and L. M. Silverberg, "Control of Structures Subjected to Seismic Excitation," *Proc., ASCE,* Vol. 109 (EM2), pp.604–618, 1983.
9. H.H.W. Leipholz and M. Abdel-Rohman: *Control of Structures,* Martinus Nijhoff Publishers, 1986.
10. T. Kobori, "Active Seismic Response Control," *Proc. 9th WCEE,* Vol. 8, pp.435–446, 1988.
11. Y. Yamada, H. Iemura and A. Igarashi, "Active Control of Structures under Stochastic Base Excitation with a Predominant Frequency," *Proc. of 5th ICOSSAR,* San Francisco, 1989.
12. Y. Yamada, H. Iemura, A. Igarashi and Y. Iwasaki, "Phase-delayed Active Control of Structures under Random Earthquake Motion," *Proc. 4th United States National Conference on Earthquake Engineerings,* Vol. 3, pp.447–456, 1990.

Summary of Group Discussion and Recommendations

Group 1 Assessment of State of Structural Systems

I. Participants

Italy: Faravelli, L., University of Perugia.
Gavarini, C., University of Rome, Italy National Council for Research.
Gherardini, P., National Research Council, Rome.
Nistics, N., University of Rome.
Righetti, G., D'Appolonia, Genova.
Zucchini, A., University of Genova.

Japan: Kameda, H., Kyoto University.
Shibata, H. University of Tokyo.

U. S.: Masri, S. F., University of Southern California.
Rehak, D., Carnegie Mellon University.
Shinozuka, M.(Co-chair), University of New York, Buffalo.
Wen, Y. K.(Co-chair), University of Illinois, Urbana.

Guest: Hirsch, G. Aachen University of Technology, Aachen, Germany.

II. General

The area of concentration of this group is assessment of state of structural systems with control under uncertain loadings. The major concerns are damage assessment, uncertainty in the loading and the structural resistance capacity with control, Knowledge-Based-Expert-System (KBES) and neural network technique, and finally special considerations of lifeline systems.

The problem of assessment of state of systems with control as a whole can be illustrated by the diagram shown in Fig.1 using control of response due to seismic loading as an example. The horizontal axis indicates the level and extent of control in the structure and the vertical axis indicates the state of structure in terms of degree of damage, indicated by dashed lines and associated response behavior. The solid lines depict the state of the system as a function of the extent of control for seismic excitation of various intensities denoted by I. Note that for extremely high intensity earthquakes, the demand on control in terms of power and control force is large, therefore until these demands are met the state of the system is hardly improved by the control. Also, when under such load the structure would become nonlinear and inelastic for which control algorithms robust enough for practical applications are yet to be developed. Therefore it is a major concern when the control becomes ineffective and adaptive control should be needed if the loading far exceeds the design load of the control system. The boundary for ineffective performance of the control device would be an increasing function of the control performance as indicated by the solid line intercepting the performance curves. In this regard KBES and neural network technique may be used for developing adaptive control systems and the criteria for establishing the boundary. It is felt that at the present stage of development of these technologies, the major concerns are the massive calculation required for real time response in implementing KBES and the hardware development and training for unforeseen seismic excitation time histories required for neural networks.

The problem of whether and how much control is necessary of course can not be separated from the uncertainties in the loadings and resistance of the structure and the issue of costs, i.e. life cycle cost including initial cost, maintenance cost, and cost of consequence of failure of the structure. A rational decision requires an analysis of the expected total lifetime cost. The initial cost and maintenance cost obviously increases with the level of control in the structure, however, the expected failure cost will decrease because of decrease in the probability of failure of the system and so on. Fig 2. shows such an illustration and possible benefit of the structural control as a function of the control performance. C_o is the expected cost of failure without control, a constant; C_1 is the expected cost of failure with control, a decreasing function of the control level; and C_2 is the cost of control including maintenance cost. The difference $C_o - C_1 - C_2$ is obviously the saving (or benefit) of the control. Note that it is possible to arrive at an optimal control performance by the benefit analysis.

III. Research Needs

The research needs in the general area of assessment of structural systems with control under uncertain future loadings are as follows.

1 Definition and detection of damage of systems in the inelastic and nonlinear range. It will facilitate development of algorithms for control of nonlinear structures. In addition to conventional mathematical models, neural network type of approaches need to be explored which gives approximate but "almost" correct answers in real time.

2 Development of control algorithms for excessive excitation, i.e. much larger than the design load and the nonlinear range of structural response. Such algorithms need to be tested extensively by simulations and experiments.

3 Reliability of the control systems under uncertain future loading. It is especially important for structures whose performance and reliability depends almost entirely on the proper function of the control system; e.g. passive control with base isolation and active control with little reserve structural strength.

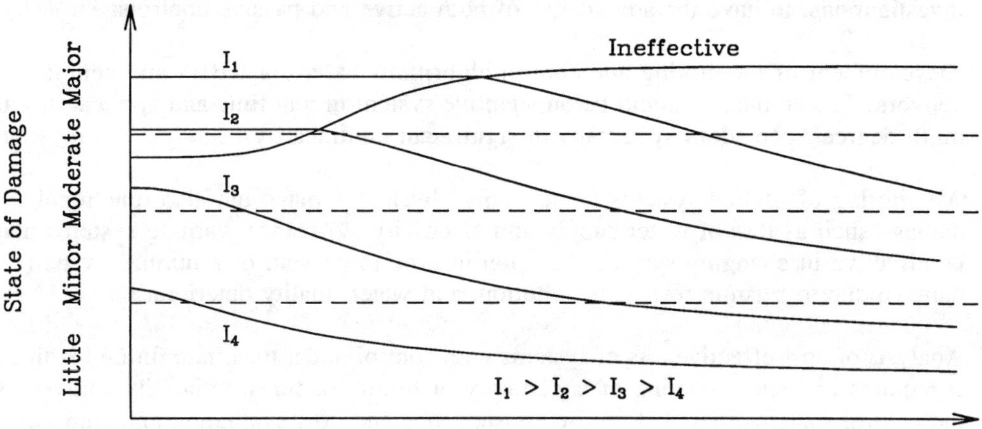

Fig. 1 State of Damage as Function of Control Performance (I = intensity of earthquake)

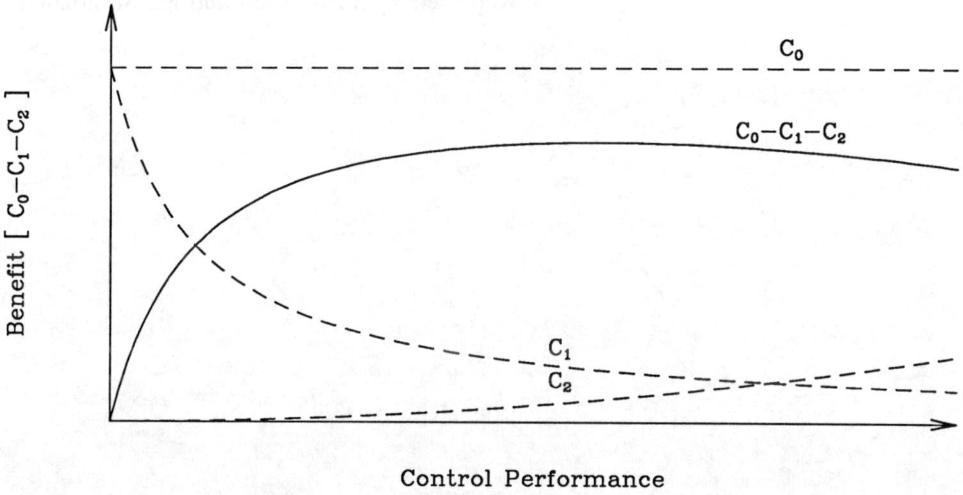

Fig. 2 Cost and Benefit as Function of Control Performance

4 Algorithm development, damage and reliability assessment of hybrid systems need to be encouraged as such system thus far has been shown, at least by analytical investigations, to have the advantages of both active and passive controls.

5 Development of monitoring and control algorithms based on KBES and neural network. The emphasis should be on adaptive system in real time and applicability to multi-degree-of-freedom systems with significant nonlinearity.

6 Monitoring of lifeline systems for not only physical damage but also functional damage such as loss of water supply and electricity. Advance warning systems may be effective in changing system operation in time to prevent or minimize system damage due to seismic force, air pollution, and water quality deterioration.

7 Analysis of cost-effectiveness of systems with control under uncertain future loadings. It requires realistic modeling of uncertainty of future loadings, reliability analysis of the control systems, detailed failure consequence and cost evaluation and study of design alternatives in terms of degree of reliance on control for structural resistance against future loadings.

Reported by Y. K. Wen and M. Shinozuka.

Group 2 Measurement of System Performance

I. Participants

Italy: A. Del Grosso (Co–chair), University of Genova

Japan: T. Inukai, Takenaka Corp.

USA: B. Wada (Co–chair)*, Jet Propulsion Lab – Pasadena
R. Claus*, Virginia Polytechnic Institute and State University
K. C. Park, partial representation*, University of Colorado

*The members did not have an extensive Civil Engineering background.

II. Scope

The discussion was limited to the technology requirements to develop measurement devices required for Civil Engineering applications; not the requirements to process, collect, store, and present the resulting data from the measurement devices. Excluded were measurement systems which did not exist because of the limitation of funds, not the lack of technology. Also, the need for one measurement device to measure all the desired data was not considered if two or more existing measurement devices could measure the required parameters.

The major consideration was on measurement devices used to establish the overall system performance rather than devices which may be required to obtain local specific information.

III. Summary

The following areas required additional research in order to provide the engineers with measurements to effectively obtain an understanding of the physical responses and to help establish new approaches to design effective Civil Systems.

(1) Long Continuous Sensors

Long continuous line type sensors with the capability to measure longitudinal and bending strain at discrete location along the long sensor are required. This type of sensors can be placed in the soil around the structure to establish the strain distribution in the soil media. Technology is also required to establish reliable techniques to attach the sensor to the media to assure the media strain of interest is being recorded. A desirable feature is the capability to change the location and/or the gage length of the strain measurement once the sensor is installed.

(2) Reliable Imbedded Sensors for Concrete

Reliable strain measurement devices are required to measure the desired strains in the concrete. Features similar to (1) are desirable.

(3) Tension Levels in Cables

The effectivity and/or the life of a structure may be dependent on the tension level in cable type elements. Techniques are required to effectively determine the tension level in cables non–destructively.

(4) Absolute Long Time Relative Displacement/Strain

Often the change in dimension (e.g., crack opening) must be monitored over a long period of time to establish the available life or the structure. The use of continuous stable instrumentation monitors is not cost effective to monitor the numerous parameters of interest. A method of determining the displacement change over a long time with periodic measurements without a continuous reference signal is required.

(5) Include More Experimentation in Education

The inclusion of more experimentation in education is required to educate engineers to make more effective utilization of experimental techniques where appropriate.

(6) Structural Design for Measurement

The approach, methodology and technology should be developed to design the structure to allow efficient incorporation of measurement devices to measure the desired system parameters of the structure.

Reported by B. Wada

Group 3 Control and Isolation of System

I. Participants

Italy: F. Casciati, University of Pavia
G. Piccarob, University of Genova
C. Bonavita, University of Genova
A. De Carli, University of Rome

Japan: S. Kawamura, Taisei Corp.
S. Kamagata, Kobori Research Complex, Kajima Corp.
H. Iemura, University of Kyoto

USA: K. C. Park, University of Colorado
A. H–S. Ang, University of California, Irvine
J. N. Yang, University of California, Irvine
H. J. Lagorio, Co–chair, National Science Foundation
T. T. Soong, Co–chair, University of New York, Buffalo

II. General

It is the consensus of this group that the workshop/seminar has been a success. Its objectives have been achieved through paper presentations, extensive discussions, and exchange of ideas. This was possible because of the following factors:

(a) Relatively small number of participants.

(b) Interdisciplinary representations from civil, electrical, and aerospace engineering.

(c) Tri–lateral format.

It is recommended that meetings of this nature be repeated at approximately two–year intervals.

It is also recommended that the group consider more active participation and organization of technical sessions on structural control in future national and international conferences. A few that come to mind include:

(a) Japan National Workshop on Structural Control, March 1992. Contact Person: S. Kamagata.

(b) 10WCEE, Madrid, 1992. Contact Person: F. Casciati.

(c) ASCE Structures Congress, Irvine, CA, April 1993. Contact Person: A. H–S. Ang.

(d) IFAC World Congress, Australia, 1993. Contact Person: A. DeCarli.

III. Assessment of the State–of–Research

It appears that the state of structural control is such that major technologies, with possible exception of improved actuators, are currently available. Analytical theory for structural applications, at least for linear systems, is also fairly well developed. Full–scale demonstration of AMD and ABS systems is in progress in Japan. Hence, significant progress has been made in the last few years.

It Europe, active control is of interest o researchers in Italy, Portugal and Spain for earthquakes, and in Germany and England primarily for Wind. French researchers have developed several interesting passive control techniques. However, in countries of low seismicity such as Italy, on–line control systems may turn out to be less attractive than off–line control due to cost of maintenance, repair, and data processing and management.

IV. Research Needs and Future Directions

It is also clear from the materials presented at the workshop that additional studies and research are needed. Recommendations for future research areas are summarized below.

(a) From the standpoint of practical implementation of structural control, it is more realistic at this stage to require control to avoid structural damage. but not complete reliance on the control system for safety against collapse.

(b) Methods for damage assessment as well as reliability need to be developed or improved for structures with control. Research in this area includes reliability of integrated software/hardware systems, reliability under exceptional loads, and issues related to uncertainties that exist in ground motion and in structure/control systems.

(c) Actuator/dynamics and technology is an area in need of further research. New types of actuators, such as electric actuators, should be developed for structural control applications. A survey of the state–of–the–art in this area perhaps can be undertaken by Japanese investigators and by U.S. industrial groups.

(d) More research on robust control algorithms, particularly for nonlinear systems, is needed.

(e) More experimental research through laboratory experiments and full–scale demonstration projects is encouraged. They can be used for comparative studies of a variety of control algorithms and actuator dynamics under different loading conditions.

(f) Hybrid control and parametric control seem to offer opportunities for improvement over passive or active systems taken individually from the standpoint of energy requirements, economy, and control efficiency.

Reported by T. T. Soong

INDEX OF CONTRIBUTORS